Information and Communication Technology for Agriculture and Rural Development

Information and Communication Technology for Agriculture and Rural Development

Editors:

R. SARAVANAN
Associate Professor (Communication and Incharge)
School of Social Sciences, College of Post Graduate Studies
Central Agricultural University, Umiam (Barapani), Meghalaya-793103

C. KATHIRESAN
Project Manager
InDG Project & Principal Investigator, Model e-Village Project
Centre for Development of Advanced Computing (C-DAC)
Hyderabad-500085, Andhra Pradesh

T. INDRA DEVI
Project Fellow, e-Arik Project, College of Horticulture and Forestry
Central Agricultural University (CAU), Pasighat-792102, Arunachal Pradesh

New India Publishing Agency
Pitam Pura, New Delhi-110 088

Published by
Sumit Pal Jain *for*
New India Publishing Agency
101, Vikas Surya Plaza, CU Block, L.S.C. Mkt.,
Pitam Pura, New Delhi- 110 088, (India)
Phone: 011-27341717, Fax: 011-27341616
Mobile : 09717133558
E-mail: info@nipabooks.com
Web: www.nipabooks.com

ISBN : 978-93-80235-88-2

Composed and Designed by NIPA

Preface

Information and Communication Technologies (ICTs) are facilitating faster sharing of information and acting as a key agent for changing agrarian situation and farmers' lives by improving access to information and sharing knowledge. Pioneering ICT experiments in India show that the rural livelihoods are greatly enhanced by access to information on improved agricultural practices, pest and diseases control, market and weather. National efforts are underway to reduce urban and rural digital divide. Nearly for the last one decade, ICT practitioners are experimenting number of innovative ICT initiatives for the agricultural and rural development in India. In this connection, as an integral part of the C-DAC sponsored research project on "Model e-Villages in North-East India", a National Seminar on ICT for Agriculture and Rural Development (ICT4ARD) was organised during 2009 at the College of Horticulture and Forestry, Central Agricultural University (CAU), Pasighat, Arunachal Pradesh State, with the objective of bringing together different ICT initiatives and its review in agriculture and rural development, involving public, private, NGOs and individuals in this field. The articles included in this manuscript focuses on; Digital divide in rural India, e-Agriculture issues, Village Knowledge Centres (VKCs), Community Information Centre (CIC), SATCOM application, e-Villages in North-East India, InDG web portal initiative, Kisan Mobile Sandesh (KMS), Dyanmic Market Information (DMI), Expert systems, Citrus e-Clinic Software, Interactive Multimedia Compact Disc (IMCD), KISSAN, Mobile Agricultural School & Services (MASS), Farmers Database creation, Pest Surveillance of Rice, Techmode Approach for Distance Learning, evaluation of Kissan Call Centres (KCCS), Web Portals and Digital Data Base in Agroforestry, Watershed modelling using GIS and Remote Sensing and ICT adoption level, e-Readiness, impact of ICT initiatives, stakeholders feedback, policy implications and recommendations were included in this manuscript. Along with the articles presented during the seminar, other few selected articles contributed by the ICT practitioners also included in this book.

We wish to express our deep sense of gratitude to Prof. S. N. Puri, FNAAS, Vice-Chancellor, CAU for his encouragement and support in making the National Seminar a grand success. We also express our sincere gratitude to Dr. V. K. Mishra, Dean and Chairman, Seminar Organising Committee for his guidance. We also thank Prof. V. Veerabhadraiah, Former Director of Extension, UAS, Bangalore, Prof. K. Narayana Gowda, Vice-Chancellor, UAS, Bangalore, Prof. P.S. Lonkar, Director of Extension, Maharashtra Animal and Fishery Sciences University, Dr. Edwin Joseph, Grand Vally State University, Allendale Michigan, USA, Mr. Dillip Pattanaik, Executive Director, IRMA-India and all other seminar participants for their participation and contribution.

We express a special note of thanks to Mrs. S. Devika, M.Sc. (Agri) for reading entire draft manuscript and suggesting corrections. We also thank all the authors for their valuable contribution.

R. Saravanan
C. Kathiresan
T. Indra Devi

List of Contributors

Abhay Wankhade, SMS (Agricultural Extension), KVK, JNKVV, Chandangaon, Chhindwara, Madhya Pradesh - 480 001, India.

Agrawal, S.K., ARIS Cell, Indian Veterinary Research Institute (IVRI), Izatnagar, Bareilly, Uttar Pradesh - 243 122, India.

Ajit Maru, Senior Knowledge Officer, GFAR, FAO, Rome, Italy.

Anandaraja, N., Assistant Professor (Agricultural Extension), e-Extension Centre, Directorate of Extension Education, Tamil Nadu Agricultural University (TNAU), Coimbatore, Tamil Nadu - 641 003, India.

Ansari, M.A., Professor (Agriculture Communication), G.B. Pant University of Agriculture and Technology, Pantnagar, U.S. Nagar, Uttarakhand - 263 145, India.

Anup Prakash Upadhyay, Department of Agricultural Communication, G.B. Pant University of Agriculture & Technology, Pantnagar, U.S. Nagar, Uttarakhand - 263 145, India.

Arun Babu, A., University of Agricultural Sciences Bangalore (GKVK) Bangalore, Karnataka - 560 065, India.

Balasubramani, N., Assistant Director (Agricultural Extension), National Institute for Agricultural Extension Management (MANAGE), Rajendranagar, Hyderabad, Andhra Pradesh - 500 030, India.

Bikash Borthakur, Division of Veterinary Extension Education, Indian Veterinary Research Institute, Izatnagar, Bareilly, Uttar Pradesh - 243 122, India.

Binoy, K., (ICT-ARD), Dhirubhai Ambani Institute of Information and Communication Technology, Gandhinagar, Gujarat - 382 007, India.

Bisht, G.S., ARIS Cell, Indian Veterinary Research Institute, Izatnagar, Bareilly, Uttar Pradesh - 243 122, India.

Borua, S., Assiatant Professor (S.G.), Department of Extension Education, Assam Agricultural University (AAU), Jorhat, Assam- 785 013, India.

Choudhury, I., Crop Inventory & Modelling Division, AFEG/RESA, Space Applications Centre (ISRO), Ahmedabad, Gujarat-380 015, India.

Das, G.K., ARIS Cell, Indian Veterinary Research Institute, Izatnagar, Bareilly, Uttar Pradesh - 243 122, India.

Dillip Pattanaik, Executive Director, IRMA-India, B-14, Phase-1, Krishna Garden, Jagamara, Bhubaneswar, Orissa - 751 030, India.

Druv Shrivastava, Subject Matter Specialist, KVK, Seoni, JNKVV, Madhya Pradesh -480 661, India

Gaurav Papnai, Department of Agricultural Communication, G.B. Pant University of Agriculture & Technology, Pantnagar, U.S. Nagar, Uttarakhand - 263 145, India.

Ghosh, S.K., ARIS Cell, Indian Veterinary Research Institute, Izatnagar, Bareilly, Uttar Pradesh - 243 122, India.

Gururaj Katti, Principal Scientist (Entomology), Directorate of Rice Research (DRR), ICAR, Rajendranagar, Hyderabad, Andhra Pradesh - 500 030, India.

Hemnath Rao Hanumankar, Professor and Director, Centre for Poverty Studies and Rural Development (CPSRD), Administrative Staff College of India, Bella Vista, Raj Bhavan Road, Hyderabad, Andhra Pradesh - 500 082, India.

Indra Devi, T., Project Fellow, e-Arik Project, College of Horticulture and Forestry, Central Agricultural University, Pasighat, Arunachal Pradesh - 792 102, India.

Indumathy, R., NRCAF Colony, National Research Centre for Agroforestry, Gwalior Road, Jhansi, Uttar Pradesh - 284 003, India.

Kathiresan, C., Project Manager, InDG Project & Principal Investigator, Model e-Village Project , Centre for Development of Advanced Computing (C-DAC), Hyderabad, Andhra Pradesh - 500 085, India.

Kishore Kumar, N., Director, Academic Programming and Standards, Agricultural Extension & Rural Sociology, Hamelmalo Agricultural College, Keren P.O.Box No. 66, Keren, Asmara, Eritrea, North East Africa.

Kumari Shweta, Department of Veterinary & Animal Husbandry Extension Education, Ranchi College of Veterinary Science & Animal Husbandry, Birsa Agricultural University, Ranchi, Jharkhand - 834 006, India.

Lakshmana Reddy, B.S, University of Agriculture Bangalore (GKVK) Bangalore, Karnataka - 560 065, India.

Lizamoni Chungkurang, e-AgriKiosk Project, College of Horticulture and Forestry, Central Agricultural University (CAU), Pasighat, Arunachal Pradesh - 792 102, India.

Lonkar, P.S., Director (Extension and Training), Maharashtra Animal & Fishery Sciences University, Seminary Hills, Nagpur, Maharashtra - 440 006, India.

Mahesh Chander, Sr. Scientist & Head, Division of Extension Education, Indian Veterinary Research Institute, Izatnagar, Uttar Pradesh - 243 122, India.

Maithili, B., Rural Development Institute, Himalayan Institute Hospital Trust (HIHT), Swami Ram Nagar, P.O. Doiwala, Dehradun, Uttarakhand, India.

Malhan, I.V., Professor, Department of Library and Information Science, University of Jammu, Jammu and Kashmir - 180 006, India.

Nalin Khare, Joint Director of Extension, Directorate of Extension, Jawaharlal Nehru Krishi Vishwavidayalaya, Jabalpur, Madhya Pradesh, India.

Narayana Gowda, K., Vice-Chancellor, University of Agriculture Bangalore (GKVK) Bangalore, Karnataka - 560 065, India.

Narmi Darang, Assistant Professor, Department of Economics, J.N. College, Pasighat, Arunachal Pradesh - 792 102, India.

Nataraju, M.S. , Professor and Head, Communication Centre, University of Agriculture Bangalore (GKVK) Bangalore, Karnataka - 560 065, India.

Nazreen Hassan, S., Department of Agricultural Extension, College of Agriculture, Kerala Agriculture University (KAU), Vellayani, Kerala - 695 522, India.

Neelam Bhardwaj, Professor, Department of Agricultural Communication, G.B. Pant University of Agriculture & Technology, Pantnagar, U.S. Nagar, Uttarakhand - 263 145, India.

Neelam Pandey, Rural Development Institute, Himalayan Institute Hospital Trust (HIHT), Swami Ram Nagar, P.O. Doiwala, Dehradun, Uttarakhand, India.

Neog, P.K., Professor, Extension Education Institute (EEI), Assam Agricultural University (AAU), Jorhat, Assam - 785 013, India.

Ninawe, A.S., Vice Chancellor, Maharashtra Animal & Fishery Sciences University, Seminary Hills, Nagpur, Maharashtra - 440 006, India.

Oraon, J., Professor, Department of Veterinary & Animal Husbandry Extension Education, Ranchi College of Veterinary Science & Animal Husbandry, Birsa Agricultural University, Ranchi, Jharkhand -834 006, India.

Pandey, A.K., Professor ,Department of Veterinary & Animal Husbandry Extension Education, Ranchi College of Veterinary Science & Animal Husbandry, Birsa Agricultural University, Ranchi-Jharkhand - 834 006, India.

Paras Kakkar, Rural Development Institute, Himalayan Institute Hospital Trust (HIHT), Swami Ram Nagar, P.O. Doiwala, Dehradun, Uttarakhand, India.

Periyar Ramasamy, D., Directorate of Extension Education, Tamil Nadu Agricultural University (TNAU), Coimbatore, Tamil Nadu - 641 003, India.

Prabhakar, I., Department of Agricultural Communication, G. B. Pant University of Agriculture & Technology, Pantnagar, Uttarakhand - 263 145, India.

Prabhakar, M., Entomolgy Division, CRIDA, ICAR, Hyderabad, Andhra Pradesh -500 059, India.

Prasad, Y.G., Entomolgy Division, CRIDA, ICAR, Hyderabad, Andhra Pradesh - 500 059, India.

Ratha Krishnan, P., Scientist (Forestry), National Research Centre for Agro-forestry, Gwalior Road, Jhansi, Uttar Pradesh - 284 003, India.

Sajeed Ali, Darjeeling Krishi Vigyan Kendra, Uttar Banga Krishi Viswavidyalaya, Kalimpong, Darjeeling, West Bengal, India.

Samita Wadhera, Research Scholar, Department of Library and Information Science, University of Jammu, Jammu - 180 006, India.

Sanjay Kumar Rewani, Department of Veterinary & Animal Husbandry Extension Education, Ranchi College of Veterinary Science & Animal Husbandry, Birsa Agricultural University, Ranchi, Jharkhand - 834 006, India.

Sanjay Kumar, Senior Scientist & Officer In-Charge, ARIS Cell, Indian Veterinary Research Institute, Izatnagar, Bareilly, Uttar Pradesh - 243 122, India.

Sarat Chandra Babu, N., Director, Centre for Development of Advanced Computing (C-DAC), Hyderabad, Andhra Pradesh - 500 085, India.

Saravanan, R., Principal Investigator, e-Arik & e-AgriKiosk Projects, Co-Investigator, Model e-Village Project, Associate Professor (Communication) & In-Charge, School of Social Sciences, College of Post Graduate Studies, Central Agricultural University (CAU), Umiam (Barapani), Meghalaya - 793 103, India.

Senthilkumar, S., Assistant Professor, Veterinary University Training and Research Centre, TANUVAS, Ramayanpatti, Tirunelveli, Tamil Nadu - 627 358, India.

Senthilkumaran, S., Director, Jamshedji Tata National Virtual Academy for Rural Prosperity, M S Swaminathan Research Foundation (MSSRF), 3rd Cross Road, Institutional Area, Taramani, Chennai, Tamil Nadu - 600 113, India.

Shibi Sebastian, e-Extension Centre, Directorate of Extension Education, Tamil Nadu Agricultural University (TNAU), Coimbatore, Tamil Nadu - 641 003, India.

Singh, S.K., ARIS Cell, Indian Veterinary Research Institute (IVRI), Izatnagar, Bareilly, Uttar Pradesh - 243 122, India.

Singh, Y.P., Scientist (SS), Indian Veterinary Research Institute (IVRI), ARIS Cell, Izatnagar, Bareilly, Uttar Pradesh – 243 122, India

Sriram, N., Assistant Professor (Agricultural Extension), e-Extension Centre, Directorate of Extension Education, Tamil Nadu Agricultural University (TNAU), Coimbatore, Tamil Nadu - 641 003, India.

Subrata Manna, Computer Programmer, Darjeeling Krishi Vigyan Kendra, Uttar Banga Krishi Viswavidyalaya, Kalimpong, Darjeeling, West Bengal, India.

Sujay Dutta, Crop Inventory & Modelling Division, AFEG/RESA, Space Applications Centre (ISRO), Ahmedabad, Gujarat- 380 015, India.

Sumanjeet Singh, Assistant Professor, Department of Commerce, Ramjas College, University of Delhi, New Delhi - 110 007, India.

Supriya Devi, Research Fellow, Model e-Village Project, College of Horticulture and Forestry, Central Agricultural University (CAU), Pasighat, Arunachal Pradesh - 792 102, India.

Surendra Pannase, Programme Co-ordinator, KVK Chhindwara, JNKVV, Madhya Pradesh - 480 001, India.

Taufiq Ahemed, Ph. D. Scholar, Department of Agricultural Communication, G.B. Pant University of Agriculture and Technology, Pantnagar, U.S. Nagar, Uttarakhand, 263145, India.

Vadivel, E., Director of Extension, Directorate of Extension Education, Tamil Nadu Agricultural University (TNAU),Coimbatore, Tamil Nadu - 641 003, India.

Vasantbhanu, K., Scientist, Andhra Pradesh Rice Research Institute, ANGRAU, Maruteru, Andhra Pradesh, India.

Vennila, S., Principal Scientist (Entomology), National Centre for Integrated Pest Management (NCIPM), Pusa campus, New Delhi – 110 012, India.

Vijayalakshmi, B.,Vertical Head (Agriculture), India Development Gateway Initiative (InDG), Centre for Development of Advanced Computing (C-DAC), Hyderabad, Andhra Pradesh - 500 085, India.

Yogiraj V. Shete, (ICT-ARD), Dhirubhai Ambani Institute of Information and Communication Technology, Gandhinagar, Gujarat – 382 007, India.

Contents

1

Digital Divide in India: Measurement, Determinants and Policy for Addressing the Challenges in Bridging the Digital Divide

Sumanjeet Singh

ABSTRACT

In the recent years, the notion of 'digital divide' has been widely researched and has attracted much debate and speculation for its economic, social and political consequences. Existing studies reveals, the gap that exists between those who have access to ICTs and those who do not create exclusion, endanger social integration and hamper economic growth. The digital divide has many dimensions and can be categorized as global, regional and national. At national level, there is no single divide, but multiple divides: for instance; within countries, between men and women, young and elderly, rich and poor and most importantly rural and urban. The present paper is mainly focused on India and tries to explore the problem of digital divide mainly in rural-urban India. In the context of present paper, digital divide essentially means teledensity, mobile and Internet divide between the rural and urban areas. The paper reveals that obstacles such as illiteracy, lack of skills, infrastructure, and investment in rural areas must be tackled if India is to diminish the gap of digital divide. At the government front, it should put thrust towards: connectivity provision, content creation, capacity augmentation, core technologies creation and exploitation, cost reduction, competence building, community participation and commitment to the deprived and disadvantaged which would definitely help in bridging digital divide.

Introduction

The Information and Communication Technology (ICT[1]) is one of the important driving forces for modern civilization. The rapid development and proliferation of ICTs has accelerated the economic and social change (Nandi, 2002), across all areas of human activity worldwide and continues to do at the rapid pace. ICTs enable interactive communication unhindered by distance, volume, medium or time and also reduce the cost of co-ordination (Fletcher *et al.*, 2000), communication and information processing (Dean, 2002; Gordon, 2002). ICTs hold great promise for development and poverty reduction in global south. In many instances, poor people have experienced benefits in the form of increased income, better health care, improved education and training, access to job opportunities (Kuhn and Skuterud, 2000; Sumanjeet, 2008; Hecker, 2001; Motohashi, 2001), engagement with government services, contacts with family and friends, enterprise development opportunities, increased agricultural productivity (Poole, 2001; Hooker *et al.*, 2001), and so on (Sumanjeet, 2009). The issue of transparency is easier to manage with ICTs, which may result in monetary savings in addition to stakeholder confidence in the development process and system (Jesus, 2003). ICTs have radically changed the way of doing business. Internet and its enabled business technologies like e-commerce have opened up vast business avenues and transformed the whole business world into a global village. Further, it is expected that ICTs will play a crucial role in the socio-economic development process, and change the pattern of people's economic models and lives. But, the potential to exploit the benefits of ICTs largely depends on the access and adoption of these technologies. In fact, the status of ICT adoption of an economy is an indicator of its potentiality to exploit the economic opportunities affordable by the new technologies, more generally, its prospects for the transition to the 'new economy'. But, as expected the adoption of ICTs vary significantly across countries.

Table 1 : Status of ICTs Adoption among Different Regions (Per 100 Inhabitants)

Regions	Fixed Telephone	Cellular Subscribers	Internet Users	Broadband Subscribers
ESCAP[2]	16.2	31.5	11.9	2.9
LLDC	6.6	13.3	4.0	0.1
LDC	0.9	9.3	0.5	11.8
SIDS	15.4	39.5	15.0	0.5
ASEAN	8.6	33.8	10.2	0.2
SAARC	3.3	15.3	5.0	0.1
Central Asia	11.1	20.1	6.3	—
Low Income	4.0	14.5	5.4	2.8

contd....

Middle Income	23.5	40.2	11.4	22.0
High Income	46.7	84.6	68.4	–
Africa	3.1	20.9	4.7	–
Latin America & Caribbean	17.7	54.2	18.2	2.7
North America	57.8	75.0	69.0	19.7
Europe	45.0	101.4	43.1	15.6
Other Asia Pacific	22.5	51.2	19.3	7.1
World	19.4	40.9	17.4	4.3

Notes : Data for the year 2006, compiled by researcher; LLDC (Least Developed Countries); LDC (Least Developing Countries); SIDS (Small Island Developing States)
Source : Statistical Yearbook for Asia and the Pacific, 2007;

Table 1 indicates that high income economies have twice as many fixed telephone lines per capita as middle-income economies, and 11 times the number in low income economies. In the low income countries and SAARC member countries, the number of fixed telephone lines per 100 people was just 4 and 3.3, respectively. In the least developed countries the number did not even reach one in 2006. At the same time, the number of mobile phone per 100 people was 14.5, 15.3 and 9.3, respectively, in low income economies, SAARC member countries and least developed countries. In high-income economies in the region, there are, on average 68 Internet users per 100 inhabitants, compared with 0.5 in LDC, 11 in middle-income economies and five in low-income economies. In 2006, there were three broadband subscribers per 100 inhabitants in low-income economies, compared with 20 in North America and 16 in Europe.

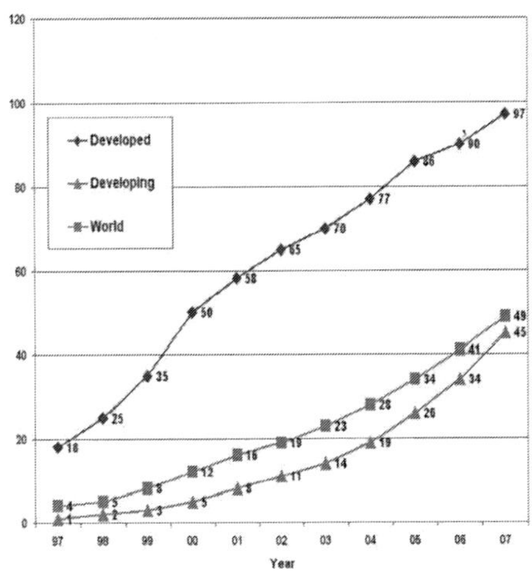

Graph 1 : Mobile Phone Users per 100 Inhabitants 1997-2007
Source: IUT, 2009

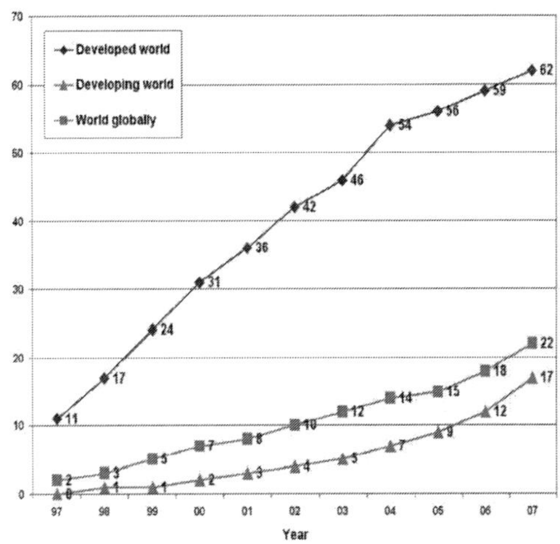

Graph 2 : Internet Users per 100 Inhabitants 1997-2007
Source: IUT, 2009

The global disparities in the access to these technologies are more clear with Graph 1 and 2. As illustrated above, the adoption level of ICTs is highest among the developed countries. In these countries, more people have access to Internet and at a higher speed and many of them have mobile phones. Developing countries are trying to keep pace with the changing technologies and adopting ICTs at faster rate, however, the overall adoption rate is much lower than the developed world. Asian countries have experienced a rise in the use of ICTs in the recent years, though the degree of expansion has varied among the countries. Internet adoption rate is very slow among developing countries.

From the above analysis, it is clear that there are millions of peoples in this world who do not have access to Internet, telephone, mobile phone *etc.* due to various reasons. Their inability to access ICTs deprives them to reap the benefits of ICTs. This has led to a serious issue *'Digital Divide'* between those who are participating in information technology revolution and those who are not (United Nations, 2006). In the recent past, the digital divide has attracted the attention of both scholars and policy makers for its economic, social and political consequences.

Methodology

From the above discussion it is clear that the concept of digital divide has many dimensions. Many technological and non-technological gaps can be considered. Added to this, there are many ways to measure the gap of digital divide. The present paper is focused on India and tries to explore the macro aspect of digital divide in India. The paper tries to explore the problem of digital divide within India. Thus, in the context of the present paper digital divide essentially means teledensity, mobile and Internet divide between the rural and urban areas. The main objective of the present paper is to study the problem of digital divide in the Indian context and address the challenges in bridging the gap of digital divide. The study also makes an attempt to identify the causes of digital divide in India. Further, in the end, an attempt has been made to suggest the policies to address the challenges of bridging the gap of digital divide in India.

1. Concept and Measurement of Digital Divide

There has been much discussion and debate about the concept of digital divide[3] and of the empirical analyses of its components (Chen and Wellman, 2004; Companie, 2001[4]; Cooper, 2002; Dewan and Riggins, 2005; Norris, 2001; Nahon, 2006). The notion of a digital divide gained attention in the 1990s with recognition that some people and institutions were not going online or were not going onto broadband. The concept of a "digital divide" between technological "haves" and "have-nots" has been a useful tool in efforts to bring greater, more equal access to powerful new information and communication technologies like the Internet. The term 'Digital Divide' is used to describe situations in which there is a market gap[5] in access to the use of ICT devices measured by, for example the number of fixed line phones per inhabitants, or the number of mobile users or the internet connections in the population. According to OECD (2001) the term 'digital divide' refers to "the gap between individuals, households, businesses and geographic areas at the different socio-economic levels with regard to their opportunities to access information and communication technologies (ICTs) and their use of Internet. It reflects the differences among and within countries". This definition is exclusively focused at national and international level. Nevertheless, digital divide exists in variety of other levels; sector, community, and individual level. For example, many communities within nation-states are far removed from the rest of the country with regard to information and communication technologies access and use. Such communities reshape the ICT to their culture and norms (Barziliani and Barziliani, 2005). Further, digital divide is also referred to as "the spiral of uneven access to and usage of information and communication technologies and the socio-economic rebound causes that have caused the emergence of information inequality throughout the world, both in and between the countries and also locally in communities. According to Hanimann and Ruedin (2007) the term 'digital divide' essentially describes three distinctive divides: a geographical digital divide (between regions and countries), a social digital divide (between social classes), and an upgraded digital divide (between technology and humans). Norris (2001), Mark, (2003) and Branko (2005) suggest that there are at least three major divides:

- A global divide between the developed and less developed worlds
- A social divide between the information rich and the information poor
- A democratic divide between those who do and those who do not use the new technologies to further political participation

Thus digital divide can be defined as economic, social or cultural deprivation generated by missing ICT access and skills. This definition goes beyond conventional definitions and it has a number of practically important characteristics. First, it explicitly spells out the three different dimensions where digital divides are important and where ICTs make a difference. In the modern knowledge and information based world, economic opportunities, such as employability, depend on ICT access and skills. ICTs, however, also play an increasingly important role in all social relationships, ranging from political participation to connecting local communities, friends and the family. Second, in the global and culturally diversified world, ICTs are also increasingly important for access to cultural resources and expression. Third, the definition also replaces traditional technology-focused characterizations of digital divide, noting that lack of technology, *per se*, is not always a problem. It is clear that technology remains inert and useless without knowledge and capabilities to use them, and when they are embedded in social without necessary human skills and competences. Technologies become real when they are combined practices.

Measurement of Gap[6] Elements for 'Digital Divide'

Element 1	Element 2
A gap in access to use ICT: Measured by the number and spread of ICTs (first order digital divide).	*A gap in the ability to use ICTs:* Measured by skills base and presence of numerous complementary assets (second order digital divide).
Element 3	**Element 4**
A gap in actual use: Measured by the telecommunications for various purpose, the number and time of online users, the number of internet hosts and the level of e-commerce, e-business and e-governance.	*A gap in the impact of use:* Measured by financial and economic returns.

Based on these four elements, many international organizations have defined development policies aimed to reduce the digital divide. However, in spite of the evolution in the concept, these principally emphasize development of a technological infrastructure. National investments and policies for the reduction of the digital divide continue to principally target connectivity development.

2. Digital Divide in India: The State of Art

ICT infrastructure is the backbone of modern society. It is the biggest enabler of change and process reforms with minimum resistance. Various studies (Lawrence and Lee, 1999; Leizero, 2000; Pastor *et al.*, 2004; Norris, 2000; Sumanjeet, 2006) revealed that good governance is not possible without the ICTs. The information and communication technologies are being increasingly used by the governments to deliver its services at the locations convenient to the citizens. Further, it brings effectiveness, efficiency and transparency in the system. Therefore, the governments around the world are busy in developing the ICTs infrastructure. India is one of the countries where telecommunication development activities have gained momentum in the past decade or so[7]. Efforts have been made from both governmental and non-governmental platforms to enhance the telecomunication infrastructure. The idea is to help modern telecommunication technologies to serve all segments of India's cultural diverse society, and to transform it into a country of technologically savvies.

Various studies (Huberman, 2001; DiMaggio *et al.*, 2001; Guillen, 2006; Servon, 2002) showed that the problem of digital divide is more visible in the developing countries like India. The gap of digital divide is significant between the rural and urban India (Dasgupta *et al.*, 2002;, Nath, 2001; Singh, 2007; Mahajan, 2003; Dutta, 2003). The major metropolises are at par with some of the developed countries, but rural areas in states like eastern Bihar and Orissa are worse off than several of the least developed countries. The problem can be more understood by studying it under three sub-sections: Teledensity divide, mobile phone divide and Internet divide between rural and urban India.

2.1 Teledensity[8] Divide

Teledensity is low throughout the countries of the South Asia. Pakistan has highest teledensity in Asia and Bangladesh has lowest. India is by far the largest South Asian country, in terms of population, economy and telecommunication network. However, there are huge disparities exist within the country, and this is evident in the uneven distribution of telecommunication access. The recent extraordinary growth in telecommunication connections in India[9], which topped 15 million per month in January 2009, has understandably grabbed the headlines. These huge numbers, however, disguise a disturbing reality, which is the enormous variation within India. Many of the less developed states have state-wise average penetration rates of below 20 per cent, including Uttranchal (10.37), Bihar (12.13), Assam (13.67), West Bengal (13.78), Orissa (14.28), Uttar Pradesh (15.58), Andaman & Nicobar Islands (17.94) and Madhya Pradesh (19.54). Teledensity in some states like Jharkhand (3.49), Chhattisgarh (4.18) and North East-II (8.71) are less than 10 per cent (Table 2).

Table 2 : State-Wise Urban and Rural Teledensity in India (In %)
(As on 29/02/2009)

Circle/States	Teledensity		
	Rural	Urban	Overall
Andaman & Nicobar Islands	13.89	24.77	17.94
Andhra Pradesh	10.15	72.44	27.34
Assam	3.85	72.46	13.67
Bihar	3.1	89.13	12.13
Chhattisgarh	1.31	14.27	4.18
Gujarat	15.57	58.15	32.34

contd....

Haryana	16.06	57.67	29.45
Himachal Pradesh	30.5	118.64	39.9
Jammu & Kashmir	7.35	59.4	20.99
Jharkhand	1.14	11.38	3.49
Karnataka	11.14	73.38	33.68
Kerala	25.5	97.46	43.98
Madhya Pradesh	4.96	58.34	19.54
Maharashtra (-) Mumbai	11.66	55.16	26.18
North East -I	6.65	89.45	26.32
North East - II	3.06	27.47	8.71
Orissa	6.66	53.64	14.28
Punjab	24.83	80.63	46.85
Rajasthan	12.07	57.98	22.98
Tamilnadu (-) Chennai	15.37	56.8	34.01
Uttaranchal	5.03	24.43	10.37
Uttar Pradesh	6	50.36	15.58
West Bengal (-) Kolkata	7.02	55.43	13.78
Kolkatta	#	57.43	62.3
Chennai	#	100.13	101.62
Delhi	#	107.96	107.96
Mumbai	#	81.41	81.41
India	**9.03**	**64.48**	**25.34**

Source: TRAI, 2009

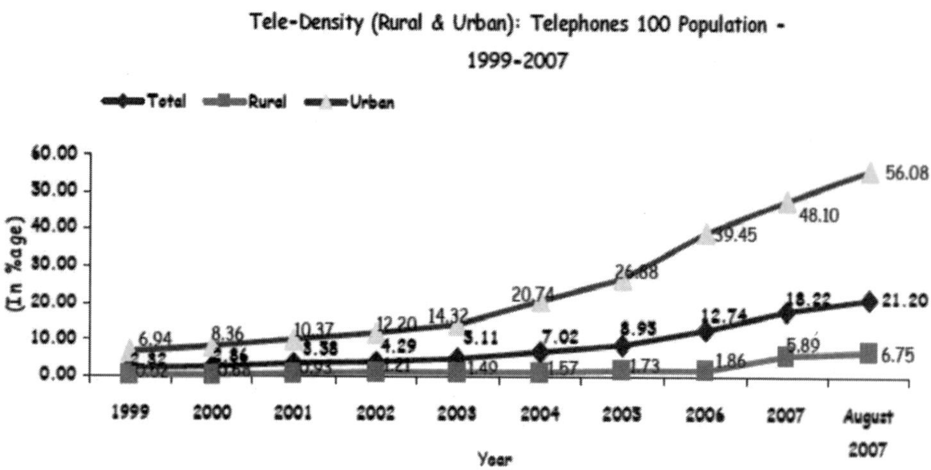

Graph 3 : Urban & Rural Teledensity (Per 100 Inhabitants 1997-2008)
Source: ICRA

More perilous, however, is the inequality between rural and urban India (Graph 3). Despite several policy initiatives to promote rural penetration, growth in teledensity continues to be skewed in favour of urban India.

2.2 Internet Divide

Internet came to India in the mid 1990'sThe Videsh Sanchar Nigam Limited (VSNL) introduced Internet in India *via* dial up in the 6 cities on August 14, 1995. At that time, there was limited Internet access only in a few major cities, all in the hands of the government. VSNL, the agency responsible for Internet activities, and the DoT (Department of Telecommunications) provided an agonizingly erratic connectivity, with miserly bandwidth and that too for few phone lines. Connection rates ran as low as five per cent (for every 20 dialups you might get connected once) and users were frequently cut off. And the rates for this pathetic level of service were among the highest in the world. Domestic users paid about $2 per hour, and lease lines, for the few companies that could afford them, ranged over $2000 per month for a 64 Kpbs line. By the end of 1998, after three years of government monopoly, there were barely 150,000 Internet connections in India. But, today the government monopoly is largely over. Many small to large Internet Service Providers have set up shop, triggering a price war and an improvement of service. Users are now estimated at over 49 million and, with a growth predicted to reach 1 billion in the next five years (Graph 4 and 5).

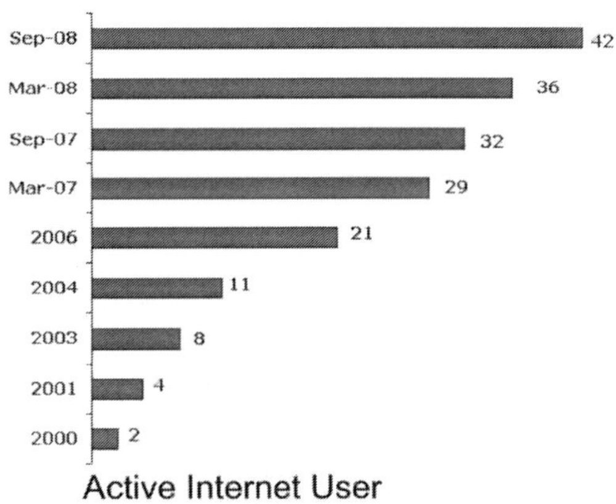

Graph4: Active[10]InternetUsersinIndia
Note: Figures in Millions
Source: IAMAI

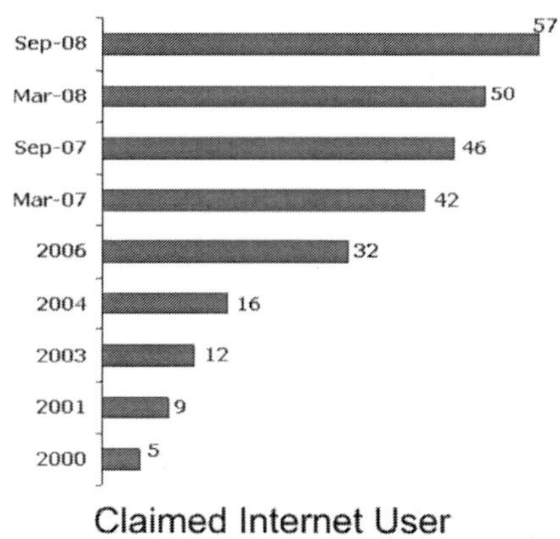

Graph5: ClaimedInternetUsers inIndia
Note: Figures in Millions *Source:* IAMAI

Small Internet kiosks have set up even in small towns, and the governments, both State and Central are pushing for growth in the Internet sector. Internet is the new buzzword. But data shows that Internet seems to be on a move in the metros of India and not the buzzword for small towns too. Today almost 70 per cent of the total Internet users in India are from the top seven cities-Mumbai, Delhi, Bangalore, Hyderabad, Chennai, Kolkata and

Pune. Only 30per cent access is happening from all other cities put together. Added to this disparity among the rural and urban Internet users is also very high (Table 3).

Table 3 : Urban-Rural Internet Users in India (In Millions)

Internet User-ship in India (Rural-Urban)	2005	2006	2007	2008
Urban-Internet using individuals (Regular)	17.63	21.95	25.17	30.03
Urban-Internet using individuals (Occasional)	5.20	1.65*	5.15	10.31
Urban-Internet using individuals (Total)	**22.83**	**23.60**	**30.32**	**40.34**
Rural-Internet using individuals (Regular)				5.06
Rural-Internet using individuals (Occasional)				4.00
Rural-Internet using individuals (Total)				9.06
All India-Internet using individuals (Regular)				35.09
All India-Internet using individuals (Occasional)				14.34
All India-Internet using individuals (Total)				49.40
All Urban internet users as % of Indian Population	7	7	9	12
All Rural internet users as % of Indian Population				1.2
All India internet users as % of Indian Population				4.5

Note: *May have been underestimated because of relatively less deeper coverage of SEC 'D' and 'E'in the land survey.
Source: Juxtconsult India Online, 2008

In rural India only 1.2 per cent people have Internet access whereas it is 12 per cent in the urban India. Urban users continue to dominate Internet use contributing to 40.34 million of the 49.40 million odd users. 30.3 million urban people are using Internet on regular basis where only 5 millions are in rural areas. In September, 2007, the number of active Internet users in urban India was 30 million. Disparities within the Indian states are huge. States like Delhi, Maharashra and Tamil Nadu have highest number of Internet subscribers followed by Kerala, Karnataka and West Bengal. Assam, North East-II, Orrisa, Andaman and Nicobar and Uttranchal have lowest Internet subscriber States of India.

2.3 Mobile Divide

Far from being a lifestyle product, Mobiles have now become a necessity. India has seen a huge spur in mobiles in the past five years and it has penetrated even to the rural areas of India to a good extent. With entrants of CDMA[11] (Code Division Multiple Access) like Reliance communications and Tata Indicom the call rates have been reduced and usability has been increased. The launch of the revolutionary offer of a mobile phone along with a connection by Reliance for Rs.500 boosted the mobile sales to sky soaring limits. Over and above with the STD rates and the call rates reducing day-by-day and new competitors are entering, the mobile market has flung open high growth prospects.

Globally in terms of mobile subscriptions, India is the world's second largest wireless market after China. At the end of June 2008, the total wireless subscribers (GSM[12], CDMA & WLL (F)) base was 286.86 million (Graph 6).

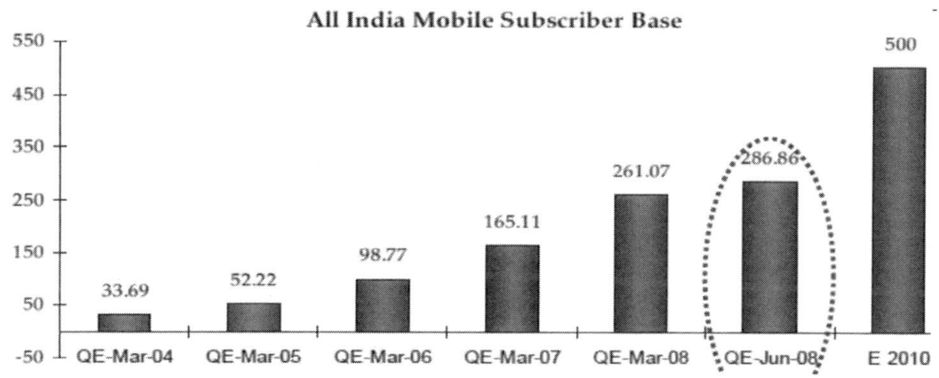

Graph 6 : All India Mobile Subscriber Base

Note : QE stands for Quarter End; E stands for Estimated; Figures in Millions.

Source : TRAI

A total of 8.94 million wireless subscribers were added during the month of June 2008 as against 8.62 million wireless subscribers added during the month of May 2008. As a result the overall tele-density rose to 28.33 per cent by end of June 2008 as against 27.59 per cent in May 2008. This growth of the sector can be clearly attributed to the favorable and improved regulatory structure, declining handset prices and innovative pre paid tariff structure. It is expected that by the end of the year 2010, India will have 500 million mobile phone subscribers and the break up could be: out of the 500 million subscribers, 60 million would have video capabilities, 10 million music, 200 million radios, 250 million camera capabilities, while 250 million will have net connections.

Thus, India has become the fastest growing mobile market in the world. The only country with more mobile phones than India is China. But, there is huge divide between mobile phone users in urban and rural areas. Still, rural mobile penetration is pretty low-just 4.92 per cent (Table 4), though it has touched double digits in some prosperous States like Punjab, Kerala and even Himachal Pradesh. However, overall rural penetration remains very low compare to the 43.88 per cent mobile density in urban areas.

Table 4 : State Wise Mobile Users (Urban and Rural) in India

Top 15 States in Terms of Rural Mobility	Rural Mobile Connections (in Million)	Total Rural Population (in Million)	Penetration in Rural Areas (in %)
Punjab	2.24	10.83	20.69
Himachal Pradesh	1.0	5.85	17.09
Kerala	2.66	25.03	10.63
Haryana	1.66	16.27	10.20
Gujarat	3.20	34.42	9.31
Tamil Nadu	2.8	32.86	8.52
Andaman & Nicobar Islands	0.02	0.27	8.27
Rajasthan	3.27	48.66	6.72
Maharastra	3.79	59.67	6.35
Karnataka	2.25	36.56	6.15
Andhra Pradesh	3.27	59.27	5.52
Jammu & Kashmir	0.43	8.24	5.21
West Bengal	3.01	62.48	4.81
Orissa	1.28	33.06	3.88
Uttar Pradesh	4.60	147.00	3.13
Total	**35.51**	**580.47**	**6.12**
All India (Total)	**39.46**	**802.00**	**4.92**

Source: Voice and Data, 2008

But, it is expected that the gap will narrow down in the coming years. Mobile industry players are eyeing rural India as their new area of opportunity. Cellular service providers seem to be answering the call of the world as they are entering the so far ignored rural market. Although a huge market in the urban segment remains tapped, most of the cellular operators have turned towards Rural India to broaden their base and reach. So the real growth is expected from this geography in near future. The low population density in rural areas has necessitated more towers of higher altitudes raising the costs further. Rural India, experts say, that the next phase of mobile phone growth would come from the hinterland.

3. Determinants of Digital Divide in India

From the above discussion, it is clear that there is huge gap of digital divide between rural and urban India. However, this gap varies from State-to-State. In some States like North East, Uttarakhand, Bihar, Jharkhand, Orrisa and Andaman and Nicobar Island the gap of digital divide is really very significant, in some other states it is narrow like Punjab, Maharashtra and Kerala. Added to this, the gap also varies from technology-to-technology. Some States are not able to adopt even one technology but others have adopted very efficiently. In some states overall adoption of technology is high, but, adoption rate in rural areas is very low. In some cities like Delhi, Mumbai, Kolkata, Bangaluru, Hyderabad, Noida *etc*, ICTs adoption rate is very high, whereas, in some other cities like, Patna, Lucknow, Ahmedabad (even these are the capital of States) *etc*, adoption rate is very low. The undeniable fact is that there is noticeably problem of digital divide in India. It is a massive problem and a very complex problem too. It is not just about people who have access and those who do not; it is not just about haves and have-nots. It is about people becoming knowers and know-nots; and doers and do-nots; those who can communicate with the rest of the world and those who can not. In this light it is also important to find out some of the important reasons behind the digital divide in India. There are many reasons, which can be held responsible for digital divide in India. Compared to many developing and developed countries, India's capacity to bridge the digital divide is very poor. Table 5 clearly indicates India's ICTs adoption capacity at international level.

Table 5 : India's ICTs Adoption Capacity among Different Countries (Ranking)

Countries	Internet Users*	Broadband Subscribers**	Computers In Use*	Communication Technology	IT Skills	Computers Per Capita
India	55	51	12	31	10	55
China	52	41	2	47	54	51
Singapore	13	21	41	1	2	19
Hong Kong	–	–	29	3	–	–
Japan	7	14	3	17	22	21
Korea	12	3	7	20	18	18
Malaysia	29	42	21	30	23	32
USA	10	17	1	12	4	2
UK	14	12	5	24	25	10
Israel	15	15	31	15	3	17
Germany	19	20	4	6	24	13
France		13	6	22	–	20
Denmark	28	1	–	–	–	–
Sweden	1	–	–	–	1	1
Finland	–	–	–	–	13	–

Note: *Rank for 2007; **2006; Internet Users and Broadband Subscribers per 1000 inhabitants
Source: Voice and Data; IMD World Competitiveness Year book, Various Issues

It is clear with the table 5 that where India stand at international level as far as various capabilities to measure the gap of digital divide is concerned. In case of broadband subscriber, India has lowest rank of 51 among the selected countries, same in the case of computer per capita and Internet users. As India is a multicultural, multi-language and multi-religion country with complex socio-economic conditions, there are six main difficulties in serving rural communities, each one of which has appeared insurmountable: poverty, unemployment, age and education. A Nation Online Survey (NTIA, 2002) found computer and Internet use correlate with poverty[13] (family income), employment status, and educational attainment. Even age[14] is also a major factor (Lenhart, 2000). Rural India has more than 70 per cent of Indian population. 75 per cent of the poor are in rural areas, most of them are daily wagers, self-employed householders and landless labourers.

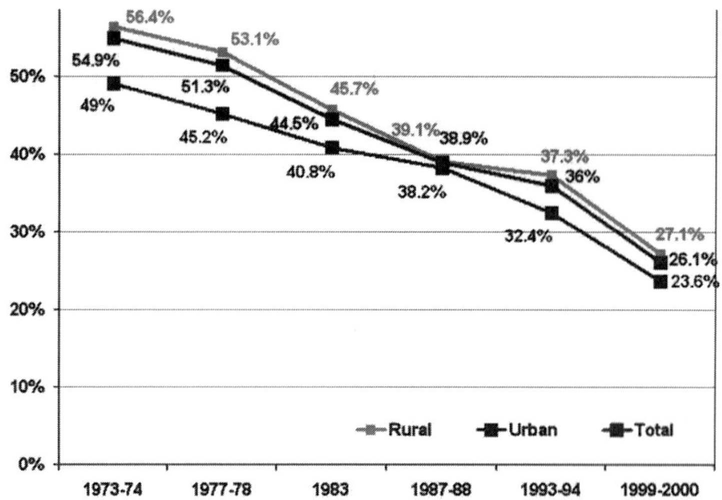

Graph 7 : People Below Poverty Line in Rural-Urban India
Note: Per cent of population living below poverty line, which is 356.35 rupees or around $7 a month in rural areas.

The Planning Commission of India uses its own criteria and has estimated that 27.5 per cent of the population was living below the poverty line in 2004–2005, down from 51.3 per cent in 1977–1978, and 36 per cent in 1993-1994. Again unemployment rate is high in the rural India than the urban. Unemployment rate in agriculture sector is 62 per cent (Planning Commission, 2007) whereas, it is only 16 per cent in manufacturing sector, 10 per cent in the services sector and 12 per cent in other job sectors.

The better educated are statistically more likely to have and use connected PCs. In particular those with college degrees or higher educated, are ten times more likely to have access. Literacy rate is high in the urban India (79.9 per cent) whereas, it is only 58.7 per cent in the rural India. Overall literacy rate in India is 64.8 per cent. There is also a huge difference in male (75.3 per cent) and female literacy rate (53.7 per cent).

Another important reason of digital divide in India is knowledge divide[15]. Knowledge divide is directly related with digital divide. More educated people[16] with computer knowledge and English language are able to access new technologies. Rural India had 368 million literate people out of whom only 63 million were found to be English-speaking as on March 2008. Given the low levels of literacy in rural India and very low levels of English speaking population and computer savvy population[17], IMAI , 2008 made a clear case of content and applications in local languages in order to ensure higher and faster adoption of internet in rural area. Internet use is primarily associated with a large section of the English-knowing urban population-though they account for only 31 per cent of the total urban population of 250 million-as many, as 84 per cent of them are PC literate. Internet penetration extended to only 0.6 per cent of the population in rural areas, with the number of active Internet users estimated at 3.3 million.

Electricity is the most basic requirement for using ICTs and many studies established the relationship between the level of electrification and digital divide. Rural India has low electricity coverage (Table 6). Almost 10 per cent villages of India have no electricity. Some areas may get agricultural power-two hours in the morning and evening-but even this is exceptional. Added to this the cost of electricity is very high. In this situation, one can not even think about using computers and Internet.

Table 6 : Reported Status of Rural Electrification (As on March 2004)

States	Electrified Villages (%)	Electrified Households (%)	States	Electrified Villages (%)	Electrified Households (%)
Punjab	100.00	91.90	Madhya Pradesh	97.43	70.00
Haryana	100.00	82.90	Rajasthan	98.38	54.70
Gujarat	100.00	80.40	Chhattisgarh	94.00	53.10
Maharashtra	100.00	77.50	West Bengal	83.63	37.50
Tamil Nadu	100.00	78.20	Orrisa	80.15	26.90
Kerala	100.00	70.20	North-East	75.32	33.20
Andhra Pradesh	100.00	67.30	Uttar Pradesh	58.73	31.90
Himachal Pradesh	99.38	94.80	Bihar	50.00	10.30
Karnataka	98.91	78.50	Jharkhand	26.00	24.30

Source: Planning Commission, 2005

The digital divide is not simply an issue of access, but also of obstacles to use ICTs. Various studies (Kennedy *et al.*, 2003; Winter and Huff, 1996; Spender, 1997; MacKenzie and Wajcman, 1985) revealed that even when women and men have equal access to the internet either through home, work or school, they may not have the opportunity to access the Internet or engage in a wide variety of uses. Women have been online less than men (Kennedy *et al*, 2003). They have been online for fewer months and when they do go online, they spend less time. The gender issue is highly relevant in the developing countries like India. Women have less access than men in India due to various social and personal factors.

Added to these, the growing population, insufficient funds, affordability[18], and delays in implementation of government policies and programmes have been some of the challenges that have lead to unequal development in the society, which is responsible for digital divide.

4. Policy for Addressing the Challenges in Bridging the Digital Divide

Despite India taking significant steps towards acquiring competence in information and technology, the country is increasingly getting divided between people who have access to technology and those who do not. India has around half a million software developers and is second only to the US, but 300,000 Indian villages do not have a phone connection, 70 per cent of the Indian population have no access to any form of technology, 26 per cent of the population lies below the poverty line and 35 per cent illiterate. There are only 5 PCs per 1000 people, 9 mobile lines per 1000 and 37 fixed lines per 1000 people in India, which is extremely low when compared to China. The challenge in front of the Indian government is to bring down this digital divide and ensure development and adoption of technology. The government has made encouraging steps to bridge the gap of digital divide in India. Most of the initiatives[19] are directed towards bridging the gap of rural and urban digital divide. The Indian government has passed Information Technology Act, 2000 to make e-commerce and e-governance a success story in India. Recognizing the potential of ubiquitous Broadband service in growth of GDP and enhancement in quality of life through societal applications including tele-education, tele-medicine, e-governance, entertainment as well as employment generation by way of high-speed access to information and web based communication. The government has announced Broadband Policy in October 2004. In the field of international telephony, India had agreed under the GATS to review its opening up in 2004. In an effort to strike a balance between a purely market driven system of allocation which would tend to concentrate on main cities and on the largest and wealthiest customers and a social consideration reducing the rural-urban disparity and improve teledensity in rural areas, the Government of India had designed a Universal Service Obligation (USO) fund[20]. The USO as a tax on service providers to facilitate redistribution of resources to 'unprofitable' rural areas. Complementing this is the Access Deficit Charge[21] (ADC) regime put in place by the Telecom

Regulatory Authority of India to help service-providers to bridge the gap between cost of supply and provision of access, especially in rural areas. Since the costs of expansion and maintenance of rural networks are high compared with dense urban areas, the USO funding was extended for mobile services and rural infrastructure in 2006. This was expected to provide incentives for private operators to break into the rural market. Recently, the regulator Telecom Regulatory Authority of India[22] (TRAI), in its recommendations on growth of telecom services in rural India, had prescribed infrastructure sharing as the solution for improving rural tele-density. In 2007, the Telecom Regulatory Authority of India (TRAI) further lowered the Access Deficit Charge (ADC) rates to push mobile growth in rural areas. With all these development in place, the ministry of telecommunication went ahead with a bid call for subsidy for rolling out the mobile networks in 2.5 lakh villages. Recently Indian government has decided to push WiMAX technology[23] especially for broadband connectivity in rural areas.

The Government has announced the guidelines for Mobile Number Portability (MNP) Service Licence in the country on 1st August 2008 and has issued a separate Licence for MNP service w.e.f. 20.03.2009. The government has in a pioneering decision, decided to auction 3G & BWA spectrum[24]. The broad policy guidelines for 3G & BWA have already been issued on 1st August 2008 and allotment of spectrum has been planned through simultaneously ascending e-auction process by a specialized agency. Apart from the 123.51 million fixed and WLL connections on March 2009 provided in the rural areas, 57167 uncovered VPTs have been provided as on March 2009. Thus, 85 per cent of the villages in India have been covered by the VPTs. More than 3 lakh PCOs are also providing community access in the rural areas. Further, Mobile Gramin Sanchar Sewak Scheme (GSS) - a mobile Public Call Office (PCO) service is provided at the doorstep of villagers. At present, 2772 GSSs are covering 12043 villages. Also, to provide Internet service, Sanchar Dhabas (Internet Kiosks) have been provided in more than 3500 Block Headquarters out of the total 6337 Blocks in the country. The target of 80 million rural connections by 2010 has already met during year 2008 itself. USO subsidy support scheme is also being utilized for sharing wireless infrastructure in rural areas with about 19,000 towers by 2010.

From the above discussion it is revealed that most of the government initiatives is directed towards developing the sound telecom infrastructure in the rural area. But, this is only aspect of addressing the problem of digital divide in India. Added to this, government should adopt the following policy to bridge the gap of digital divide in India:

- First, as our earlier analysis suggests that relationship between education and ICT is critical. International evidence suggest (Pluss, 2004;

Rheingold, 2000; Smith and Peter, 1998) that education is strong complement to use of technologies like Internet and that the relevant education levels are secondary and tertiary levels as they are expected to upgrade the national capacity for adaptation and innovation[25]. Like many other developing countries, the main emphasis of Indian government is to boost primary education that yields the higher rate of social returns. But, to bridge the gap of digital divide government should introduce some innovative policy measures to encourage students to go for further education in the rural areas. The expenditure, which Indian government is doing for rural education, is far below the level of satisfaction (Table 7).

Table 7 : Expenditure on Education (Rural vs. Urban)

Components of Education	**Monthly Household Expenditure (In Rs. At 2004-05 Prices)**			
	Rural		**Urban**	
	1993-94	**2004-05**	**1993-94**	**2004-05**
Books and Journals	8.9	12.6	21.3	21.7
Newspapers & Periodicals	1.7	2.9	16.7	20.5
Library Charges	0.2	0.1	0.6	0.1
Stationary Articles	7.2	14.9	15.5	21.8
Tuition Fees	7.5	29.1	67.7	118.3
Private Tutor	5.4	10.3	25.8	41.7
Others	6.1	2.8	13.7	5.5
Total Education	**37.0**	**72.7**	**161.3**	**227.7**

Source: Consumption Expenditure Data from 50th and 61st Rounds of NSS

Table 7 indicates that there is huge difference between expenditure on education for rural India and urban India. Prime Minister of India also realized the need to increase the level of investment in education in rural India. Nevertheless, one does not need to rely exclusively on the government for promoting secondary and tertiary education. Many individuals who would like to take advantage of ICT opportunities are economically well off and may not need government financial assistance. For others, improving the availability of education loans from the financial institutions can be a major hold in financing their educational expenses. In addition to formal education, there is a need to promote technical education in rural India and among the women of India. Information technologies should be introduced when (and only when) they constitute the most effective available way of meeting basic human needs and fulfilling fundamental human rights. ICTs can have a positive role in development. But, ICTs are neither a panacea nor necessarily the first line of attack in combating poverty, misery, and injustice. The utility of ICTs must always be judged against the role they can play in meeting core human needs.

- There is need to promote technologies which are best suited for the rural India. For example, to bridge the gap of digital divide in its real sense there is need to increase PC penetration. The reason is very simple; mobile cannot do everything a PC can. But, mobile are cheaper, more portal and their extended battery life is suited to regions where access to electricity is lacking or non-existent. The infrastructure needed to connect wireless devices to the Internet is easier and less expensive to build. There are also no learning curve, no literacy barrier and no technical-support challenges to overcome. There are no costly and burdensome applications to load, maintain and update. Thus, mobile is best suited for the rural people. Therefore, not only government but also private players should encourage mobile penetration in India.
- The main barrier in Indian rural society is the fact that people do not associate the benefits of the Internet and other communication technologies with their personal needs, believing that "computers are not for them". As a result they behave very passively towards the ICTs. In order to bring 'passive people' to the Internet and ICTs, a broader understanding of the 'ICTs for everyone' notion has to be promoted, the motivational barriers being the main barrier to passive people. To remove this barrier, media productions such as T.V. serials *etc.* enjoying great popularity among 'passive people' to be exploited. Only after the broad social barriers have been removed, more specific services will become more attractive to the current non-users.
- There is need to develop innovative strategies to address constraints the world's women face in their access to and use of ICTs. For women, content is directly linked to use. If women are to be able to make use of the Internet for income-generation, education or advocacy, there must be more relevant content. This includes both substance and languages. Women are producers as well as consumers of information and knowledge and this is an important area of support.
- Like many other developing countries, in India as well great attention has been paid to ICT training in the recent years. Government schools across the country are making use of the computer literacy programmes funded by giant transnational companies like Microsoft Corporation and IBM. In an ironical twist, it is the world's largest technology companies who have taken it upon themselves to help bridge the "digital divide" in India. Microsoft Corp, the world's largest software maker,

has signed MoUs with nine governments around the country, and has committed $20 million to promote computer literacy among disadvantaged kids in rural areas. The Confederation of Indian Industry (CII), and the Global Leaders of Tomorrow (GLT) of the World Economic Forum (WEF), have launched an initiative called 'Shiksha India' in December 2001 during the India Economic Summit primarily to bridge the digital divide and promote better quality education in Indian Schools. There are many more examples. It is expected that these project will undoubtedly play a great role in overcoming the lack of interest and will help in bridging digital divide. However, the problem that is likely to arise with the elderly population is that the newly emerged interest may wane if one sees barriers rapidly appear, particularly if the language problems inhibits coping with the user interfaces of computers and one's relatively poorer ability to memorize hampers learning the necessary operations. Psychological obstacles may be the unwillingness "to reveal one's lack of skills". Therefore, the effectiveness of the training programmes will have to be evaluated (to make sure that the trainees have become convinced Internet users) and improvements made wherever necessary.

- Promotion of telecommunication infrastructure in the rural India is the most important condition for bridging the rural-urban digital divide and Indian government can play a very significant role in creating the IT infrastructure in the rural India. A special expenditure should be marked for bridging the digital divide among the regions. It is clear from international experience (Rowan *et al.*, 1998; Wahid *et al.*, 2006; Mwesige, 2004; Mosse and Sahay, 2003) that bridging the digital divide is impossible without additional expenditure from the national budget. India should learn from China. In China, government has not only invested heavily in the creation of IT infrastructure, but also in the universal telecommunication access in rural and remote areas. To bridge the widening Internet connections gap between rural and urban areas, the Chinese government has launched 'every village has a phone, 2004' and 'Gold Farm Engineering, 1994' projects which promoted telephone access and Internet applications in the rural areas.

- Linguistic issues are of major policy importance in India, given the dominance of English language in software and Internet. The desire to promote cultural diversity is one reason behind interest in linguistic issues, but so is the avoidance of social exclusion among non-English speaking population of India. Hence, these matters are critical to various aspects of digital divide. How can they be addressed? One roundtable example came from New Zealand, where language especially Maori-is a prominent feature of political, cultural and educational life. The New Zealand web portal for teachers is bi-lingual, with the Maori contents presented before the English version. As far as provision for early childhood learning is concerned, there is Maori ICT network across different centers, characterized by strong focus on family and inter-generational learning. It would be useful to compile similar examples from other communities and countries in policy framework.

Further, there is need for greater resources and investment by the private sector; as funds from governments and developed countries in implementing the ICT projects have been on the decline due to budget constraints and global economic slowdown. Last but not the least, with the effective deployment of ICT applications in the core areas of education, healthcare, and connectivity for redressing the grievances of the people in the countryside, the digital divide can be narrowed as wireless and satellite links have made them economical and affordable.

Conclusion

The unprecedented development of ICT, its applications, and the emergence of a global information society are changing the way people live, learn, work and interact. Enhanced access to knowledge is rapidly becoming a potent tool for empowering the people and communities in their quest for new opportunities, dignity and a better life. The divide between technology's haves and have-nots threatens to exacerbate the gaps between the rich and poor, within and among countries. Unfortunately, in India all people have access to the Internet and ICT, and an amazingly large number of people especially from the rural areas does not have abilities to use the ICTs in a proper way and, therefore can not draw the advantages from its usage. The issues of digital divide is posing a herculean task before the Government of India to provide the maximum benefits to the stake holders. However, some IT experts believe that digtial divide is vanishing myth from India. But, the present paper cleary highlights the problems of digital divide in rural India. Added to this, the paper revealed that the problems of digital divide also exist within and among the varioius States of India. While some people are rich and have many resources, others do not. The educational system of India also has been slow to achieve the set target framed by various commissions and committees and schemes launched from time to time. The lack of sound ICT strategies and policies in India is the main cause of these troubles. Added to this, inadequate Internet and telephone connectivity to India's rural areas, where more than 70 per cent of India's population lives, is a key challenge for a number of government agencies, NGOs, and multilateral aid agencies. The corporate sector too is discovering that bridging this digital divide could translate into new market opportunities. Therefore, obstacles such as illiteracy, lack of skills, and infrastructures in rural areas must be tackled if India is to diminish the gap of digital divide. At the government front, it should put thrust towards: connectivity provision, content creation, capacity augmentation, core technologies creation and exploitation, cost reduction, competence building, community participation and commitment to the deprived and disadvantaged which would definitely help in bridging digital divide. It is evident that many projects have failed to bridge the gap between rural and urban India. In order to increase the benefits of its projects for women, who constitute a huge mass of people; efforts have to be made by the Government to bridge the "gender digital divide". There is also need to strengthen the capabilities of local communities and organizations to create, communicate and exchange their knowledge, through the use of ICTs.

As far as future of digital divide is concerned in India, it is really very difficult to predict it. But, it is true that the present situation in India is not alarming (if not highly satisfactory). The gap of digital divide is getting narrower. It is expected the government policies and public private partnership will help in bridging the digital divide. But, it is not possible to completely bridge the gap of digital divide in India, as gender, age, culture, language, sex, *etc.* are all fundamental components that often affect daily activities and experiences including the virtual world.

Acknowledgement

While bearing full responsibility for any mistakes, I wish to thank Prof. L. N. Dahiya; Prof. S. D. Vashistha and Richard Nyangosi for reading the earlier versions of this chapter and making a number of helpful comments and constructive criticisms. I also thank the comments and suggestions of anonymous referees, which were also of great importance. However, I am solely responsible for all the remaining errors and inadequacy.

End Notes

1 Information and communication technologies is an umbrella term which is currently used to refer to a wide ranges of services (telephony, FAX, Internet etc.), applications (such as distance education and management information systems etc.), and technologies (anything from old technologies such as Television to new technologies such a cellular phones), using various types of equipment and software, often running over telecom networks.

2 The Economic and Social Commission for Asia and Pacific (ESCAP) of the United Nations is the regional development arm of the United Nations for the Asia-Pacific region. With a membership of 62 Governments, 58 of which are in the region, and a geographical scope that stretches from Turkey in the west to the Pacific island nation of Kiribati in the east, and from the Russian Federation in the north to New Zealand in the south, ESCAP is the most comprehensive of the United Nations five regional commissions.

3 Just about a year prior to the turn of the century, the concept of the digital divide was dramatically catapulted onto the world stage with disturbingly striking statements such as: "80 per cent of the world's population have never made a phone call;" "There are more telephones in Manhattan than in all of rural Asia;" and "There are more Internet accounts in London than in all of Africa." About almost the same time that the dotcom frenzy had reached its height, concern was also being expressed in different circles that the information revolution was in fact bypassing the vast majority of people inhabiting planet Earth.

4 According to Compaine " the digital divide refers to the perceived gap between those who have access to the latest information technologies and those who do not".

5 Such gaps have narrowed and in some cases, even reversed over time, but other disparities have risen. This suggests that: the digital divide is a dynamic concept, which evolves over time; older technologies tend to be more evenly diffused than newer ones. For example, TV sets are more evenly distributed than 3G mobile phones; there is not a single divide, but multiple divides: for instance, within countries, between men and women, between the young and elderly, different regions *etc.*, and the main factor underlying these divides is differences in wealth, between countries and between individuals. While disparities in wealth continue to exist, the digital divide will persist.

6 As one can deduce, the concept of the digital divide has changed over time. In the beginning, it basically referred to connectivity problems (gap in access to use of ICTs). Later, it began to introduce the concern for the development of capacities and skills required to use ICTs (capacity-building and education), and finally, there is also reference to the use of integrated resources in the technology. Thus, the concept of the digital divide basically focuses on three areas: Infrastructure, capacity building and focus on resource usage.

7 After 1995, the government set up TRAI (Telecom Regulatory Authority of India) which reduced the interference of Government in deciding tariffs and policy making. The DoT opposed this. The political powers changed in 1999 and the new government under the leadership of A.B Vajpayee was more pro-reforms and introduced better liberalization policies. They split DoT in to two- one policy maker and the other service provider (DTS) which was later renamed as BSNL. The proposal of raising the stake of foreign investors from 49% to 74% was rejected by the opposite political party and leftist thinkers. Domestic business groups wanted the government to privatize VSNL. Finally in April 2002, the government decided to cut its stake of 53% to 26% in VSNL and to throw it open for sale to private enterprises. TATA finally took 25% stake in VSNL. This was a gateway to many foreign investors to get entry into the Indian Telecom Markets. After March 2000, the government became more liberal in making policies and issuing licenses to private operators. The government further reduced license fees for cellular service providers and increased the allowable stake to 74% for foreign companies. Because of all these factors, the service fees finally reduced and the call costs were cut greatly enabling every common middle class family in India to afford a cell phone.

8 The number of landline telephones in use for every 100 individuals living within an area. A teledensity greater than 100 means there are more telephones than people.

9 Whether the country has adequate teledensity commensurate with its development can be best judged by reference to a table prepared by the United Nations some years ago in which the co-relation between the per capita income and teledensity was worked out. According to the table, for a per capita income of $1000 an annum, a teledensity of 3 per cent is considered adequate. So, India has adequate teledensity, as we have not reached the per capita GDP of $1000 (it is about US$ 400).

10 Active Internet users are those who have used the Internet at least once in the last one month-this is an internationally accepted benchmark for enumerating Internet users.

11 Short for Code-Division Multiple Access, a digital cellular technology that uses spread-spectrum techniques. Unlike competing systems, such as GSM, that use TDMA, CDMA does not assign a specific frequency to each user. Instead, every channel uses the full available spectrum. Individual conversations are encoded with a pseudo-random digital sequence. CDMA consistently provides better capacity for voice and data communications than other commercial mobile technologies, allowing more subscribers to connect at any given time, and it is the common platform on which 3G technologies are built.

12 Short for Global System for Mobile Communications, one of the leading digital cellular systems. GSM uses narrowband TDMA, which allows eight simultaneous, calls on the same radio frequency. GSM was first introduced in 1991. As of the end of 1997, GSM service was available in more than 100 countries and has become the de facto standard in Europe and Asia.

13 Although Indian economy has grown steadily over the last two decades, its growth has been uneven when comparing different social groups, economic groups, geographic regions, and rural and urban areas. Between 1999 and 2008, the annualized growth rates for Gujarat (8.8%), Haryana (8.7%), or Delhi (7.4%) were much higher than for Bihar (5.1%), Uttar Pradesh (4.4%), or Madhya Pradesh (3.5%).Poverty rates in rural Orissa (43%) and rural Bihar (41%) are among the world's most extreme. 80% of Indians live on less then half a dollar a day.

14 The Oxford's research finds that age is the primary factor determining Internet usage patterns in developed Western countries, not money. Almost 98 per cent of students in Britain regularly use the Internet, while only 22 per cent of British retirees' surf the Web. All youngsters, whether or not they are numerate or illiterate, appear capable to access the Internet. Professor Richard Rose, the lead researcher on the project, indicates that the idea of wealth-based divide is wrong. It is all about age and not so much about social class. Rose's observed pattern of Internet usage is similar across Western Europe. Russians under 30, for example, are 10 times more likely to surf the Web regularly than Russians over 60.

15 Usually the concept of digital divide is used to relate the technological aspect of knowledge gap.

16 Literacy rate (2001) in Kerala was 90.86 per cent against 47.00 per cent in Bihar. Life expectancy at birth (2001-2006) is 71.61 for males and 75 for females in Kerala. In Bihar, it is 65.66 for males and 64.79 for females. Infant mortality (per 1,000 live births, 2002 provisional data) was only 10 in Kerala against 61 in Bihar. Birth rate (per 1,000, provisional) is 16.9 in Kerala against 30.9 in Bihar. Death rate (per 1,000,

provisional) is 6.4 in Kerala against 7.9 in Bihar. The differences clearly indicate that literacy is the key to improvement in quality of life.

17 According to I-Cube 2008, a survey jointly undertaken by IMRB International and Internet and Mobile Association of India, there were 3.3 million active internet users in rural India as on March 2008.

18 "Unfortunately telecom networks are designed for people who can afford to pay around US$35 in monthly bills, and very few people in the rural areas can afford that," said Ashok Jhunjhunwalla, a professor at the prestigious Indian Institute of Technology (IIT).

19 For example Microsoft has announced a slew of initiatives to accelerate IT literacy and enable e-governance to bridge the digital divide in the country. The low-cost device, developed jointly by the Indian Institute of Technology (IIT), Chennai, and Indian Institute of Science (IISc), Bangalore, for the Human Resource Development (HRD) ministry, is aimed at bridging the digital divide and making access to computer literacy affordable to the masses.

20 The Universal Service Support Policy came into effect from 1st April, 2002. The guidelines for universal service support policy were issued by DoT and were placed on the DoT website www.dot.gov.in on 27th March 2002. Subsequently, the Indian Telegraph (Amendment) Act, 2003 giving statutory status to the Universal Service Obligation Fund (USOF) was passed in December 2003. The Fund is to be utilized exclusively for meeting the Universal Service Obligation by providing access to telegraph services to people in the rural and remote areas at affordable and reasonable prices. The USO Fund was established with the fundamental objective of providing access to 'basic' telegraph services. Subsequently, an Act has been passed on 29th December, 2006 as the Indian Telegraph (Amendment) Act 2006 to amend the Indian Telegraph Act, 1885 to enable provision of all types of telegraph services.

21 The Telecom Regulatory Authority of India (TRAI) has phased out Access Deficit Charges (ADC) on domestic calls and slashed ADC on international calls to 50 paise from the present Rs. 1, effective April.

22 TRAI's mission is to create and nurture conditions for growth of telecommunications in the country in a manner and at a pace, which will enable India to play a leading role in emerging global information society. One of the main objectives of TRAI is to provide a fair and transparent policy environment, which promotes a level playing field and facilitates fair competition. In pursuance of above objective TRAI has issued from time to time a large number of regulations, orders and directives to deal with issues coming before it and provided the required direction to the evolution of Indian telecom market from a Government owned monopoly to a multi operator multi service open competitive market.

23 WiMAX stands for Worldwide Interoperability for Microwave Access. It is a telecommunications technology providing wireless data over long distances in a variety of ways, from point-to-point links to full mobile cellular type access. It is based on the WirelessMAN (IEEE 802.16) standard. WiMAX is a highly scalable, long-range system, covering many kilometers using licensed spectrum to deliver a point-to-point connection to the Internet from an ISP to an end user. WiMAX can be

used to provide a wireless alternative to cable and DSL for broadband access, and to provide high-speed data and telecommunications services. WiMAX can also be used to Connect many Wi-Fi hotspots with each other and also to other parts of the Internet.

[24] The 3G will allow telecom companies to offer additional value added services such as high-resolution video and multi media services in addition to voice, fax and conventional data services with high data rate transmission capabilities. BWA will become a predominant platform for broadband roll out services. It is also an effective tool for undertaking social initiatives of the Government such as e-education, telemedicine, e-health and e-Governance. Providing affordable broadband, especially to the suburban and rural communities is the next focus area of the Department. BSNL & MTNL have already been allotted 3G & BWA spectrum with a view to ensuring early roll out of 3G & WiMax services in the country. They will pay the same price for the spectrum as discovered through the auction. While, Honbl'e Prime Minister launched the MTNL's 3G mobile services on the inaugural function of 'India Telecom 2008' held on 11th December 2008, BSNL launched its countrywide 3G services from Chennai, in the southern Tamil Nadu state on 22nd February 2009.

[25] Much of the Internet based information is textual and in English. In India, significant portion of the rural population is either illiterate or has an education no higher than the elementary level. Therefore, a large segment of the Indian rural population may not be able to access and comprehend the web-based information.

References

Barziliani, Nohan and Barziliani, G., 2005. Cultured Technology: The Internet and Religious Fundamentalism. *The Information Society*, Vol. 21, No.1, pp 25-40.

Branko, Milanovic., 2005. *Worlds Apart: Measuring International & Global Inequality.* Princeton University Press: Princeton.

Chen, W. and Wellman, B., 2004. *Charting and Bridging Digital Divides: Comparing Socio-Economic, Gender Life, Stage and Rural Urban Internet Access in Use in Eight Countries.* AMD Global Consumer Advisory Board (GSAB).

Companie, B. M., 2001. *Digital Divide: Facing a Crisis or Creating a Myth?.* MIT Press, MA: Cambridge.

Cooper, M., 2002. Does the Digital Divide Still Exists? Bush Administration Shrugs, But Evidence Says 'Yes'. Consumer Federation of America, Consumer Union, The Civil Right Forum, Washington, DC. http: www.chnm.gmu.edu/digitalhisotry/links/pdf/introduction/0.26a.pdf.

Dasgupta, S., Lall, S. and Wheeler, D., 2002. Policy Reform, Economic Growth, and the Digital India: Policy Review. *Working Paper of the World Bank, No. WPS-2567*, Washington, DC.

Dean, Parham, 2002. Productivity Gains: Importance of ICTs. *Agenda*, Vol. 9, No. 3, pp 195-210.

Dewan, S. and Riggins, F. J., 2005. Digital Divide: Current and Future Research Direction. *Journal of Association for Information Systems*, Vol. 6, No. 2, pp 298-337.

DiMaggio, Paul, Eszter Hargitti, W. Russell, Newman and John, P. Robinson., 2001. Social Implications of Internet. *Annual Review of Sociology*, Vol. 27, pp 307-336.

Dutta, Subrat, 2003. Impact of Information Communication Technology on Society. *Yojana*, Vol. 47, No.2, p 24.

Fletcher, Ameila, Simon, Gaysford and Adele, 2000. OFT Study on e-Commerce and Competition. http://www.out-law.com/page-954.

Gordon, R., 2002. Does the 'New Economy' Measure up to the Great Inventions of the Past?. *Journal of Economic Perspective*, Vol. 14, No. 4, pp 49-74.

Guillen, Mauro, 2006. Explaining the Global Digital Divide. *Social Forces*, Vol. 84, pp 681-708.

Hanimann, Thomas and Ruedin, Etienne, 2007. Digitale Gräben oder Digitale Brücken? Chancen und Risiken für Schwellenländer, http://eprints.rclis.org/11455/

Hecker, E. Daniel, 2001. Employment Impact of Electronic Business. *Monthly Labor Review*, May.

Hooker, N. H; J. Heilig and Ernst, S., 2001. What is Unique about e-Agribusiness?. IAMA World Food and Agribusiness Symposium, Sydney: Australia, June.

Huberman, Bernardo, 2001. *The Law of Web: Patterns in the Ecology of Information*. MIT Press: Cambridge.

Jusus, M. Frias., 2003. The Importance of ICTs for Developing Countries. *Interdisciplinary Science Review*, Vol. 28, No. 1, pp 10-14.

Kennedy, Tracy, Barry, Wellman and Kristine, Klement., 2003. Gendering the Digital Divide. *IT & Society*, Vol. 1, No. 5, pp 72-96.

Kuhn, Peter and Mikal, Skuterud, 2000. Internet and Traditional Job Search Methods, 1994-1999. Paper presented to the IRPP and CERF conference on Creating Canada's Advantages in an Information Age, May.

Lawerence, Steve and Lee, Giles, 1999. Accessibility of Information on the Web. *Nature*, Vol. 400, pp 107-109.

Leizerov, Sagi, 2000. Privacy Advocacy Groups Versus Intel: a Case Study of How Social Movements are Tactically Using the Internet to Fight Corruptions. *Social Science Computer Review*, Vol. 18, pp 461-483.

Lenhart, Amanda, 2000. Who's not online: 57% of those without Internet access say they do not plan to log on. Available: http://www.pewinternet.org/reports/toc.asp?Report=21 (Accessed on September 21, 2000).

MacKenzie, D. and Wajcman., 1985. *The Social Shaping of Technology: How the Refrigerator Got Its Hum*, Milton Kenynes, Open University Press, UK.

Mahajan, Subrat, 2003. Impact of Digital Divide on Developing Countries with Special Reference to India. *SERALS Journal of Information Management*. Vol. 40, No. 4, pp 328–329.

Mark, Warschauer, 2003. *Excellent Technology and Social Inclusion: Rethinking the Digital Divide.* MIT Press: Cambridge.

Mosse, E. and Sahay, S., 2003. Counter networks, communication and health information systems: A case study from Mozambique. In: Korpela *et al.*, (Eds.) *Organizational Information Systems in the Context of Globalization.* Kluwer Academic Publishers, Boston/Dordrecht/London.

Motohashi, K., 2001. Economic Analysis of Information Network Use: Organizational and Productivity Impact on Japanese Firms, Research and Statistics Department, METI Tokyo, Japan. 22-23.

Mwesige, P.G., 2004. Cyber elites: A survey of Internet café users in Uganda. *Telematics and Informatics*, Vol. 21, No. 1, pp 83-101.

Nahon, K. Barzilai, 2006. Gap and Bits: Conceptualizing Measurement for Digital Divide/s. *The Information Society*, Vol. 22, pp 269-278.

Nandi, B., 2002. Role of Telecommunications in Developing Countries in the 21st Century, 14th Biannual Conference, International Telecommunication Society, Seoul.

Nath, V., 2001. Employment and governance through ICTs: Women's Perspectives. *The International Journal of Information and Library Review*, Vol. 33, pp 317-339.

Norris, Pippa, 2000. *Information Poverty and the Wired World.* The Harvard Journal of International Press Politics, Vol. 5, pp 1-6.

Norris, Pippa, 2001. *The Digital Divide, Civic Engagement, Information Poverty and Internet Worldwide.* Cambridge University Press: Cambridge.

OECD, 2001. *Understanding the Digital Divide*. OECD, Paris.

Pastor, Satorras, Romualdo and Alessandro, Vespignani, 2004. *Evolution and Structure of the Internet: A Statistical Physics Approach.* Cambridge University Press: Cambridge.

Pluss, Martin, 2004. Digital Divide and the Role of Education. http://plu.wikispaces.com/file/view/dd_education_04.pdf.

Poole, B., 2001. How Will Agricultural e-Markets Evolve?, Paper Presented at the USDA Forum, Washington DC, February 22-23.

Rheingold, Harold., 2000. *The Virtual Community: Homesteading on the Electronic Frontier.* Cambridge, MA: MIT Press.

Rowan, Brian, Andrew, W. and Miracle, Jr., 1983. Systems of Ability Grouping and Stratification of Achievement in Elementary Schools. *Sociology of Education,* Vol. 56, pp133-44.

Servon, Lisa., 2002. *Bridging the Digital Divide: Technology, Community and Public Policy.* Blackwell: Oxford.

Singh, Neena, 2007. Bridging the Digital Divide in India: Challenges and Opportunities. *World Libraries*, Vol. 17, No. 1.

Smith, Mark and Peter, K., 1998. *Communities in Cyberspace.* Routledge: New York .

Spender, D., 1997. The Position of Women in Information Technology-or Who Got There First and With What Consequences. *Current Sociology,* Vol. 45, No. 2, pp 135-147.

Sumanjeet, 2006. e-Governance: An Overview in the Indian Context. *Indian Journal of Political Science,* Vol. 67, No. 4.

Sumanjeet, 2008. Impact of e-Commerce on Economic Models: Little to Lose; More to Gain. *International Journal of Trade and Global Markets,* Vol. 1, No. 3, pp 319-337.

Sumanjeet, 2009. Social Implications of Electronic Commerce. *Journal of Social Studies,* Accepted for publication.

United Nations., 2006. *Encyclopedia Britannica.* http:search.eb.com/ eb/ article-250455.

Wahid, F., Furuholt, B., and Kristiansen, S., 2006. Internet for development? Patterns of use among Internet café customers in Indonesia. *Information Development,* Vol. 22, No. 4, pp 278-291.

Winter, D. and Huff, C., 1996. Adapting the Internet: Comments from a Women-Only Electronic Forum. *The American Sociologists,* Vol. 27, No. 1, pp 30-54.

2

e-Agriculture in India : Agriculture Knowledge Communication with the Application of Information and Communication Technologies

I.V. Malhan and Samita Wadhera

ABSTRACT

Agriculture is an important sector of economy that ensures food security of the nation. About 58 per cent of the Indian population is engaged in Agriculture and planners are concerned to increase agriculture productivity. The National Knowledge Commission has also looked into various problems of this sector and made some recommendations from the knowledge perspective. This paper includes review of existing plans and programs for improvement of knowledge accessibility to Indian farmers. Article also discusses how the Information Communication Technology (ICT) applications, especially, the hand held devices like mobile phones can improve knowledge accessibility in the agriculture sector and describes how the ongoing programs of mobile telephony are helping Indian farmers.

Introduction

Agriculture occupies a key position in the Indian economy because of its contribution to over all economic growth through supplies of food, raw materials and exports. It is a source of livelihood for a majority of the population and provides a large market for non-agricultural goods and services. Agriculture in India is a prime occupation for a major segment of population and a vast majority of rural population depends on this prime occupation. To plough the land initially animals were used that was replaced by tractors and now tractors are being replaced by laser technology but in India still the first two methods are used to plough the land. India being a developing country is far away from developed countries in the production and application of agricultural knowledge.

The National Knowledge Commission has shown the serious concern about the deep crisis in the Indian agriculture as it contributes just 18.5 per cent of the National GDP though the proportion of the population employed in the agriculture is as much as 58 per cent. In comparison to the Non-Agriculture sector, the per capita GDP and average living standard in agriculture sector are quite low. Thus, the appropriate application of knowledge in this sector can lead to the transformation of India into a knowledge economy. Further, attending agriculture sector on top priority basis and implementation of recommendation of various commissions and working groups set by Government of India can provide a competitive edge to the Indian farmers in the International market.

The National Knowledge Commission (NKC) recommends the following concrete steps for the revitalization of knowledge generation and application in agriculture:

A. Knowledge Generation

1. Modernise and stimulate agricultural research institutions, co-ordinate research and make research support more flexible.
2. Improve the organisation of agricultural research.
3. Direct more research to neglected areas.
4. Provide more effective incentives for researchers.
5. Change the curricula in agricultural universities to ensure greater relevance.
6. Exploit the opportunities and meet the challenges of Intellectual Property Rights.
7. Encourage and regulate private agricultural research.

B. Knowledge Applications

1. Make knowledge applications in agriculture community-driven and farmer-led.
2. Redesign the existing support systems.
3. Document and disseminate successful experiences and good practices.
4. Improve the capacities and expertise of extension workers.
5. Restructure Agricultural Technology Management Agency (ATMA) to make it more decentralized, participatory and locally responsive.
6. Enhance and regulate the role of private players in agricultural extension delivery.
7. Create a web-enabled Knowledge Bank on all aspects of agriculture.

The NKC has further pointed out that these changes are both necessary and desirable in order to encourage the creation and dissemination of useful and relevant knowledge in agriculture (Pitroda, 2009).

The present era is based on knowledge economy and the major argots of 21st century are information revolution and information management. Information technology has removed the geographic barrier and has made the society highly information intensive and networked one. The society has moved from agricultural society to knowledge society and with the changing times a number of new technologies have been introduced which help to perform the task efficiently than ever before. The creation of new technologies has shown its impact in each and every field and help people of different fields to handle their work in a better manner. These technologies have brought a number of new jobs and business opportunities which lead to higher incomes and prosperity by providing the more efficient and effective mechanisms of commerce. Information and Communication Technology (ICT) has a lead and edge over other technologies. With the passing era, communication has become so fast and reliable that one can take well-informed and timely decisions. ICT helps to disseminate the knowledge derived from the huge network of education and research institutions. Each and every aspect of the life is changing in the electronic form like e-Business, e-Governance, e-Library, e-Agriculture, and so on. We are living in a world where all spheres, including those most resistant to change, cannot escape getting influenced by the developments in the ICTs.

e-Governance

The National e-Governance plan (2003-2007) of Indian Government seeks to lay the foundation and provide the impetus for long-term growth of e-

Governance within the country. The plan seeks to create the right governance and institutional mechanisms, set up the core infrastructure and policies and implements a number of Mission Mode projects at the centre, state and integrated service levels to create a citizen-centric and business-centric environment for governance. In 2005, the World Bank signaled its willingness to increase funding further (if required) for a range of e-Governance initiatives in India as part of the first phase of the country's National e-Governance Plan (NeGP). NeGP Vision - All Government services accessible to the common man in his locality through a One-stop-shop (integrated service delivery) ensuring convenience, efficiency, transparency and reliability (NeGP, 2008).

e-Governance network backbone provided by the National Informatics Centre and support given to various Government bodies like Central Government, State Government, UT Administrations, District Administration and others is a major initiative for e-Governance at the national level. India portal is providing services in multi-lingual content. If these technologies become widely available and affordable for the vast majority of our citizens, their impact would not remain limited to the services that are provided by the Government. With the help of e-Governance the government can achieve efficiency in task, transparency in work and ensure active participation of citizens. It helps to move towards good governance, work in the welfare of citizens and create citizen's awareness and empowerment. e-Governance also contributes to nation's economic growth.

Today many e-Governence projects simply layer technology over top of bureaucracy and offer little advantage to citizens. Government must transform their processes to be truly successful in e-Governence (Shirur, 2007). The comments of the Shivraj Singh Chouhan, Chief Minister of Madhya Pradesh, India while inaugurating the 10th National e-Governance Conference at Vidhan Sabha Bhawan premises in the capital city of Bhopal about the e-Governance is quoted as "Greater public awareness about e-Governance is necessary for providing corruption-free and more efficient administration. Besides adequately training and equipping government officials with the knowledge of computers, it is also essential to enhance awareness among people about this. e-Governance needs not only computers but also necessary infrastructure including power. Language is a critical issue involved in e-Democracy and English continues to remain a hurdle in rural areas. Earlier, most of the government websites were in English. Directives have now been issued to make them bilingual" (Khan, 2007).

A number of products and services of e-Governance supported by NIC are: AGMARKNET, BHUIYAN, ePOST, JUDIS, VAT, Examination Results Portal, Gyandoot, Passport website and Rural Bazaar.

e-Agriculture

Implementation of ICT within agriculture creates hope within the community. Agriculture portal is to be a one-stop-shop for farmers, government, business, and other institutions. "A multi stakeholder working group e-Agriculture was set up in 2006 and the e-Agriculture Community of Expertise was officially launched as a global initiative to enhance sustainable agricultural development and food security by improving the use of ICT in the sector during September, 2007. It is a platform for the community, a virtual meeting and sharing place where users can share experiences, opinions and good practices related to agriculture" (Ritu and Subir, 2009). A number of international agencies are working with hand-to-hand to empower the project of e-Agriculture. Some of the agencies are FAO (Food and Agriculture Organization of the United Nations), Consultative Group on International Agricultural Research (CGIAR), International Association of Agricultural Information Specialists (IAALD), Inter-American Institute for Cooperation in Agriculture (IICA), The National Agricultural Library (NAL), The Technical Centre for Agricultural and Rural Cooperation (CTA), The Global Forum on Agricultural Research (GFAR), The International Fund for Agricultural Development (IFAD) and so on (e-Agriculture, 2008).

A mission was initiated with the aim to provide knowledge connectivity to every village of India and the mission was termed as "Mission 2007" as the deadline date of mission was 15th August, 2007 which mark the 60th anniversary of the words "India's tryst with destiny" called by the first Prime Minister of the Country Pt. Jawaharlal Nehru.

In the 11th Plan some of the major challenges are described and Agricultural Crisis: Regaining Agricultural Dynamism is on priority. During the 10th Plan period the actual growth of agricultural GDP including forestry and fishing is below 2 per cent and in 11th Plan the target is 4 per cent (NeGP, 2008).

The Common Services Scheme is the front end delivery points for Government and other sectors to rural citizen of India and e-Agriculture is one of the remarkable examples. The NIC began initiatives by laying the roadmap for IT in agriculture way back in 1995. DACNET was an e-governance initiative executed by the NIC to facilitate agriculture on-line and implement e-governance solutions in the 40 directorates and 132 field offices of Department of Agriculture and Co-operation. The demand had been made to earmark at least 6 per cent of annual agriculture budget to develop IT in agriculture department. DACNET project acts as a facilitative platform to inform multiple stakeholders on how agriculture can be developed in India support them in improving productivity and help them in participating in the market, including global markets. With this, the latest prices of about 300 agricultural products will be available on the Net and will be updated everyday (DACNET News, 2005).

The project, Market driven Agricultural Initiative through IT enabled Agri Business Centres in Kerala enables farmers and other stakeholders through Agri-Business Centres to interact with Agricultural Service Providers in the Private, Government and Non Government sectors. It also provides a web-based solution enabling the small and medium farmers as well as owners of large landholdings. About 550 e-kendras have been leveraged to reach the masses of farming community and other stakeholders in Agriculture sector. The project shall be driven by a state level resource cell and district level implementation cell. Project services envisaged shall include Information/ Transaction/ Payment Platforms and expected benefits to member farmers, member Agricultural Input Providers and member Agricultural Activists, NGOs, Government organization.

Government Schemes for Promoting Agriculture

Agriculture sector needs well functioning markets to drive growth, employment and economic prosperity in rural areas of the country. Currently, AGMARKNET covers 2800 market nodes and 300 commodities and provides information in ten different languages. In order to increase the institutional credit for agriculture "Farm Credit Package" announced in June, 2004. Prior to this Kisan Credit Card Scheme was introduced in August 1988 for short- and medium-term loans. Government of India has announced a package for Short term Rural Cooperative Credit Structure in January, 2006 and NABARD has been appointed as an implementation agency. Jammu and Kashmir is not the part of this package. Krishi Vigyan Kendras were set up by Indian Council of Agricultural Research to promote the awareness among the farmers about the latest techniques and technologies. But due to lack of coordination these programmes have not achieved the target. Farmer Knowledge Centers are one of the ways to work at the village level and implemented successfully to provide the adequate Agriculture content accessible to the farmers and supported by a number of agencies. Department of Agriculture and Cooperation has launched Kisan Call Centres on January 21, 2004 to utilize significant advancement in telecom infrastructure. Set up of farm schools and farmer field schools should be promoted to improve dissemination of technology by demonstrating advanced agricultural production practices on farmers' fields in the village.

New initiatives taken by the Government of India as per the Economic Survey 2007-08 are as under:

1. National Food Security Mission: This is a centrally sponsored scheme launched by Department of Agriculture and Cooperation, Ministry of Agriculture. In this financial year (2008-09) Rs. 1100/- crores was allocated for the mission.
2. Rashtriya Krishi Vikas Yojana: This Yojana was approved by the Government on August 16, 2007.
3. National Policy for Farmers, 2007: This policy has been approved by the government as per the recommendations of National Commission on Farmers and state governments. The areas covered under this scheme are;
   - Economic well being of the farmers
   - Asset reforms
   - Water use efficiency
   - Encouragement of new technologies
   - Establishment of national agricultural bio-security system
   - Seeds and soil health
   - Support services for women
   - Credit and insurance
   - Set up of Gyan chaupals
   - Appropriate social security scheme for farmers
   - Implementation of minimum support price
   - Enlargement of food security basket

Private Agencies Initiatives for Promoting the Agriculture

India is predominately rural society and 2/3rd of the population is rural based. According to an estimate (FAO database of 2006), agriculture is the main source of income for 58per cent of the Indian workforce.

IFFCO Kisan Sanchar Limited (IKSL) is a joint venture of Bharti Airtel and Indian Farmers Fertilizers Co-operative Ltd. to provide the rural telephony. These facilities empower through the relevant and updated information in time and in turn boost the rural economy and increase the rural employment. In the first phase, services were provided in Uttar Pradesh, Punjab, Haryana, Rajasthan, Madhya Pradesh, Chhattisgarh, Bihar, Jharkhand, Orissa, West Bengal, Tamil Nadu and Himachal Pradesh. In the second phase, services extended to Gujarat, Maharashtra, Goa, Karnataka, Andhra Pradesh and Kerala (Efytimes, 2008).

In an interview, Shri. Sunil Bharti Mittal, Chairman and Managing Director Bharti Airtel Ltd. shared the view that with this service the farmer receive five free voice messages every day to provide relevant information on mandi prices, animal husbandry inputs, farming techniques, weather forecasts, dairy farming, rural health initiatives, fertilizer availability *etc.* In addition to this a dedicated helpline managed by agriculture experts will respond to queries from farmers.

Uttar Pradesh Agricultural Marketing Board (Mandi Parishad) and Bharat Sanchar Nigam Ltd. has jointly started a voice based scheme project known as Mandi on Mobile. With this service one can get the information about the latest mandi rates of 108 items ranging from grains, pulses, fruit and vegetables (Cell Bharat, 2008).

Kerala State Information Technology Mission with the co-operation of National Institute of Smart Government, Hyderabad, Kerala Department of Agriculture and Indian Institute of IT and Management, Kerala has launched a toll free agricultural trade call centre (1800-425-1661) and the revamped e-krishi web site (http://www.e-krishi.org). It provides information on a number of topics related to agriculture including agricultural prices and trends. It creates a sense of community and spirit of unity among farmers in the state (The Hindu-Business Line, 2007).

e-choupal also unshackles the potential of Indian farmer who has been trapped in a vicious cycle of low risk taking ability - low investment - low productivity - weak market orientation - low value addition - low margin - low risk taking ability. This made him and Indian agribusiness sector globally uncompetitive, despite rich and abundant natural resources (One choupal, 2009).

e-krishi creates a network of farmer community which has access to information on market demand, prices, good agricultural practices and quality agricultural inputs. This is achieved through a technology-enabled robust transaction platform that facilitates all their offline activities. e-Krishi has also been instrumental in the setting up of farmer groups at the panchayat level named Bhoomi clubs - meet once in a month and discuss issues related to agriculture in general and marketing in particular. It has also helped the farmers to know about hundreds of Government schemes in the agriculture sector and take benefits from those schemes. It has helped in promoting good agriculture practices due to better information dissemination and awareness (Kumar, 2009).

Nokia's launch of life tools service for farmers which will fetch them instant information related to their need and choice. The service helps the farmers to get easy access on agrarian, commodity rates, weather and allied information through their mobile phone. Life Tools service was formally launched across India on June 12, 2009 after successful launching of the scheme in Maharashtra earlier in April 2009. The service in Maharashtra will be provided in association with Maharashtra State Agricultural Marketing Board (MSAMB) where Nokia has signed a MOU with the MSAMB. This service in Maharashtra brought livelihood gains through relevant information such as market rates for farmers' produce, greater awareness on market conditions, tips on weather, news, crop advisory, learn English and General Knowledge. It also enabled the consumers to fill their information gaps by being better informed, save time and money, and improve their confidence and social standing. This complete Nokia Life Tools solution will be available on the newly launched Nokia 2323 classic and Nokia 2330 classic devices and will be later expanded to other Nokia devices. It has range of three primary services - Agriculture Services (plug the information gaps and needs of farmers *via* their mobile devices, by providing information on seeds, fertilizers and pesticides, market prices and weather; temperature, rainfall, wind conditions), Education Services and Entertainment Services. This service will be expanded to select countries across Asia and Africa later in 2009 and beyond (Yahoo! Finance, 2009).

French Institute of Pondicherry with Rice Wheat Consortium for Indo-Gangetic Plains, India, CIRAD of France, Netherlands based Wageningen University in a collaboration initiated a software named as OSCAR (Open Source Simple Computer for Agriculture in Rural Areas) used to eliminate the unwanted plants/weeds from the land. "The software has varied illustrations for precise results. The user can get all the information on a particular weed and a set of instructions on how to eliminate it. The information can be accessed in three languages Hindi, English and Bengali. The software has a database of 50 different weed species found in the Indo-Gangetic plain and would be updated with 250 more varieties of such unwanted plants found in the region" (Rediff News, 2006).

Similarly, Farm Trac and Farm Site are the software developed by the US Company Farm Works which help the farmers to improve productivity by providing the information of the use of resources and plan for the future. "Farm Trac is a GPS-integrated field record-keeping package written for MS Windows that allows farmers to keep notes on field history and chemical usage, calculate direct costs incurred at the field level and plan future farming operations. Farm Site adds GPS compatibility to Farm Trac and provides state-of-the-art positioning and a limitless number of layers, including soil types and soil tests, details of previous activities and the outline of future plans" (ICT Update, 2006).

Constraints

Adoption of new technology involves change. "In contrast to this the per capita income of the farmers are 1/5th of that of an average Indian. Little knowledge about the facts led to the over exploitation of the resources as they did not know the actual information about the: Kind of crops grown in the field, How to grow the crops, What planning they need to do with respect to area and soil conditions, what are the market dynamics to get the fruitful results. The production of agriculture crops, besides the weather induced fluctuations, significantly depends on the availability of inputs like fertilizers, irrigation, certified seeds, credit support and appropriate price signals" (eAgriculture India, 2007).

Indian government is experiencing difficulties in disseminating information regarding novel farming techniques and best practices in a country which is multilingual and divided into nearly 300 agro-climatic zones.

Insufficient agriculture infrastructure, support facilities and institutional capacities, lack of awareness regarding suitable agricultural methods among the farmers, inadequate use of public-private partnerships, lack of common platforms, insufficient use of ICT for agricultural purposes, quality standardization, land monitoring, watershed planning, yield forecasting, cropping pattern analysis, analytical modeling, *etc.* are the major constraints for the implementation of any new project.

One of the most common reasons for the failure of programmes and schemes is the faulty and incomplete design. Care and attention must be taken to formulate programmes, projects and schemes in a more systematic and professional manner.

Obstacles in implementation of ICT in the field of agriculture are:

1. *Geographical distances:* The farmers are located in the far-flung areas and in majority of the cases the road connectivity is not well developed. In this condition the connectivity with the farmers become the major hindrance for providing the relevant information to them.
2. *Lack of key persons:* Key persons play a major role in the success of any project but in the implementation of e-Agriculture this is missing. It has been observed that the appropriate human resources with right kind of attitude are missing altogether and the information from one end to other end is not communicated properly. Rs 281/- crores was sanctioned in the session 2008-09 for agricultural extension programmes to disseminate frontline technologies and Rs. 350/- crores for agricultural education to improve the quality of human resources, upgrade infrastructure and educational reforms. But, it is strange to observe that

the link between the lab and land shows variation. The link from lab to land is weak and from land to lab is missing altogether.

3. *Lack of investment:* Ministry of Agriculture has spent an average of Rs. 8132/- crores in 2007-08 and has in hand Rs. 9660/- crores for 2008-09 (Economic Survey, 2007-08) but still this amount did not fulfill the requirements of different departments functioning under this ministry. Private partnership initiatives in the field of agriculture are very limited due to risk factors. Government has taken an initiative by introducing National Agriculture Insurance Scheme to promote the involvement of private agencies but still the investment has not reached the desired level. Rs. 644/- crores has been allotted in the mission for the ongoing financial year.

4. *Lack of communication:* Communication gap has created a barrier between the government agencies and the farmers. Due to this gap the information about the major government programmes is not circulated among the end users and the latter thus can't extract the maximum benefits from those projects.

5. *Different languages:* India is multilingual country. Almost every State has its own local language and the most of the websites are in English/ few sites are in Hindi. Barriers of languages have also become an obstacle in the implementation of the ICT in agriculture. Different services for farmers have initiated their working in different languages but still this gap require to be filled by providing a local language interface at some places.

6. *Lack of integrated services:* Integration is necessary between the different agencies of the Government and the Farmers so as to get the projects implemented in a better way. It also helps in sustainability assessment of local needs, backward-forward information, extension services and research. Government has sanctioned Rs. 257/- crore for National Agriculture Innovative Project for 2008-09 (Economic survey, 2007-08) to increase the involvement of end-users by providing innovative ideas for the agriculture development.

7. *e-illiteracy:* More than 50 per cent of the literate people are e-illiterate and what can be said for illiterate people. Majority of farmers are e-illiterate as they don't know how to use the ICT. Due to the lack of networking, access to internet is poor and again the information about the ongoing projects and helpful programmes remains unknown to many.

Solutions

Agricultural productivity can be enhanced to a considerable level even if the existing knowledge and best practices are communicated to most of the Indian farmers. Extension departments are not able to reach upto a large number of farmers and hence knowledge gaps continue to exist. Two different farmers having the same size of land may obtain different economic value from their farms because of different levels of agricultural knowledge. If we require to increase the agricultural productivity to next desired level we must initiate measures to accelerate the process of knowledge accessibility and bridge the knowledge gaps. Policy makers should promote innovation, competition and fastest diffusion and dissemination of best practices and latest agricultural technologies.

Twenty four hours agricultural television channels, development of interactive media kits, extension of NICNET upto key villages can help to bridge knowledge gaps in agriculture. Mobile vans loaded with the communication infrastructure such as Internet, Interactive media kits may available in villages to help farmers to get access to latest ideas and information in the field of agriculture. Mobile telephone network is fast expanding in India including the rural areas. This infrastructure should be fully exploited to help Indian farmers for accessing agriculture knowledge and sharing their ideas and experiences.

Conclusion

A number of efforts have been made from time to time to improve agriculture productivity and ameliorate the agriculture knowledge infrastructure so that the best practices and state of the art need based knowledge is made accessible to Indian farmers. However, agricultural developmental process is not taking place upto the desired level because of knowledge gaps besides other reasons. There is no dearth of agricultural knowledge but there is lack of adequate outreach services for knowledge transfer to farming communities. Besides inadequate communication from Lab to Land, there is also lack of communication regarding experiences gained at farmers' field to labs. Wide spread illiteracy among farmers is one of the potent reason for this situation.

Agricultural development process cannot wait till farmers are educated. We have to look at ways how agricultural knowledge is made accessible to farmers over the given situations and conditions. We must revamp our school curricula so that school children find it interesting and amusing and compatible to social ecosystem. Agriculture should be included as a subject in the school curriculum and computer education should be an important part of agricultural education system which is one of the recommendations of e-Agriculture India 2007 conference. It is well said that rather than G2C there should be G for C (eAgriculture India, 2007).

The application of ICTs for information communication in rural India provides a means for faster communication of information of agricultural best practices, weather information, flood alarms, market prices, soil specific and region specific agricultural knowledge. Interactive multimedia may also be used to make farmers understand various techniques and processes for pursuing better productivity and gainful agricultural practices. The NICNET has excellent backbone for e- agriculture and e-governance and this infrastructure is required to be adequately used for communication of agricultural knowledge to key players who may further disseminate knowledge and information to farmers. Private sector companies are also extending their networks in the rural areas with the twin objective of helping the farmers and expanding their business and trade in the rural India. The ITC's e-choupal network in Indian villages is helping the farmers with relevant agricultural information along with trading agricultural commodities offering the best deal for farmers.

The mobile telephony is the fastest growing communication facility in India. Presently every fifth Indian has a mobile telephone. Mobile companies have set up their infrastructure in rural India and some companies have started information services for farmers. Such handheld devices help to access information just in time and just in place and reach upto difficult terrains like hills, deserts, deep valleys and islands. As the technologies advance, mobile telephony may play a crucial role to transfer knowledge from agricultural knowledge centres to individual farmers and may help to establish linkages among farmers for more effective sharing of information and ideas.

References

Cell Bharat, 2008. Mandi on Mobile – Live Commodity Updates in Mobile Phone. http://cellbharat.com/913/mandi-on-mobile%e2%80%99-live.(Retrieved on December 26, 2008).

DACNET News, 2005. 172 Directorates under DACNET. Spatial India.

eAgriculture India, 2007. Enabling Agriculture. www.eIndia.net.in/eagriculture/index.asp (Retrieved on December, 25, 2008).

e-Agriculture, 2008. http://www.e-agriculture.org/partners.html (Retrieved on December 26, 2008).

Economic Survey, 2007-08. Agriculture and Food Management. Government of India. 1-169. http://indiabudget.nic.in (Retrieved on December, 25, 2008).

Efytimes (2008). Bharti Airtel to Offer Affordable Telephony to Farmers. http://www.efytimes.com/efytimes/fullnews.asp?edid=26319 (Retrieved on December 26, 2008).

ICT Update, 2006. TechTip: Farm Management Software Helps Farmers Improve Productivity. http://ictupdate.cta.int/en/layout/set/print/Regulars/Rester-en-contact/TechTip-Farm-man......html (Retrieved on December 25, 2008)

Khan, Danish, A., 2007. e-Governance Awareness Must for Corruption-Free, Efficient Administration. Event Diary, 37-38. www.egovonline.net (Retrieved on December 25, 2008)

Kumar, Ajay., 2009. e-Krishi in Kerala. *Information for development*. Vol. VII No. 2, pp.15-17.

NeGP, 2008. The National e-Governance Plan. e-Governance, 1-3. http://www.it.iitb.ac.in/~prathabk/egovernance/national_egov_plan.html (Retrieved on December 27, 2008)

One choupal, 2009. http://www.echoupal.com/frontcontroller.ech (Retrieved January 31, 2009)

Pitroda, Sam, 2009. Letter to Prime Minister to India on 2nd March, 2009 by Chairman National Knowledge Commission.

Rediff News, 2006. Software to Help Farmers Log out Weeds. http://in.rediff.com/money/2006/apr/25weeds.htm (Retrieved on December 27, 2008)

Ritu, Srivastava and Subir, Dey, 2009. e-Agriculture.org: Revisiting e-Agriculture.org. *Information for development(i4d Magazine)*. http://www.i4donline.net/articles/current-article.asp?Title=e-Agriculture.org&articleid=2177&typ=eagriculture (Retrieved on January 31, 2009)

Shirur, Prachi, 2007. Leveraging ICT for Improved Government Services. Event Diary, 30-33. www.egovonline.net (Retrieved on December 25, 2008)

The Hindu-Business Line, 2007. Kerala e-Krishi Portal, Agri Trade Call Centre Launched. http://www.blonnet.com/2007/03/20/stories/2007032004901100.htm (1-2 Retrieved on December 27, 2008)

Yahoo! Finance, 2009. Nokia Life Tools Launched Across India. http://finance.yahoo.com/news/Nokia-Life-Tools-Launched-prnews-15508515.html?.v=71 (Retrieved on June 15, 2009).

3

ICTs in the Perspective of Rural India: A Review

Gaurav Papnai, Anup Prakash Upadhyay, Taufiq Ahemed and Neelam Bhardwaj

ABSTRACT

The term rural development also represents improvement in quality of life of rural people in villages. As per Chambers (1983) "Rural Development is a strategy to enable a specific group of people, poor rural women and men, to gain for themselves and their children more of what they want and need." ICT is an integral part of development strategies of both developing and developed countries. It has great potential to bring in the desired social transformations by enhancing access to people, services, information and other technologies (Dutton *et al.*, 2004). ICT applications can enhance poor people's opportunities by improving their access to markets, health and education. Furthermore, ICT can empower the poor by expanding the use of government services and reduce risks by widening access to micro finance (Cecchini and Scott, 2003). The uses of ICT for development are actively promoted for economic development, job-creation, rural development and poverty-alleviation. The term 'ICTs' can be interpreted as including a wide range of media, 'new ICTs' is used to denote "the use of computers and communication systems between computers" (CTA 1999:4). The most significant use of new ICTs is the World Wide Web, which enables people to access information on millions of other computers. Traditional media and new ICTs have played a major role in diffusing information to rural communities, and have much more potential. Studies on information systems serving rural communities have focused on specific sectors such as agriculture or health, instead of covering the rural community needs in a holistic manner. Rural information systems must involve rural communities and local content must be of prime importance (Mchombu 1993). Traditional media have been used very successfully in developing countries and rural radio in particular has played a major role in delivering agricultural messages. New ICTs, have the potential of getting vast amounts of information to rural populations in a more timely, comprehensive and cost-effective manner, and could be used together with traditional media. The rural poor are the engines of agricultural production in developing countries. Any bottlenecks to the general livelihood of the rural poor - health, disaster, education, lack of infrastructure, lack of financial services, and many others - will have significant impacts on Agricultural production at the household, regional and national levels. ICT interventions that improve the general livelihoods of the rural poor may also yield significant Agricultural development investments on the part of rural families. However, a full range of rural livelihood assets and strategies should be examined to enable us to better understand the relationships between ICTs, agriculture and poverty reduction. A variety of fairly straightforward rural ICT interventions may have greater impact on Agricultural production and post-harvest activities than those that are strictly focused on agriculture. Agricultural extension playing a broader role in helping to harness ICTs, extensionists will be better able to harness information and communication technologies (ICTs) to improve rural livelihoods If they equip themselves with analysis and arguments to bolster the case.

Introduction

In the rural context, development involves use of physical, financial and human resources for economic growth and social development of the rural economies (Burkey, 1993). The term rural development also represents improvement in quality of life of rural people in villages. As per Chambers (1983) "Rural Development is a strategy to enable a specific group of people, poor rural women and men, to gain for themselves and their children more of what they want and need." Singh (1999) defines Rural Development as "A process leading to sustainable improvement in the quality of life of rural people, especially the poor".

Jhunjhunwala *et al.*, (2006) underlined that the business model for rural development should be based on collective partnerships and must incorporate the traditional knowledge available within local community. Incorporation of traditional knowledge systems (TKS) would also ensure involvement and ownership of the rural beneficiaries themselves. In the Indian rural context, there are already several encouraging examples such as "Honey Bee Network" which is a database of grass root innovations and technologies and serves as an effective solution to problems of local development.

Garai and Shadrach (2006), conclude that there is an urgent need to recognize the role of local knowledge in sustainable development. They also argue that interactions between communities based local bodies and development worker need to be enhanced to ensure success of the development process. Since communities are the closest to grassroots' problems, they are the best judge to evaluate technology alternatives and provide innovative solutions for the problems of their respective areas. This "from the inside out" and "bottom up" perspective to technology has been supported by several socialists (see for instance Lee, 2001). Community participation has been pointed out by Yerramsetti (2005) as a key component of success of telecentres in ushering development and social change in rural areas.

The information revolution is another intervention with the potential to ensure that knowledge and information on important technologies, methods and practices are put in the right hands. The relevance of this revolution is supported by Balit (1998) who pointed out that the least expensive input for rural development is knowledge. Knowledge and information are basic ingredients of food security and are essential for facilitating rural development and bringing about social and economic change. According to Albert Waterson, as quoted by Cohen (1987), the purpose of rural development is "to improve the standard of living of the rural population – is multi-sectoral including agriculture, industry, and social facilities". Rural communities require information inter alia on supply of inputs, new technologies, early warning systems (drought, pests, and diseases), credit, market prices and their competitors. The success of the Green Revolution in Asia and the Near East indicates that giving rural communities access to knowledge, technology and services will contribute to expanding and energising agriculture.

ICT and Rural Development

ICT is an integral part of development strategies of both developing and developed countries. It has great potential to bring in the desired social transformations by enhancing access to people, services, information and other technologies (Dutton *et al.*, 2004). ICT applications can enhance poor people's opportunities by improving their access to markets, health, and education. Furthermore, ICT can empower the poor by expanding the use of government services, and reduce risks by widening access to micro finance (Cecchini and Scott, 2003). The uses of ICT for development are actively promoted, for economic development, job-creation, rural development and poverty-alleviation.

The term 'ICTs' can be interpreted as including a wide range of media, 'new ICTs' is used to denote "the use of computers and communication systems between computers" (CTA, 1999). Electronic mail is the most commonly used new ICT and has caused a cultural revolution in the way individuals and organisations interact, in terms of time, cost and distance. The second most significant use of new ICTs is the World Wide Web, which enables people to access information on millions of other computers.

Traditional media and new ICTs have played a major role in diffusing information to rural communities, and have much more potential. There is need to connect rural communities, research and extension networks and provides access to the much needed knowledge, technology and services (Forno, 1999). Studies on information systems serving rural communities have focused on specific sectors such as agriculture or health, instead of covering the rural community needs in a holistic manner. Rural information systems must involve rural communities and local content must be of prime importance (Mchombu, 1993). Traditional media have been used very successfully in developing countries, and rural radio in particular has played a major role in delivering agricultural messages. Print, video, television, films, slides, pictures, drama, dance, folklore, group discussions, meetings, exhibitions and demonstrations have also been used to speed up the flow of information (Munyua, 2000). New ICTs, however, have the potential of getting vast amounts of information to rural populations in a more timely, comprehensive and cost-effective manner, and could be used together with traditional media.

Indian model relies heavily on the Internet for the transfer of information, the local Digital Doorway focuses on the provision of educational and life skills material and both local and global information in an environment where Internet is not yet available. The kiosk serves as both a tool for computer literacy development and as a digital library/information centre, covering a range of subjects. Snapshots of the Wikipedia and Project Gutenberg form the mainstay of the information resources menu, with additional information packages on computers, crafts, agriculture, health and safety, literature, small businesses, accounting, budgeting, financial literacy and numeracy. PDF files on employment and finding work, empowerment and governance, the constitution, conservation, safety, hygiene and numerous other applications focus on life skills development. Users are able to create personal accounts on the Digital Doorway to which they can save documents and a USB port allows transfer of documents to a flash drive for later printing.

Although the Internet is not a panacea for food security and rural development problems, it can open new communication channels that bring new knowledge and information resources to rural communities (Bie, 1996). Traditional communication channels have been used successfully but these have been monologic and have not allowed for much interaction with users. For example, Radio has been very effective for disseminating information to all types of audiences, but broadcasting times are sometimes not appropriate for most people. But radio could be linked to the Internet, and a few initiatives have been started on this concept, such as the project Internet Radio in Sri Lanka.

Some examples of areas where ICTs could play a catalytic role in developing rural areas include:

Decision-making process: Sound decision making is dependent upon availability of comprehensive, timely and up-to-date information. Food security problems facing developing countries demonstrate the need for informed researchers, planners, policy makers, development workers and farmers. Information is also needed to facilitate the development and implementation of food security policies. e-mail and the Internet could be used to transmit information to and from rural inaccessible areas.

Market outlook: Farmers could promote their products and handle simple transactions such as orders over the web while payment transactions for the goods can be handled off-line (O'Farrell *et al.*, 1999). It has been shown to be cheaper and faster to trade online than on paper-based medium, telephone or fax. Electronic-commerce could, therefore, enable entrepreneurs to access global market information and open up new regional and global markets that fetch better prices and increase farmers' earnings.

Empowering rural communities: ICTs can empower rural communities and give them a "voice" that permits them to contribute to the development process. With new ICTs, rural communities can acquire the capacity to improve their living conditions and become motivated through training and dialogue with others to a level where they make decisions for their own development (Balit, 1998). Giving rural people a voice means giving them a seat at the table to express their views and opinions and become part of the decision making process. The approach should be participatory and could lead to improved policy formation and execution.

Targeting marginalized groups: Most rural poor people lack the power to access information. ICTs could benefit all stakeholders including the civil society, in particular youth and women (UNDP, 2000). Other disadvantaged groups that could be targeted include the disabled and subsistence peasants.

Creating employment: Through the establishment of rural information centres, ICTs can create employment opportunities in rural areas by engaging telecentre managers, subject matter specialists, information managers, translators and information technology technicians. Such centres help bridge the gap between urban and rural communities and reduce the rural-urban migration problem. The centres can also provide training and those trained may become small-scale entrepreneurs.

ICT Initiatives in India

1. *Akshaya – Mallappuram, Kerala:* In 2001, a joint project between local bodies (gram panchayats) in rural areas, municipalities in urban areas and private entrepreneurs in Mallappuram district of Kerala was started to bridge the digital divide by providing community access to computers and the Internet. Five hundred and sixty-five Community Technology Centres (CTCs) have been implemented in the district. Akshaya operates PPPs in establishing the CTC in remote villages.

2. *Anand Milk Collection Centres – Anand, Gujarat:* The Anand Milk Union Limited (AMUL) with more than 578,000 members was the first cooperative dairy to be established in Gujarat. It was established in 1956, and now AMUL collects 0.8 million litres of milk from 1,003 milk societies every day. AMUL introduced an electronic automatic milk collection system in 691 milk collection centres, which reduced the time required for collecting the milk. The system weighs the milk and measures its fat content at the time of delivery to the centre, and this has enabled immediate payments to the farmer, thereby eliminating the previously opaque procedures that often left the farmer short-changed and with little recourse for query.

3. *Bhoomi – Bangalore, Karnataka:* The Department of Revenue, Government of Karnataka, has computerized 20 million land ownership records of 6.7 million farmers in the state. Each record is available online from 177 taluka kiosks at a cost of Rs. 15 per record. The project has been widely acclaimed as possibly the most successful ICT project for land records in the country.

4. *Computer-Aided Administration of Registration Department (CARD) – Hyderabad, Andhra Pradesh:* In this project, 214 registration offices have been computerized since 1998. This project provides services such as encumbrance certificates, valuation certificates, market value search, *etc.*, on user charge bases. The time taken for registration of documents has been reduced from one day to 15 minutes. More than 4 million documents have been registered, and 2.16 million encumbrance certificates and 3.73 million registration slips have been issued since 1999.

5. *Community Information Centres (CICs) – Gangtok, Sikkim:* This project was started by the Department of Information Technology, Government of India with the technical support of the National Informatics Centre

(NIC). This pilot project started in 30 blocks of seven North Eastern states in 2001. Presently, all 487 blocks in 79 districts of the eight States, including Sikkim, are provided with one CIC at block level. The CICs provide e-governance, e-health, e-education and e-business opportunities.

6. *e-Choupal – Ujjain, Madhya Pradesh:* This web-based initiative of Indian Tobacco Company's international Business Division in Central India caters to soya growers for information, products and services required in soya farming. The kiosks facilitate the supply of high quality farm inputs and purchases of soya at the doorsteps of the villagers. This project was started in 42 villages of Ujjain district and around 1,800 kiosks in Madhya Pradesh and has around 3,300 kiosks in Central India. The kiosks also handle dealerships of various commodities like Hero cycles and Eicher tractors. This feature has provided extra benefits to the villagers in terms of minimizing their cost on travel expense.

7. *e-Seva – Hyderabad, Andhra Pradesh:* The project was started as a pilot in the twin cities of Hyderabad and Secunderabad, and was thus called TWINS (Twin Cities Integrated Network Systems). It was started at the cost of Rs 10 million, fully funded by the Government of Andhra Pradesh. The project provides registration of birth and death certificates, and vehicles and learners' driving licenses.

8. *Fast, Reliable, Instant and Efficient Network for Disbursement of Services (FRIENDS) – Thiruvananthapuram, Kerala:* This one-stop service centre uses computers to provide public services such as payments of electricity bills, examination fees, motor vehicle tax, building tax, property tax, water bills and telephone bills. This project is operational in all 15 district headquarters of the state. The services are provided on a user-charge basis and government officials operate the counters at the FRIENDS centres.

9. *Gramdoot – Jaipur, Rajasthan:* Aksh Optifibre Ltd is India's second largest manufacturer of optic fibre cables. The company has provided an integrated hardware and software solution for connectivity in the Gramdoot project. Gramdoot provides e-governance through broadband services to 200 gram panchayats in Jaipur district. The project also provides cable connections to rural households on which 32 television channels are telecast. High-speed non-dial-up Internet access at 70 Kbps is available to 200 villages. Land records, prevailing market rates of agricultural commodities, Hindi e-mail facilities, application for

certificates and online grievance reporting opportunities are also provided.

10. *Gyandoot–Dhar, Madhya Pradesh:* This project started as a comprehensive community network in Dhar district with 40 ICT-equipped information kiosks. More than 24 public services are installed, including land records, agriculture commodity rates, grievance opportunity, and applications to government departments. These are charged to users. All the kiosks are either community financed or privately owned. Wireless in Local Loop (WLL) technology is used. This project has been operational since, January 2000. It has been replicated in more than 45 districts in India.

11. *India Agriland – Nellikuppam, Tamil Nadu:* In this project, EID Parry, a 212-year-old private company, working in the field of sugar production, caters to 100,000 sugarcane growers. The company has partnered with N-Logue Communications Pvt. Ltd. in 48 kiosks and has been in operation since, 2003. The project disseminates market and commercial information to farmers and provides them with direct access to their markets. Information includes crushing details of sugarcane, payments due to farmers, local news, weather forecast, information on cultivation and farming techniques, e-mail, *etc.* These services are provided at user charges ranging from Rs 5 to Rs 10 per service. Information kiosks also collect soil samples for testing and sale seeds, sugar, tea and candies.

12. *Janmitra – Jhalawar, Rajasthan:* This project is a joint initiative of UNDP and the Government of India, operational since, 2002. It has been implemented with the help of district administration Jhalawar, Department of Information Technology (Government of Rajasthan) and RajComp, a state agency. A rural intranet provides e-governance, e-education, e-health and e-commerce services to the villagers through 30 Community Information Centres (CICs); 21 departments are connected to the server through dial-up connectivity and 13 departmental offices are on a Local Area Network (LAN) with the server. The kiosks also function as stamp vendors, petition writers, computer education providers and desktop publishing (DTP) providers.

13. *Mahitishakti – Panchmahal, Gujarat:* In this project, about 80 telecentres have been set up in Panchmahal district to cater to the information needs of villagers. The network provides more than 200 online forms of different government schemes. It also provides updated sanctions of development works from the District Rural Development Agency and the District Planning Board, along with some Geographical Information System (GIS) functions. The information is available in the local language (Gujarati), and the project has been operational since, 2001.

14. *N-Logue Telecentres – Madurai, Tamil Nadu:* N-Logue Communication Pvt Ltd, a commercial offshoot of Indian Institute of Technology (IIT), Madras, has pioneered its own version of WLL technology. In Madurai district of Tamil Nadu, 30 telecentres have been set up, which provide a link between the doctors at Madurai Medical College and the villagers. Besides telemedicine, several other web-based services are also provided to the villagers.

15. *Self Employed Women's Association (SEWA) – Ahmedabad, Gujarat:* SEWA is a large primary trade union working since, 1972 for women workers in the informal sector. SEWA's two main goals are full-employment and self-reliance. SEWA started using the SatCom (Satellite Communication) Programme in 1998 with its receiving terminals in nine districts and transmitting terminal in Gandhinagar. SEWA has started computer training for semi-literate women workers. Many of its milk cooperatives are using computerized milk collection software.

16. *TARAhaat – Jhansi, Uttar Pradesh:* The project was started by Development Alternatives (with the help of 12 project partners) in four districts of North India as a business model to cater to the unserved rural markets. The project provides services like TARAbazaar (e-bazaar), TARAvan (mobile kiosks), TARAguru (e-education), TARAdhaba (cybercafé), TARAreporter (news), TARAdak (e-mail), TARAvendor (e-commerce) and TARAcard (e-greetings). This project provides connectivity to franchisee kiosks through C-band satellite, VSAT or dial-up modem, according to the infrastructure available.

17. *Vidyal Information Service Provider (VISP) – Tiruchirapalli, Tamil Nadu:* Activists for Social Alternatives (ASA) has been working in five districts of Tamil Nadu in rural microfinancing since, 1993. It has 2,000 women's credit and thrift groups and has 60,000 women members. In May 2003, ASA launched VISP in six villages. The project provides services such as prices of agriculture commodities, horoscopes, rural market place, matrimonial services, educational services, grievance opportunity, government forms, *etc.*, by using the software developed by Drishtee Foundation. The kiosks also provide services like web-browsing, DTP, data entry job work, net-to-phone and basic computer education.

18. *Warana Wired Village – Kolhapur, Maharashtra:* Seventy villages in Kolhapur and Sangli districts of Maharashtra have been linked through a Wide Area Network (WAN) using dial-up connectivity and VSAT technology. This project aims to provide benefits to members of the sugar cooperative and the villagers. More than 12 public services, such as measuring the content of carbohydrate in sugarcane, payments due to

farmers, land records, *etc.*, have been introduced. This project was started in 1999 by the Government of India, the Government of Maharashtra and Kolhapur Sugar Cooperative on a cost-sharing basis.

19. *Bellandur Gram Panchayat Computer System:* The Bellandur Gram Panchayat computer system was begun in response to community members' requests for increased public projects. The Gram Panchayat (village administration), in order to finance such projects, installed a computer system for tax-revenue management. The system offers e-Governance services, consisting of bill payment and accounting for the Panchayat. The project's location within the Gram Panchayat building means that users may also access additional Panchayat services. Bellandur offers few informational and transactional services. The informational services cover basic government information (such as contacts and procedures) and local news. These are also available through signboards, paper documents, and local television broadcasts (which also broadcasts Panchayat meetings).
20. *Boodikote Jagruthi Resource Center:* The Jagruthi Resource Center, set up in 2002, primarily addresses a wide range of informational needs, in line with its objectives. Informational services offered include basic agriculture, education, news and health information, information on managing Self Help Groups (SHGs), and legal and financial information concerning loans and insurance. The center benefits from co-location and linkage with Namma Dhwani, the community radio, which is popular in Boodikote, through sharing of traffic and information. A number of transactional services are offered, including explanations of procedures and printout of forms for loans, insurance and applications for utility setup, matching buyers and sellers of goods and services locally, and entertainment. Among the e-Governance services, the center provides form printouts and an opportunity to submit grievances to the local government.
21. *HP iCommunity in Kuppam:* Kuppam CIC's broader goal is to provide all ICT services. Informational services make up the bulk of Kuppam's offerings, covering information on government programs and contacts, agricultural and veterinary information, information on employment, education, and health and local news. SHG management information and computer training are also offered. Kuppam's transactional services include email, sale of insurance, horoscopes, and digital photographs. Entertainment services, web browsing, games, and watching movies, are popular.
22. *M.S. Swaminathan Research Foundation Info-Village Knowledge Centres :* MSSRF's Info-Village emphasizes information services. Much of this

information is derived from the Internet and broadcast in innovative ways, such as the public address system at fishing villages or the siren that awakes fishermen when it is time for them to begin the fishing day. Local language newspapers and signboards outside the centers are also an effective way to spread knowledge. People in the Info-Village communities report a high level of satisfaction. Some transactional services are offered, but there are few e-Governance services. Communication is the most used transactional service, followed by loan applications. Entertainment and SHG accounting services are available.

23. *n-Logue Chiraag Kiosks:* n-Logue is India's largest operator of for-profit rural kiosks, under the "Chiraag" brand, focusing on fee-based transactional services. Chiraag kiosks provide a substantial number of informational services using content that is localized and stored on site. Basic informational offerings include agriculture, education, health, government programs, and local news. N-Logue's aims include providing e-Governance services, but the absence of adequate digitized content such as downloadable certificates, limiting to e-Governance services. Still, submission of grievances and passport and visa services are available.

Lessons from ICT Projects

1. *Absence of useful content:* With some exceptions, such as Bellandur, the absence of a compelling ("killer") application rather than inadequate technology, location, electric power or management, is the reason for sparse usage. Most rural ICT projects are, at best, in the same position as n-Logue's Chiraag kiosks, which, despite having an inventive technology infrastructure, are unable to attract users. Users appear to be more concerned although largely unable to identify the specific programs they miss. Low content relevance is probably a combination of unfulfilled promises from government, language and other localization issues, poor awareness (among both users and providers) and inadequate understanding of user needs. Less skilledoperator selection and inadequate training and lack of local listening.

2. *Lack of awareness among users:* In many cases, users were unaware of the full range of available services and concluded that the computer was either for a specific use, such as bill payment (AP-Online) or accounting (Bellandur), or a toy (some of the n-Logue sites). For example, the agricultural advice system at Kuppam is vastly underutilized because farmers do not know that they can use it to interact with an agricultural extension officer. Similarly, in Gyandoot, many were unaware of the e-Governance services available, and viewed the kiosks simply as centers

for telephony. Some sites did not allow customers to use the computer, thus reducing the benefits of experimentation and, potentially, causing low usage.

3. *Less qualified operator selection and inadequate training:* The centre operator needs to understand the basic technology, such as how to navigate the Internet or to maximize the use of the available tools, and marketing. One Kuppam Community Information Center (CIC) was tucked quietly into a retail shop, where it received few users and earned only Rs. 200/ month in revenue. In both cases, usage was correlated with the operator's level of ingenuity and effort.

4. *Segments of the population are sometimes excluded:* Several projects discriminate among the local population, usually unintentionally. For example, the e-Choupal and Warana initiatives serve farmers, thus excluding the female and landless population. The e-Choupals are located in individuals' homes and the home-owner control the access. Authors of this article visited more than one chaupal in Madhya Pradesh in which members of the backward castes were not allowed to enter the house in which the e-Choupal was located. The dearth of female kiosk operators(though some initiatives like n-Logue and Gyandoot encourage women to become operators) also discourages female clientele. Early on, several MSSRF Knowledge Centres located in the private houses were closed precisely because they were exclusive (now the organization, as a policy, locates centres only in public buildings). By contrast, Bellandur is an example of a project that, by design, serves all villagers.

   It is impossible for a site to serve every villager and it could be argued that a viable starting point is key, even if it discriminates by plucking "low-hanging fruit", such as the richer farmer. The question is whether an alternative approach, while focused on a single service, can address the broad scope of poverty, gender discrimination and lack of basic social services. Our findings show that, while usage is generally low across all the projects, levels of satisfaction are as high with projects, such as MSSRF Knowledge Centers, that are broad-based as those that serve wealthier segments.

5. *Undefined expectations in sharing project management:* Several projects began as partnerships or involved contractual relationships among several parties. For some, failure was often attributed by one party to other parties that believed that their responsibility was complete. The failures involved both content and technological services. For example, in the Warana project (after the initial project phase), content maintenance was neglected, and technology frequently broke down.

Village operators attributed these failures to external providers' inability to provide skills that they did not and were not expected to have. This problem may be tied to the absence of a program to build-up rural capabilities prior to exit, thus compromising the project.

6. *Government resistance to e-Governance:* Government plays a very important role in rural areas as a provider of social and economic services and subsidies, and as tax collector. If adequately provided, e-Governance services could be the "killer application" that most initiatives lack. However, the government has failed to deliver e-Governance as promised. The problem appears to be less severe where the government has a financial or reputation stake. For example, AP-Online has been able to provide bill-paying services due to the government's financial involvement. On the other hand, n-Logue and Kuppam initially intended to serve as an interface to the government, particularly to provide official government documents, such as birth and death certificates, caste certificates, income certificates, and land records. Their expectations were based on promises from high government officials on the provision of electronic content. But this was an unsustainable basis and both projects could not provide e-Governance services because the government failed to provide digitized content.

   An alternative may be to begin with e-Governance projects that are localized. This is because village officials are more likely to be accountable than more remote officials who work out of the state or district headquarters. The Bellandur project may, therefore, be a useful indicator of sustainable e-Governance. Of course, there are no easy solutions. But the experiences described above do suggest the need to develop an innovative approach for the provision of e-Governance across all the Panchayati Raj institutions, *i.e.*, not just at the village level, but at the block and district levels as well.

7. *Lack of a realistic financial model:* For all the projects, the costs of deployment are very high relative to revenue. The major deployment cost is building out the technology infrastructure. Despite claims by some site managers, usage at most sites is simply not high enough to cover costs in the foreseeable future. This is true even for sites with high user satisfaction.

8. *Widespread infrastructural and hardware problems:* Both power shortages and connectivity are major issues. In seven of the nine initiatives, we witnessed significant power outages. The outages were most prevalent in Madhya Pradesh (eChoupal and Gyandoot) where some kiosks had power for fewer than four hours a day. While all of the projects do have Uninterrupted Power Supplies (UPS) as backups, and some even have

backup solar power, these backup supplies are ineffective against such massive grid power failure. In some locations, the UPS systems are broken or last for only ten minutes. The solar panels provide energy for only a few hours more. Battery-operated devices that require intermittent charging may be a solution.

However, most such devices, such as laptop computers, are currently designed for portability rather than to overcome the shortage of grid power, thus raising costs to unviable levels. Since most computing and transmission equipment is inherently low-power consuming, technological solutions to the problem of power outages may exist. Connectivity is also an issue. For example, some Warana centers we visited (which had purchased connectivity from n-Logue), the authors of this article witnessed connectivity failure for two days at the time of their visit. Failing hardware also poses a problem at some sites. For example, MSSRF machines, which are donated secondhand systems, are often unreliable and require costly maintenance that can offset the costs saved through buying second-hand. While making use of secondhand computers is an innovative idea, such an approach unduly burdens projects with maintenance costs. Warana faces a similar issue, systems at that site are now six years old and require significant maintenance.

9. *The projects have not adequately transferred capabilities to rural areas:* A key issue for sustainable rural development is the transference of capabilities to rural areas.

Success Factors of ICT Projects

1. *Use of the Internet as a carrier of informational services :* While few users or kiosk operators know how to search the Internet, sites that provide useful information such as local news and agricultural information are accessed online, thus supplementing local content. Warana's village booths, for example, carry only the software necessary for sugarcane management, but users use the Internet for accessing sites of local interest. Strategies to leverage the Internet better are needed for several reasons. First, as noted, the Internet enables users to access information not available onsite, especially when updated information is needed, such as the latest weather report. Second, it enables content developed for a given village to be used in villages. For example, if an NGO develops digital content to train users to monitor infant health, it may be more useful if that content is available over the Internet so that such training can be accessed more widely, such as by the NGO in contiguous villages or even, perhaps, by other NGOs in distant locations. Third, access to the Internet facilitates

transactions, including e-Governance. Some sites, such as AP Online and eChoupal, use the Internet for transactional services.

2. *e-Governance services attract the most users:* The villager is typically dependent on a range of government services: information on employment in government projects; customized information such as land records and birth certificates; approvals, such as for 'below poverty line' status, and grievance redress; health, education, entitlement and other social services; mandatory services: taxation, updating land and population databases; exchange services: postal, banking and utility services. e-Governance, though sparsely provided, is the most popular service, perhaps owing to higher transparency and convenience. For example, at AP Online centers, users paying utility bills save travel over long distances at a small additional cost. Likewise, Bellandur's users benefit from higher transparency, which leads to less corruption.

3. *Path-dependency (Initially positive outcomes enhance acceptance and vice versa):* Providing one type of service effectively in order to capture a receptive audience, and then using that audience to capture more users and offer related services is at present the best method to increase usage and provide needed services (Moore, 2002). Bellandur's success in providing a transparent bill-paying service raised usage of related services, such as obtaining information on licenses and accessing public records of meetings. On the other hand, one of Kuppam's first projects (which offered e-Governance services such as certificates) failed, and appears to have affected its usage generally.

4. *Community involvement adds value to informational services:* The MSSRF and Boodikote initiatives are examples of community-based organizations that offer valuable informational services. In the fishing villages surrounding Pondicherry, for example, the MSSRF Knowledge Center saves lives by sharing expected wave patterns for the day. Because of its relationship with community members, the Boodikote center is made aware of and is able to respond to local problems such as the shortage of nurses at the local Primary Health Centers. This is not surprising, since communities know their needs best.

5. *Partnerships for content development add value:* Since content requires domain knowledge, it is perhaps a truism that content development would require partnerships with experts. Those recognizing this have done well. For example, n-Logue offers videoconferences and web-chats across a region with experts on health, agriculture, animal husbandry, education, government, and financial matters. HP's Kuppam CIC partners with financial institutions to provide banking and insurance

services. MSSRF's Info-Villages are linked to many NGOs and local CBOs, such as the local coconut farmers association, the local dairy cooperative, and the Rajiv Gandhi College for Veterinary and Animal Sciences. The need for partnerships in content development raises a problem: NGOs and government providers of ICT services at the local level have traditionally not looked to partner with each other or the private sector.

6. *Intermediaries improve usage:* At many sites, adults were hesitant to use computers. Hence, intermediaries can be important. The kiosk operator plays this role, often supplemented by children initially playing games. This was noticed, for example, in Gyandoot, an area with low levels of education and literacy. As we have noted earlier, however, the inability of end-users to use computers directly can be detrimental to overall usage of the kiosk.
7. *Attention to maintenance adds value:* At sites where the infrastructure often failed, such as at Gyandoot, either due to power shortages or poorly maintained computers, users were less willing to use ICT. By contrast, n-Logue and e-Choupal manage infrastructure well with correspondingly higher usage. n-Logue has three engineers at each regional site, usually located within 25 km of a kiosk. Additional engineers are located at zonal project sites, each of which oversees three access centers. At e-Choupal, engineers are located at each hub (where farmers sell their goods), each responsible for 80 choupals. Though e-Choupal in Madhya Pradesh suffers from the same electricity shortages as Gyandoot, the presence of solar backups and the engineers' contribution to ensuring connectivity helps it to garner more users than the Gyandoot project.

Points to be Considered to Rejuvenate the ICT Projects

Some general problems that have been experienced in a number of ICT-based rural projects and initiatives include:

1. *Policy considerations:* In most developing countries, especially in Africa, the formulation and implementation of policies in the ICT sector is still very rudimentary and calls for an integrated set of laws, regulations and guidelines that shape the generation, acquisition and utilization of ICTs. Most countries lack policies and strategies that facilitate the harnessing of new ICTs for rural development and where policies have been formulated, proper implementation plans are needed (Fillip, 2000). In addition, review strategies are often lacking. At present, the regulations are rigid and telecommunication tariffs and import duties on ICT equipment are high. The situation is compounded by lack of political good will.

2. *High telecommunication costs in some developing countries:* The cost of basic Internet remains a strong deterrent in many developing countries such as Angola, Kenya, Cuba and India. Although market liberalization has led to the entry of several private sector ISPs, service provision is through government phone companies, whose service is inadequate in terms of robustness, low bandwidth, congestion and noisy lines. Others make international calls to ISPs in the United States or elsewhere in Europe, which renders the service rather expensive.

3. *Infrastructure:* The telecommunication and electricity infrastructure in developing countries is lacking or is poorly developed in rural areas. Satellite and wireless technologies are now in use in some developing countries, but these are largely developed around urban cities and even here, the infrastructure is often inadequate. There are problems of low bandwidth and there is a need for strengthening the Internet backbone.

4. *Lack of local content and language barrier:* Information available through ICTs is mostly in English, which the majority of developing country rural communities cannot read. There is a marked shortage of relevant material in local languages that responds to their needs and this calls for "significant investment and support for local content" (O'Farrell *et al.*, 1999). Bartecchi has described one example in India – the Village Earth Project and its Consortium for Sustainable Village Based Development. The project works with NGOs and individuals to help connect rural villages and develops appropriate content using ICTs for individuals or organizations in rural areas. The project has a library of books, microfiche and CD-ROMs that provide appropriate information resources but most of these materials are in English, and may not be appropriate for every region or culture. Thus, there is need to develop more local content in appropriate languages (David Bartecchi, personal communication 1999[3]). Various local, national, regional and international organisations have useful information resources (statistical, bibliographic, factual and full text) of relevance to rural communities, packaged as print, audio tapes, radio and television programmes, videos, CD-ROMs and on the World Wide Web. For many of these, there is lack of bibliographic control and there are no commonly agreed standards.

5. *High rate of illiteracy in rural areas:* "Illiteracy is a fundamental barrier to participation in knowledge societies" (Mansell and Wehn, 1998). A large proportion of the rural population of developing country nations, majority of who are women, is illiterate and most pictographic and audio-visual information usually has some text that goes with it. This means that these individuals are disadvantaged and lack the basic skills required

to harness the benefits of ICTs. The assistance of intermediaries may thus be required.

6. *Gender insensitivity:* In a developing country, men and women play different productive and community roles in rural development and have different needs and preferences. Women produce more than half the world's food (World Bank, 2000; FAO, 1998) and face many problems in addressing food security and rural development. These include weak extension services, non-adoption of technologies, low status and therefore non-involvement in decision- and policy-making, varied and heavy workloads, poor access to credit; and lack of access to education and training (Munyua, 2000). When new technologies are introduced, they are seen as a domain for men, and women have often been left out of initiatives associated with new ICTs. Rural women, however, have wisdom and indigenous knowledge that is rooted on culture, traditions, values and experience. Their methods of communication and information exchange should thus be harnessed and be complemented with new ICTs (FAO, 1998). Women also spend most of their income on family welfare and have a greater impact on increasing agricultural productivity and improving the quality of family life.

   The youth are another special group that deserves special mention. In Kenya for example, out of the present 28 million people, about 50% are youth, below 18 years of age. The youth have been given little opportunity to contribute to rural development issues, despite their numbers, fresh and innovative ideas. The School Net model has been used in some developing countries (South Africa, Uganda,) to connect the youth to the Internet to enable them to move into information and knowledge based societies and develop young talent.

   Women, youth and the disabled will, therefore, require special treatment (affirmative action), and should be deliberately integrated into all ICT projects and initiatives through gender sensitive project development and implementation (FAO, 1998). It is only by doing so that a critical mass of information aware people, and leaders of tomorrow can be developed to ensure that their countries realise food security and rural development goals.

7. *Inadequate human resources:* To ensure more meaningful participation in rural development, and to pave way for the creation of a critical mass of people that effectively harness ICTs in developing countries, training and capacity building must be an integral part of all ICT projects. It has been observed that "a critical factor in meeting the challenge of ensuring food security in Africa is human resource development through

knowledge building and information sharing" (Forno, 1999). Users of ICTs have to be trained in the use, application and maintenance of ICTs before they become confident and comfortable enough to use them.

Most of the staff managing ICT-based projects lack adequate training that would enable them to take full advantage of the new technologies. There is need to invest in training and advisory services for information intermediaries, telecentre staff, frontline workers and women's groups. Norrish (1999) has pointed out the need to identify the best training approaches for rural communities targeting different user categories and different technologies. Such training could be done through conferences, workshops or training of trainers' courses. Introductory and sensitising workshops could be organised for different categories of users and these local experts could provide ongoing on-line support. People in rural communities of developed countries are already reaping the benefits of electronic distance education and developing countries could offer similar services through selected centres.

Focus should be on such skills as how to use ICTs through practical and participatory approaches. The 'Self Employed Women's Association' in India, for example, has trained rural women in the production and use of video to generate income, disseminate new skills and to advocate for changes in policy (Balit, 1998). Some key players in the training activity have been FAO, CTA, IDRC, IFAD, UNESCO and ministries of agriculture. Traditional media will remain important and should continue being used alongside new ICTs. "It is through the use of a variety of media and their integration with local communication networks that more people throughout Africa can be heard and can be reached" (FAO 1998).

8. *Sustainability of projects:* Most projects established with external funding face major challenges after the project period has ended. Sustainability of these projects should be considered right from the outset and, wherever possible, should have government, private sector and community support. Users should also pay for services but the cost will depend on how much they can afford. There are as yet few examples of success in attaining such sustainability, and there is an urgent need for viable models to be developed and tested.

Conclusion

The rural poor are the engines of agricultural production in developing countries. Agricultural production and post-harvest activities account for the primary livelihood assets and strategies available to the rural poor. Any bottlenecks to the general livelihood of the rural poor - health, disaster, education, lack of infrastructure, lack of financial services, and many others - will have significant impacts on Agricultural production at the household, regional and national levels. ICT interventions that improve the general livelihoods of the rural poor may also yield significant Agricultural development investments on the part of rural families.

ICTs can be harnessed to have a direct impact on enhancing both rural livelihoods and the Agricultural sector in which the rural poor are immersed. However, ICT project investments driven by the "technology of the day", or by approaches that neglect to analyze how ICTs can truly enhance rural livelihoods, are to be avoided. A full range of rural livelihood assets and strategies should be examined to enable us to better understand the relationships between ICTs, agriculture and poverty reduction.

A variety of fairly straightforward rural ICT interventions may have greater impact on Agricultural production and post-harvest activities than those that are strictly focused on agriculture. This is especially true of ICT interventions focused on extension of various financial services, provision of basic telephone access, and improved multi-stakeholder dialogue and louder rural Agricultural voices in the national policy and program context.

If extension personnel equip themselves with analysis and arguments to bolster the case for Agricultural extension playing a broader role in helping harness ICTs, they will be better able to harness information and communication technologies (ICTs) to improve rural livelihoods.

References

Balit, S., 1998. Listening to Farmers: Communication for Participation and Change in Latin America. In: *Training for Agriculture and Rural Development*, 1997-98. FAO, Rome Italy. pp. 29-40.

Bie, Stein, 1996. Development Communication and Internet: Introduction. FAO.

Burkey, Stan. 1993. *People First: A Guide to Self-reliant, Participatory Rural Development.* Stan Burkey, 29-39. London and New-Jersey: Zed Books Ltd, 1993.

Cecchini, Simone and Scott, Christopher, 2003. Can Information and Communications Technology Applications Contribute to Poverty Reduction? Lessons from Rural India. *Information Technology for Development*, Vol. 10, Issue 2 (2003): 73-84.

Chambers, Robert, 1983. *Rural Development: Putting the Last First*, Longman, London.

Cohen, J.M., 1987. *Integrated Rural Development: The Ethiopian Experience and the Debate.* Scandinavian Institute of African Studies, Uppsala, Sweden.

CTA. 1999. Information and Communication Technologies: A Remarkable Revolution. *SPORE,* No. 79 February 1999. pp. 4-5.

Dutton, William H, Sharon Eisner Gillett, Lee W McKnight and Malcolm Peltu. 2004. Bridging Broadband Internet Divides: Reconfiguring Access to Enhance Communicative Power. *Journal of Information Technology.* 19(1) (2004) : 28-38.

Fillip, B., 2000. *Distance Education in Africa: New Technologies and New Opportunities.* Japan International Cooperation Agency. USA. 8 pp.

Forno, D.A., 1999. Sustainable Development Starts with Agriculture. In: *Sustainable Agriculture Solutions: The Action Report of the Sustainable Agriculture Initiative.* Novello Press Ltd, London. UK. pp. 8-11.

FAO. 1998. Knowledge and information for food security in Africa: From traditional media to the Internet. 22 pp. FAO Research, Extension and Training Division. Communication for Development Group, Extension, Education and Communication Service (SDRE).

Garai, Atanu and Shadrach. B., 2006. Processes and Appropriation of ICT in Human Development in Rural India: Bridging the Research and Practice Gaps. In: Atanu Garai and B. Shadrach (Eds.), *Taking ICT To Every Village,* 1-35. New Delhi: Oneworld South Asia, 2006 (accessed in February, 2006 from http://www.dgroups.org/ groups/oneworld/ OneWorldSA/docs/TICTEIV_pdf.pdf) .

Jhunjhunwala, Ashok, Anuradha Ramachandran and Sangamitra Ramachander. 2006. Connecting Rural India: Taking a Step Back for Two Forward. *Information Technology in Developing Countries,* Vol. 16, No. 1 (February, 2006), Telecommunications and Computer Networks Group, Madras, IIT-Madras. http://www.iimahd.ernet.in/ egov/ifip/ feb2006/article1.htm (Accessed on December, 2006) .

Lee, R., 2001. Community Development and the Internet, Brussels: Research Centre 'Communication for Social Change' (CSC). Paper prepared for presentation at the International Conference on Information Technology, Communications and Development in Kathmandu, Nepal in November 2001.

Mansell, R. and Wehn, U. (Eds). 1998. *Knowledge Societies: Information Technology for Sustainable Development.* United Nations Commission on Science and Technology for Development. Oxford University Press. New York, USA.

Mchombu, K., 1993. Information Provision for Rural Development: A Final Report on Phase One of the INFORD Research Project. IDRC. 17 pp.

Moore, Geoffrey A., 2002. *Crossing the Chasm.* New York: Harper Collins, 2002.

Munyua, H., 2000. Application of Information Communication Technologies in the Agricultural Sector in Africa: A Gender Perspective. In: Rathgeber, E. and Adera, E. O. (Eds). *Gender and Information Revolution in Africa,* IDRC/ECA. pp. 85-123.

Norrish, P., 1999. Advancing Telecommunications for Rural Development through Participatory Communication.Richardson, D. 1999. The Internet and rural development.

O'Farrell, C., Norrish, P. and Scott, A., 1999. Information and Communication Technologies (ICTs) for Sustainable Livelihoods: Preliminary Study. April – Nov 1999.

Singh, Katar. 1999. *Rural Development: Principles, Policies and Management,* (Second Edition), Sage Publications, New-Delhi.

Yerramsetti, Srinivas. 2005. Role of Information Technology for Rural Development – A Case Study of Rural e-*Seva* Project in West Godavari District, M. Phil Dissertation at Jawaharlal Nehru University (JNU), New Delhi: 126.

World Bank. 2000. Agricultural Knowledge & Information System: Gender - Topic Brief. http://wbln0018.worldbank.org/essd/susint.nsf/4e7577b391e880c2852568ad00790893/3dae7c6ac8a33cad852568b000167fef?OpenDocument.

UNDP. 2000. http://www.undp.org/info21/index5.htm and http://www.undp.org/info21/pilot/pi-sa.html

http://www.fao.org/sd/Cddirect/Cdre0055b.htm

Warana Wired Village Project, Warananagar.25 Jan. 2005,

<http://www.warana.org/home.asp>.

Taarahat.com. 14 Oct. 2003 <http://www.tarahaat.com/tara>.

Soyachoupal. 31 Jan. 2005. <www.soyachoupal.com>.

Naidisha. Government of Haryana. 12 Oct. 2003 <http://naidisha.nic.in/>.

Kuppam iCommunity. 25 Jan 2005. <www.kuppamhpicommunity.stph.net>.

Information Village Research Project, Union Territory of Pondicherry.

http://www.mssrf.org/special_programmes/ivrp/ivrpmain.htm

eSevaOnline. 10 Oct. 2001. Government of Andhra Pradesh. 12 Oct 2003.

http://www.esevaonline.com/

Drishtee. 12 Oct. 2003 www.drishtee.com

Community Information Centers. 12 Mar. 2003. Government of India. 13 Oct. 2003. http://portal.unesco.org/ci/ev.php?URL_ID=1259&URL_DO=DO_TOPIC&URL_SECTION=201&reload=1066684392

www.iimahd.ernet.in/egov/documents/ict-and-egovernance-for-rural-development.pdf

APOnline. 25 Jan 2005. www.aponline.gov.in

www.odi.org.uk/networks/agren/papers/agrenpaper_135.pdf

www.caribank.org/titanweb/cdb/webcms.nsf/AllDoc/5AD04502737E512D0425741E006B2739/.../...

www.anancy.net/documents/file_en/WD8034.pdf

www.departments.agri.huji.ac.il/economics/gelb-how-11.pdf

www.uncapsa.org/Flash/flash0208.pdf

www.archiv.rural-development.de/fileadmin/rural-development/volltexte/2005/05/ELR_dt_18-19.pdf

4

Cyber Extension: Info-tech Empowerment of Farmers of India

M.A. Ansari and I. Prabhakar

ABSTRACT

Recent advances in Information and Communication Technologies (ICTs) have changed the way we live, learn and interact. They have transformed the way knowledge is produced, processed, retrieved and transferred between different stakeholders. The innovative ICT applications are dominating and driving the development scholarship and thinking in agriculture sector. The increasing availability and use of the Information and Communication Technologies offers the Extension System and all its stakeholders/partners an undeniable opportunity to redefine the way it provides information and interacts with farmers. ICTs are not just a new tool for information dissemination; it is empowering and changing the very culture of society's information-seeking behavior. And Extension must change, too. This change opens the door to serve a more diverse audience, build a broader constituency, and make objective information and services available from the existing national agriculture research and extension system. Historically speaking, the information and communication support for transfer of agriculture technologies during the last five decades has been mainly conventional. The existing system of agricultural extension is a moribund superstructure wherein the grassroots-level key functionary (*i.e.* Extension Personnel) suffers from low perceived competency, high levels of de-motivation and frustrations, systemic operational inefficiency besides fiscal constraints. Consequently, a new paradigm of agricultural extension is fast emerging in both developing and developed countries and expanding in all directions. This involves harnessing the potential of recent advances in the field of Information and Communication Technology (ICT) for transfer of agricultural products and technologies. Cyber Extension refers to the systematic use of the power of online networks of computer, computer based communications and digital interactive media to facilitate the dissemination of information to the intended uses. The aim of this paper is to present an enabling framework of cyber-extension that will attempt to harness the potential of ICTs and facilitate delivering of agricultural information at farmers' door-step, advice and consultation for high-tech agriculture and precision farming. Rural Knowledge Centres (RKCs) established under India's "Mission 2007 : Every village a knowledge centre" has been the focal point of nationwide efforts to mobilize the private and public sectors, including NGOs, Civil society organisations, mass media and academia, to apply ICTs to cater the Information, Communication and Training needs of rural India. The prevailing technology environment can provide ample opportunities for skill up-gradation of rural youth for possible diversification in agriculture sector and catalyze community based development initiatives through value addition in agricultural products supply chain and triggering entrepreneurial activities in villages through SHGs.

Introduction

The World Bank report in 2000 articulated that "more than natural resources, more than cheap labour, more than financial capital, Knowledge is becoming the key factor of production". Access to information and improved communication is a crucial requirement for sustainable agriculture development. Modern communication technologies, when applied to conditions in rural areas, can help improve communication, increase participation, disseminate information and share knowledge and skills (Anandaraja *et al.*, 2005). Recent advances in Information and Communication Technologies (ICTs) have changed the way we live, learn and interact. They have transformed the way knowledge is produced, processed, retrieved and transferred between different stakeholders. The innovative ICT applications are dominating and driving the development scholarship and thinking in agriculture sector.

The increasing availability and use of the Information and Communication Technologies offers the Extension System and all its stakeholders/partners an undeniable opportunity to redefine the way it provides information and interacts with farmers. ICTs are not just a new tool for information dissemination; it is empowering and changing the very culture of society's information-seeking behaviour. And Extension must change, too. This change opens the door to serve a more diverse audience, build a broader constituency, and make objective information and services available from the existing national agriculture research and extension system.

In order to transform Indian agriculture into information driven, market-led and competitive business enterprise, the role of ICTs can't be over emphasized. The report of the "Task Force on India as Knowledge Super Power" by the GoI in 2001 emphasized the necessity of developing the capacity to generate, absorb, disseminate and protect knowledge and exploit it as powerful tool to drive societal transformation. ICTs are seen as a means of achieving such a transformation.

The people working in agriculture sector are least equipped with required knowledge and skills to deal with imperatives of a globally competitive agrarian economy. Besides, the existing Transfer of Technology mechanisms and extension programmes, mostly run by the government, are no longer relevant and able to meet the increasing demands of the farming community. They are woefully inadequate and ill-equipped to act as bridges between research system and client system (Farmers).

Information is Crucial

In order to have a competitive edge in a globalised agrarian economy, Indian agriculture, especially agricultural extension system, would have to reinvent itself to meet the demands of precision farming and high-tech agriculture. This would require upscaling of competencies of extension workers as well as bring radical changes in their outlook, perceptions, motivation level and attitude towards their obligated responsibilities. This is a gigantic task requiring Herculean efforts to spruce-up the infrastructure, innovative approaches to agriculture extension and streamline operations management in an effective and efficient manner.

Historically, the information and communication support for transfer of agriculture technologies during the last five decades has been mainly conventional. The existing system of agricultural extension is a moribund superstructure wherein the grassroot-level key functionary (*i.e.* Extension Personnel) suffers from low perceived competency, high levels of de-motivation and frustrations, systemic operational inefficiency besides fiscal constraints. Consequently, a new paradigm of agricultural extension is fast emerging in both developing and developed countries and expanding in all directions. This involves harnessing the potential of recent advances in the field of Information and Communication Technology (ICT) for transfer of agricultural products and technologies. Recognising the potential of ICTs, to transform rural communities, both socially and economically, the Government of India is committed to extending the telecommunications infrastructure and expanding access to telecommunications services to all parts of the country.

Cyber Extension

Cyber: According to the Oxford dictionary, the literal meaning of the word "cyber" relates to information technology, internet and virtual reality, *i.e.* cyber space.

Extension: Extension stands for "the action or the process of enlarging or extending something".

Cyber Extension, thus, can be defined as the systematic use of the power of online networks of computer, computer based communications and digital interactive media to facilitate the dissemination of information to the intended users. It includes the effective use of information and communication technologies, national and international Information Networks, Internet, Expert Systems, Multimedia Learning Systems and computer based training systems to improve information access to the farmers, extension workers, research scientists and extension managers (Sharma, 2004).

In fact, cyber extension is an ICT enabled approach of extension linking farmers to key stakeholders. The new system does not intend to replace the existing system of extension; rather, it will augment its capacity through enlarging its reach and providing two-way instant interactivity/ connectivity between different stakeholders. The continuing expansion of telecommunication infrastructure and its ever-increasing penetration into the new rural/ remote areas will facilitate and reinforce agriculture extension services. Cyber extension system will bring increased access to new information services and provide greater control over the transfer of technology. It will empower the farmers in improving their decision making abilities and reduce their dependence over extension workers. It will provide door-step delivery of information and communication services right at the click of a button.

Cyber Extension System: Getting Started

Cyber Extension System attempts to harness the potential of ICTs and facilitate electronic delivering of agricultural information, advice and consultation to the target clients/farmers for high-tech agriculture and precision farming. This has many advantages over the traditional system of extension which are as follows:

1. *Access to Extension System: Anytime Anywhere Anyone:* The key attribute of Cyber Extension system is its non-stop and continuous availability – 24 Hours, 365 days a year – to the target users/ farmers. Besides, the

Cyber Extension System would eliminate the time and distance barrier that are associated with traditional system of extension. The farmers and other stakeholders can get the latest information on any particular problem and can get the expert advice from the best expert/ scientist from any part of the world without the need for moving from their place of residence.

2. *Saves Time, Money and Effort:* Electronically updated versions of information can be loaded into computers, hosted on websites or e-mailed to the prospective farmers and others as and when desired. This way, it will save money and time to reach the end-user instantly.
3. *Eliminated Time Lag in Diffusion/ TOT:* Through ICTs, we can disseminate the information about new technology instantly and simultaneously *via* videoconference or teleconferencing. Besides, the needs of capacity building of lower level extension personnel can also be addressed through online learning capsules, thereby eliminating the need for Monthly/ Zonal workshops for briefing the extension workers.
4. *Information Rich and Interactive:* The extension literature can be converted in the forms of multimedia interactive modules (on CDs) which will make learning interactive as well as entertaining. The information dissemination can be enriched through animated online learning which can be centrally delivered as well monitored.
5. *No Distortion:* The information disseminated through traditional system of extension gets distorted while in transit through many layers of extension personnel. However, in ICT-enabled systems of e-Extension, the number of actors will be reduced. Besides, the information can be packaged in such a way (*eg.* in CD-ROM) that it dose not get changed.

Proposed Model of Cyber Extension

The increasing availability of ICTs and its use offers an undeniable opportunity to redefine the way extension activities are undertaken. In the 21st century, extension needs to become the resource of choice and discover how to respond to the multifarious information needs of the client farmers in high tech agriculture and precision farming. Because of communication and education role of our extension system, it is incumbent on us to design a system that will serve the interests of farmers nationally as well as regionally; and enable Indian agriculture to survive in a global agrarian economy.

The goal of the proposed Cyber Extension system is to plan and implement a web-based information and education network leading to info-tech empowerment of farming community. The key elements of such a system would include the following components:

1. Knowledge Warehousing through World Wide Web

The Web, also referred as World Wide Web, is the largest and richest source of knowledge which can effectively meet the requirements of ICT enabled agricultural information system. Its massive digital holdings covering all aspects of World agriculture, natural resources and food systems enable farmers to locate needed information to improve yields, plan for weather contingencies, access research, visualise precision data, manage finances, buy inputs and outputs and monitor prices in local, regional, national or international markets.

The internet is not just a new tool for information dissemination; it is changing the very culture of society's information seeking behaviour. And, for this very reason, the Extension must also change so as to be in tune with the farmers.

2. Reaching the Unreached

The prevailing technology environment in the contemporary society has the potential to engage the farmers in remote rural areas through placement of Village Information Kiosks. These information kiosks – which can be made interactive - can be linked through Internet to information highways for accessing the latest and best technology.

3. Agriclinics

The Agriclinics can serve as platform for serving the multi-dimensional needs of the farmers through engagement. It can provide consultancy advice, agri-inputs as well as cater to the general information needs of farming/village community through effective networking. The young and enterprising agriculture graduates could be motivated to open the self-sustaining Agriclinics at block level which could have some of the essential facilities such as soil testing, a plant clinic or provide market intelligence or weather forecast for the curious and demanding farmers.

3. Krishi Vigyan Kendra (KVK)

The KVKs at district level can provide back-stopping, for Agriclinics, through its research and scientific experts who are engaged in location specific research. They can serve as a resource person for problem solving, motivating farmers for adoption of new and improved agriculture technology. There are 589 KVKs across the country serving the farming community. Indian Council of Agricultural Research (ICAR) plans to provide e-linkage to all the KVKs that would serve as resource back-up for the extension personnel and be in direct touch with the farming community.

4. Agriculture Universities (AUs)

There exists a vast network of 47 AUs besides a number of ICAR research centres and stations. The AUs can be inter-linked to form an Information Gateway which any farmers could access from anywhere for seeking advice on specific problem. This common platform can also be used for content generation.

Enablers of Cyber Extension System

1. *ICT Infrastructure :* A state-of-art ICT infrastructure needs to be created for connecting key stakeholders which include ICAR, SAUs, ZRSs, KVK, *etc.* Any basic ICT infrastructure should have at least four essential elements for harnessing the true potential of ICTs: (i) Assured availability of electricity up to village level, (ii) Telephony *i.e.* both fixed, cellular / mobile network as well as wireless systems (iii) Teledensity, *i.e.* penetration of telephone network across the entire spectrum of community, even in rural and remote areas., and (iv) ICT hardware as well as Software to provide full gamut of e-services. Besides providing the required ICT infrastructure, all out efforts should be made to develop and promote the user-friendly software. Although the use of ICTs is increasing day by day, it is yet to cross the threshold level to harness the true potential of ICTs for development.
2. *Information Packaging :* A onetime digitisation of all the extension literature needs to be done. Crop specific multimedia interactive CDs needs to be developed which could be distributed at village level.
3. *Capacity Building :* Developing competencies of all extension stakeholders in the use of ICTs so that they could access and retrieve the information from multiple sources.
4. *Services Provided :* Networking with e-governance initiatives, integration of services provided along with market intelligence and price related alerts should form the core of services provided at KVKs/Agriclinics/ rural kiosks. Besides, links to all the general interest information/services should be part of the mandate of village information kiosks so that they are self-sustainable.

Some of the guiding principles for an ICT enabled e-Extension, according to ADEC (2002) are listed below:

1. The system must be electronically integrated. This will require a system with multiple access points for users who could be farmers, extension managers/ administrator/ workers, scientists or policy makers spread over a larger geographical area.

2. The system must provide easy and seamless access to useful information. This would require an open and flexible navigation system.
3. The system must be based on openness. This means allowing multiple points of view in development efforts. Communities of interest are a key to eventual success.
4. The system must be evolutionary in nature. This will include a process that involves simultaneous experiments, thinking about various exploratory options, and taking some calculated risks.
5. The system must be flexible, nimble and quick and standards- based.
6. The system must address the needs of users and actively engage with various communities of interests in mission driven, organised and entrepreneurial fashion.
7. The system must be objective and comprehensive; and must identify new communities.
8. The system would be the outcome of dedicated virtual teams working together across borders and boundaries, focuses on rapid response to pressing educational needs of farmers emphasising quality, sound education approaches and appropriate use of information and communication technology.
9. The system is competitive and involves access anywhere, anytime and anyone who needs it.

Conclusion

This framework of Cyber Extension system should attempt to harness the potential of ICTs and facilitate electronic delivering of agricultural information, advice and consultation for high-tech agriculture and precision farming. Rural Knowledge Centre (RKC) being established under India's "Mission 2007 : Every village a knowledge centre" should be the focal point of nationwide efforts to mobilise the private and public sectors to apply ICTs to cater the Information, Communication and Training needs of rural India. The prevailing technology environment can provide ample opportunities for skill upgradation of rural youth for possible diversification in agriculture sector and catalyse community based development initiatives through value addition in agricultural products supply chain and triggering entrepreneurial activities in villages through Self Help Groups (SHGs).

References

ADEC. 2002. The Time is Now- e-Extension. *Draft paper of ADEC Think Tank*, Atlanta, Georgia.

Anandaraja, N., Chandrakandan, K., H. Phillip and C. Karthikeyan. 2005. Empowering Indian Farmers through Information Technology Gadgets (Personal Communication).

Ansari, M. A. and Parveen Khatoon. 2007. e-Extension Strategy for Accelerating Agriculture Development in India. *Indian Media Studies Journal,* Vol. 2, No. 1-2, January-December

Anonymous, 2007. e-Extension: A National Strategy. www.asred.msstate.edu/e-extension/extension.htm (Accessed on 5th January, 2007)

Chattopadhyay, B.N., 2004. ICT for Agriculture Development (Unpublished). *Handout circulated to Summer School participants at G.B. Pant University of Agriculture and Technology*. Pantnagar. Sept. 2004

FAO. 1993. Potential of Micro Computers in Support of Agricultural Extension, Education and Training. Rome: FAO.

King, D. A. and Boehlje, M.D. 2000. Extension; On the Brink of Extinction of Distinction? *Journal of Extension*, Vol. 38, No. 5.

Samanta, R. K. 1993. *Extension Strategy for Agriculture Development in 21st Century*. Mittal Publications, New Delhi.

Sharma, V.P., 2004. Cyber Extension: Connecting Farmer in India-Some experience. www.manage.gov.in/publications (Accessed on 1st December, 2007)

5

Village Knowledge Centres and Village Resource Centres –An Overview

S. Senthilkumaran

ABSTRACT

The Village Knowledge Centre programme was started in January 1998 in the Union Territory of Pondicherry (now Puducherry) in the east coast of South India with the support of IDRC and CIDA. In order to ensure access to all, the VKCs are located in public buildings (Panchayats, SHG's building, Community Hall, School, Farmers and Fishermen Associations, Women Self-Help Groups, Community Based Organization, *etc.,*). The information content is developed in close interaction with the local people in a bottom-up manner. On 23 August 2003, Jamsetji Tata National Virtual Academy for Rural Prosperity (NVA) was officially launched with the generous support of Sir Dorabji Tata Trust and Tata Education Trust. Since then all ICT-enabled development activities of MSSRF are carried out under the umbrella of NVA. Experiences shows that Building partnership is important for establishing connectivity and technology, creation of time-and location-specific, demand-driven content and dissemination of the same for ensuring livelihood, food and health security of the rural communities.

Introduction

M.S. Swaminathan Research Foundation (MSSRF) is a non-profit research organization. It was established during 1988. M. S. Swaminathan Research Foundation has all along been following a pro-nature, pro-poor, pro-women and pro-sustainable livelihood models of development using appropriate rural technology development and dissemination modules.

M.S. Swaminathan Research Foundation is doing research in six major thematic areas such as Coastal Systems Research (restoration of mangrove forests, alternative livelihood for fishing community, *etc.*), Biotechnology (developing salt tolerant transgenic rice varieties, testing availability of oil content in different biofuel crops/ plants, lichens diversity, *etc.*), Biodiversity (documenting endangered and medicinal plants, drafting the Protection of Plant Varieties and Farmers Rights Act 2001, providing necessary training to village community to maintain biodiversity register, maintaining seeds in the form of community based gene bank, *etc.*), Ecotechnology (promoting bio-villages - biovillage paradigm involves sustainable management of natural resources and developing sustainable on-farm and non-farm livelihoods as eco-enterprises managed by self-help groups), Food Security (promoting community based grain bank, developing urban and rural atlases to understand the factors involved in food insecurity, providing livelihood training to women farmers particularly in the farmers suicide hotspots, *etc.*) and Information, Education and Communication (addressing the issues related to sustainable development using different information communication technology tools) for knowledge empowerment of the resource-poor, largely illiterate and unskilled rural women and men.

Birth of the Village Knowledge Centre

The Foundation has been launching a series of annual inter-disciplinary dialogues on technology enabled development since, 1990, under the general theme "New Technologies: Reaching the Unreached". MSSRF held an Interdisciplinary Dialogue on Information Technology: Reaching the Unreached in January 1992 with the support of International Development Research Centre (IDRC), United Nations Development Programme (UNDP), Department of Space, Government of India, International Tropical Timber Organisation (ITTO) and Council for Advancement of People's Action and Rural Technology (CAPART). The dialogue revealed that the future of food security in the developing world, especially in South Asia, would be dependent less on resource intensive agriculture and more on knowledge intensity. The concept of Village Knowledge Centre could thus be traced to the dialogue in 1992.

The participants in the dialogue in 1992 suggested that, the generic content received from Universities, National Informatics Centres (NIC) and Remote Sensing Agencies, need to be made more demand-driven and location-specific. With the help of experts, these contents are converted into vernacular language in an understandable manner, based on the needs of the rural community. This location-specific demand-driven content is transmitted to block level knowledge centre. The Village Knowledge Centre receives this location-specific demand-driven knowledge from block level knowledge Centre. After that, this location-specific knowledge is disseminated to the rural community through different technologies.

During 1993-1996, MSSRF conducted a number of studies in about 25 villages in the Union Territory of Pondicherry and the State of Tamil Nadu. One of the studies was on estimating the reach and impact of electronic media in the rural areas. This study revealed that the reach of electronic media especially television, was reasonably high despite the prevalence of poverty in the villages surveyed. Another study revealed that farmers gather most of the information they need from the local shopkeeper, the market place, and the suppliers. Considerable amount of information transaction takes place among the rural poor households and this indeed acts as a primary mode of information dissemination.

MSSRF believes that Information and communication technologies could play a major role in environmentally sustainable rural development, not only in taking knowledge to the poor but also helping them to achieve food security and social justice.

Based on the outcome of these surveys, MSSRF developed the concept of Knowledge System for Sustainable Food Security and an operational plan for translating knowledge into action. This involved spelling out different components of the knowledge system for food security at individual level (availability, access and absorption), different eco-farming systems (soil health care, water harvesting management, crop and pest management, energy management, post harvest management, crop and animal components of farming systems and, information, skill, organization, management and marketing empowerment) and different databases (eco-jobs, household entitlement database, self-employment, information related to women and children).

Village Knowledge Centre (VKC)

The Village Knowledge Centre programme was started in January 1998 in the Union Territory of Pondicherry (now Puducherry) in the east coast of South India with the generous support of IDRC and CIDA. In order to ensure access to all, the VKCs are located in public buildings (Panchayats, SHG's building, Community Hall, School, Farmers and Fishermen Associations, Women Self-Help Groups, Community Based Organization, *etc.*). The information content is developed in close interaction with the local people in a bottom-up manner.

VKCs are connected with hub centres through a hybrid wired and wireless network. The VKCs provide information on agriculture, health, employment, weather, education, government entitlements, microenterprises training, *etc.* In each case, the community or local partner provides an accessible rent-free building, electricity and volunteers. In turn, the project provides all the needed equipment, training and helps in collecting data. Gender concerns are central to the project. More than half of the volunteers are women. This has positively reflected in the increase in the number of women users. And, the VKCs are inclusive in the sense that no one is denied access on the basis of caste, creed, religion, *etc.*

An important feature of this programme is the strong sense of community ownership that the village communities have developed towards the village centres. Almost in all villages, the knowledge centre were demanded by the local people. MSSRF did not launch any centre on its own. The MSSRF acts as a catalyst, and facilitator to promote knowledge empowerment to the rural women, and men through the VKCs and VRCs.

This project has won the Motorola Dispatch Solution Gold Award (1999) and Stockholm Challenge Award (2001). Many journals have covered this project. Scientific American selected this project (2004) for special mention when it chose Prof. M. S. Swaminathan as one among the 50 scientific leaders of the world who are making a difference.

Jamshedji Tata National Virtual Academy for Rural Prosperity (NVA)

Based on the VKC experience, MSSRF organized a concept and operational plan consultation for the launch of NVA on May 8, 2003. In this consultation, aspects such as role of data generators and providers (Universities, Research Stations, ISRO, IMD, NGOs, Financial Institutions, IGNOU, IMD, AIR, Doordarshan, Corporate Sector, Policy Makers, *etc.*), role of data managers (managing both content and connectivity) and the dissemination of locale specific demand driven content to the rural community were discussed.

On 23 August 2003, NVA was officially launched with the support of Sir Dorabji Tata Trust and Tata Education Trust. Since then all ICT-enabled development activities of MSSRF are carried out under the umbrella of NVA.

The main aim of the programme is to empower vulnerable people in order that they make better choices and achieve better control of their own development and to build skills and capacities of the rural poor with a view to enhancing livelihood opportunities. The NVA programme is to bring together the experts and grassroots level communities in two-way communication with the objective that knowledge should reach every home and hut.

ISRO-MSSRF Village Resource Centres

The Hon'ble Prime Minister of India, Dr. Manmohan Singh inaugurated this programme through video conferencing in October 2004. In his inaugural speech, the Prime Minister emphasized, "Community based vulnerability and risk related information, provision of timely early warning and dissemination of severe weather related information can lead to reliable disaster management support at the village level".

Under this VRC programme, most of the MSSRF VRCs are connected through Indian Space Research Organization's (ISRO) uplink and downlink satellite facilities. The satellite based ISRO-VRCs, aim at digital connectivity to remote villages for providing multiple services such as telemedicine, tele-education, interactive farm and fishery advisories, government schemes and entitlements, weather services and remote sensing applications through a single window. Users located at one node of this network can fully interact with others located at another node through video and audio links. Each node can further be expanded using different technologies such as notice boards, pamphlets, public address system, community newspaper, press releases, cable TV, audio/ video conferencing through wireless, telephone, meetings, mobile phone and intranet web site for dissemination of the useful and necessary information.

Three-tier Knowledge Network

To strengthen the management among VRC, VKCs and strategic partners (data providers, generators, capacity building providers, *etc.*) the NVA developed a three-tier knowledge network. The main aim is to develop a user-controlled, user-owned and user-managed network, which will help to reach the unreached and include the excluded in terms of information, knowledge and skill empowerment.

First-tier

MSSRF, Chennai has been connecting several data generators and data providers (universities, experts, financial institutions, corporate sector, technocrats, grassroots academicians, *etc.*) focussing primarily on content and capacity building. The Informatics Division of MSSRF is acting as a facilitating agency to carry out the entire project.

Second-tier or Village Resource Centre (VRC)

Normally, the Village Resource Centre is located at block level or community level or mandal level or at the centre point of a cluster of villages. The informatics division of MSSRF has been connecting with Village Resource Centres (VRCs - block level hub) through Indian Space Research Organization (ISRO) uplink/downlink satellites and transfering the knowledge/information to VRCs and vice-versa. The, locale-specific, demand driven information is disseminated from the VRCs through Village Knowledge Centres (VKCs). In a few cases, the VRC is also connected through internet (broadband: maximum 1 MB or dial up). Normally VRC is located in ground floor to give access the all sections of people (including physically challenged people) with minimum facilities such as two rooms, one a training hall (sitting capacity for about 20 people) with an open terrace for setting up wireless communication, small library of a few reference books related to that area, major livelihoods, *etc.* and a few monthly and daily local newspapers and magazines. Regarding hardware part, VRC has about have 8-10 systems, internet facility, phone, printer, scanner, web camera, digital camera, power backup and speakers. One VRC caters to the needs of 15-20 VKCs with the radius of 60-100 km based on the range of access.

Third-tier or Village Knowledge Centres (VKCs)

Village Knowledge Centres (VKCs) are normally located among a cluster of villages or panchayat level. VKCs space and electricity are provided by the partners (elected village administration, traditional village administration, farmers and fishermen associations, milk societies, youth clubs, NGOs, region based institutions such as temples and churches, village development council, *etc.*). They are also involved in the areas of selecting Knowledge Workers, spelling out needs of the communities particularly in the areas of content and training/ awareness programmes, supporting purchase of a few equipments and buying newspapers for knowledge centre's library.

Strengthening four-way Information Flow System through three-tier knowledge system

In this consultation, NVA developed an information system that connects lab-to-lab (organizing a consortium of scientific institutions and data providers), lab-to-land (symbiotic linkages between the providers of information and the users, so that the information disseminated is relevant to the life and work of the rural families), land-to-lab (considerable traditional knowledge flow and wisdom from rural and tribal families concerning sustainable management of natural resources, particularly water and biodiversity; to scientific institutions for learning, validation, conservation and propagation. Therefore, technical experts should not only learn from traditional knowledge and experience, but also take steps to conserve for posterity, the dying wisdom and the dying crops) and finally land-to-land information flow (where much scope for lateral learning among rural families exists; such learning has high credibility because the knowledge coming from fellow farm women or men would have been subjected to an impact analysis from the point of view of its economic and social relevance to the population).

Process for Setting up of VKC

- District/ Block/ Community level orientation meeting with different partners (elected panchayat leaders, traditional panchayat leaders, leaders of farmers' and fishers associations, NGOs, Self-Help Groups, *etc.*).
- Some times the villagers directly approach the VRC based on awareness of the existing VKCs through newspapers, farmer to farmer or fisherfolk to fisherfolk or village to village interactions *etc.*
- Identification of boundary partners to set up VKC based on the above steps.
- Conducting feasibility study (different caste groups and their relations and livelihood, economic status of the village, *etc.*; institution - partner organization and their functions, the functions of different community-based organizations, *etc.*; technical - availability of infrastructure, power, *etc.*).
- Concept-sharing with boundary partners (several formal meetings are conducted with boundary partners and other members who are all involved in the process of setting up of VKCs).
- Concept sharing with villagers through village level awareness programme, meetings with different sections of community and addressing Grama Sabha.
- Signing of Memorandum of Understanding with partners/ Resolution passing in the Grama Sabha.
- Collection of secondary data (from village administration office, various departments, *etc.*).
- Collection of village profiles.
- Documenting time line series.
- Identification of knowledge workers.
- Training to knowledge workers.
- Inauguration of VKC.

Finding out the Needs of the Community

NVA collects the village profiles which consist of several parameters such as name of the panchayat, name of the VKC village, total number of hamlets, different caste groups in the panchayat, educational status, occupational pattern, land use pattern, land holding classification, rainfall details, type of soil, cropping pattern, fishing pattern, infrastructure facilities, communication network, *etc.* NVA also documents time line series for important events of the village and the effects, season wise. NVA is using different participatory techniques - focus group meetings with different sections of the community, transect, time and trend analysis (identify village boundary with important land marks – mountains, lake, temple, *etc.*, past events and trends – important events and years, development intervention over a period of time, *etc.*), social mapping (demographic and socio economic profile - age, caste, education, health, land holding, occupation, different types of houses, infrastructure facilities like electricity, water facilities, ration card, assets, television, radio, landholdings, boats, nets, *etc.*), wealth ranking (identify the economic stratification and prioritize the target group – wealthiest, wealthy, average, poor and poorest), resource mapping (exploring the available resources and its current usage pattern - different types of income generating trees, irrigation resources, agriculture and fisheries related resources, other livelihood options, *etc.*), venn diagram (importance of different organizations to the village and villages existing linkages with such departments), seasonal diagram (season wise variations in agriculture, fisheries and other occupations and associated risks and the coping mechanisms, migration pattern, *etc.* - employment opportunities, prevailing diseases, wind flow and direction, rain, occupation, *etc.*), problem matrix (exploring the crucial issues its related causes and effects, adaptability and possible solutions - marketing facilities, water availability, *etc.*), house hold sample surveys, *etc.* This is usually carried out through VRC staff and social workers.

NVA also adopts different strategies such as sub-grouping the fishing communities based on specific needs (eg. sea fishermen, laborers, women vendors, inland fishermen, backwater fishermen, fisherwomen and men in mangroves, *etc.*), and sub-grouping of the farming communities based on their needs (labourers, dry land farmers, small and marginal farmers, *etc.*). The purpose of such sub-grouping is to provide focused awarness and training to each of these, identifying the prime movers in each sub-group, rapport building with different sectors and identifying representatives from different sub-groups to represent in the Village Management Committee. This in nutshell enables pooling together the varied needs of the farming, fishing and the landless rural communities. Hence, the VKCs and VRC programme are successful.

NVA also initiated GIS in need assessment particularly to get visual presentation and different layers of a particular village for different parameters such as critical facilities (common open well, hand pump, worship place, rice mill, shop, rainwater harvesting structure, *etc.*), gender disaggregation, types of houses, family size, *etc.*

This need assessment helps the VRC to create awareness among villagers about the concept and activities of VKC to involve the community from the day one in the project activities, to find out the problems of socially under privileged people, to assess the village needs, affordability and willingness to contribute, technological options, interest of the community, to develop plan for the village, to enable the VRC staff to understand the village situation and establish a rapport with the villagers, *etc.*

Three Important Aspects of VRC and VKC Programme

Through a collection of several secondary data and a well-planned need assessment, VRC and VKCs develop a location-specific demand-driven content, organize training and awareness programmes and make linkages with several leading institutions/organizations for transforming the content into action/ application.

Knowledge Workers

NVA has been selecting knowledge workers with the help of local partners for maintaining the VKCs. NVA has been continuously imparting a series of training to knowledge workers on various aspects such as concept of VKC, methodology of collection and dissemination of need-based content, managing users and visitors, how to maintain the user register, bill books and other records pertaining to VKC, how to build rapport with users and community members, gender concerns, basic hardware and software training, maintaining the user register which contains name of the user, age, education, occupation, the kind of services expected from VRC/VKC, frequently asked questions, collection of feed back in the form of case studies, identifying their skills and aptitudes, *etc.* to access the effective usage of VRCs and VKCs. VRCs also conducts quiz programme for Knowledge Workers (KWs) based on the availability of content in VKCs. Some of the Knowledge Workers have developed a few training materials related to VKC management based on their field experience and VRC trainings.

Content Generation and Dissemination

Creation and updating of relevant content to suit local needs is a key factor in the programme. The information provided should be time-and location-specific, demand-driven and relevant to the day-to-day life and work of rural women and men.

VRCs continuously debate what is content (dynamic-weather, wave height, market, potential fishing zone, *etc.*, static-government schemes, historical information, *etc.*, need based - system of rice intensification, how to control eriomite pest in coconut?, *etc.*, exogenous - how to insert turtle excluder device in nets, right to information act, *etc.*, complete modules-fish pickle, groundnut, *etc.*, questions and answers, *etc.*).

NVA is providing several demand driven location- specific content related to agriculture, animal husbandry, fisheries, government entitlements/ announcements, micro-enterprises, employment news, education, health, civic services (bus and railway timings, blood bank details, *etc.*), rural yellow pages (details of agricultural equipments for rental, motor repair details, fishing net availability in different shops, fertilizer shops, *etc.*), *etc.*, NVA also regularly conducts several awareness and capacity building training programmes for converting knowledge into action.

Village Resource Centres establish linkage with number of institutions, organizations, experts and government departments regarding the content, addresses and linkages of other related institutions and capacity building programmmes. These linkages are in the form VRC level meetings, face-to-face in their institution or field, phone, email, *etc.*

VRC collects and generate the content based on season based queries (agriculture, fisheries, health, *etc.*), community newspaper readers opinion, user register analysis, findings from village profile, thematic wise segregation, need based user meetings, government announcements/ entitlements, employment news from local companies, education queries, area specific, target group (women, children, youth, *etc.*), *etc.*

VRC also collects and disseminates important phone numbers such as government departments, fire services, hospitals, colleges, police stations, travel agencies, schools and colleges, *etc.*

VRCs also collect the local information from several local newspapers, journals and web.

Dissemination

NVA is using different technologies (both modern and traditional) such as notice boards, user meetings, knowledge workers, mobile phone, cable TV, pamphlets, wired and wireless public address system, K YAN-PC (it contains PC, projector, TV tuner card, DVD player, amplified speakers and modem), Very High Frequency (VHF), WiFi based video conferencing such as 802.11 b and 802.11 g, satellite based video conferencing, offline CDs for both audio and video, SMS server, Internet Radio, mobile vans - soil and water testing van, GSM based wireless public address system, community newspapers in local languages and providing answers through telephone and advisories through video conference to disseminate the location-specific demand driven information to the community.

Through Wireless/Wired Public address system, VKCs are disseminating audio content such as wave height, weather, employment news, when the local ration shops and fair shops distributed sugar, rice, kerosene, disaster preparedness announcements, local news, ration card application, 100 days employment schemes, voter's ID, other NGOs' awareness programmes, missing materials information, distribution of eggs in the mid day meal programme, final date for paying several services - electricity, water, *etc.*, panchayat leaders decision, local meeting announcements - village meeting, women SHGs, anganwadi parents meeting, school teachers, *etc.*

Technical Aspects

Technical team from MSSRF, Chennai has been regularly providing inputs/training to the VRC technical staff in all the technical aspects such as hardware and ICT-based educational programmes such as Microsoft Unlimited Potential Programme (MUPP), Computer Aided Learning Programme (CALP) and Intel Learn Programme. Inputs are regularly provided through video-conferencing and once in every three months technical staff meeting is held at different VRCs to discuss various technical issues, to develop/upgrade VRC wise technical guidance and the issues related to ICT-based educational curricula such MUPP, CALP and Intel Learn Programme. In turn, VRC technical staff is providing basic hardware and software training to Knowledge Workers to maintain effective system maintenance at VKC level. VRC technical staff also prepares several technical manuals based on the problems faced by the VKCs.

Community Newspaper

VRCs are using both modern and traditional technologies to spread the knowledge to the community. One of the technologies involves disseminating information through vernacular newspaper. Each Village Resource Centre (VRC) is publishing its own community newspaper (fortnightly) in their local language. Reporters are mostly knowledge workers, NVA Fellows and VRC Staff. The community newspaper is distributed through knowledge workers, NVA Fellows, village leaders, school headmaster, students, youth clubs, local schools and colleges, NGOs and courier to various government departments (fisheries, police, taluk, panchayat, municipality, temples, home guard, collectorate, public relation office, education, women development corporation, superintendent, health, agriculture, forest, health, *etc.*), organizations (banks, hospitals, post office, NGOs, Rashtrabhasha Printing Press, coffee board, species board, *etc.*), mini bus drivers, farmer and fishermen associations, MUPP students, milk societies, fertilizer shops, churches, mosque, key informants and institutions (schools, central marine fisheries research institute, KVK, S J Social work college, *etc.*). This newspaper is also distributed to different villages, VKCs and other VRCs. The content is published in the form of dialogue, announcements, games, bulletin, questions and answers, *etc.*

Based on the feed back, VRCs included educational and career programme after +2, season based crop advisories (sowing to harvesting), moral stories, history of national leaders, innovations (rural), spoken English, about the district/ village sports, more local news - handicrafts exhibition, eye camps, animal husbandry camps, collector's visit, subsidies for fishing boat, *etc.*

Advisories through Video Conferencing

Apart from daily advisories, several thematic video conferencing programmes are constantly provided.

Training and Awareness Programmes

NVA regularly conducts several awareness and capacity building training programmes for converting knowledge into action.

Open Knowledge Network (OKN)

During 2002-2004, M. S. Swaminathan Research Foundation and One World International demonstrated the pilot experiment called "Open Knowledge Network" in the VRC and VKC programme. The aim of the experiment is to create a network that would connect the rural poor of the world and facilitate knowledge sharing among them at affordable costs. Under this programme, a few VKCs feed local news, advertisements, items of traditional knowledge, government announcements and suggestions every day, based on which a daily news sheet was produced and send it to another VKCs through World Space Radio technology and Internet. The VRCs and VKCs also create content under the categories of Knowledge, Classified Advertisements, Local News, Government Announcements, Suggestion Box, Market Prices, Weather and Jobs. All the knowledge-based information is collected only from village people and not from printed sources.

Under the Classified Advertisements several villagers provided the information under 'sale' and 'wanted'. This advertising provides a ready reckoner for the people in and around the village to locate the right person for the right work. For example, agriculture labourers for farm work, bi-cycle sale, who is renting out a tractor and at what price. A tailor buying a new machine can announce the sale of his old machine. This promotes the business of all categories in and around the village. In the local announcements, villagers announce district collector's visit, animal husbandry camp and local school annual festival, *etc*. This is helpful for people from other villages to attend the camps and festivals. Many villagers expressed their happiness about the suggestion box during the project period. Before this experiment very few youth clubs and self help groups sent petitions to the government departments regarding their basic needs. Unfortunately they were not sent to the right department or were not acted upon. At the time of Open Knowledge Network Project villagers discussed with project staff and identified the concerned person's name and department and sent their petitions.

In the experiment, data gathered from different villages using a specially designed format are stored as both metadata and full content. The content is stored in a server at MSSRF, and the metadata is transmitted to a server at the One World International *via* Internet. Within an hour VRCs and VKCs can download the daily newssheet either through Internet or through World Space radio. Through this network access points from India and Africa share their local news through World Space Satellite and Internet. The local news is produced in Tamil and Swahili and the Meta tags are in English.

The entry form is stored in each knowledge centre in the hard disc. It was designed to collect the local knowledge items, and store the results on the local hard disc. The village volunteer should enter his/her name, title of information item, type of information (like knowledge, local news, suggestion *etc*.), date of input, and date at which information should expire. Each knowledge centre has its own code.

During the project period, in Villianur hub (Puducherry VRC) the incoming information is first checked - an important step for quality control and the avoidance of malicious or misleading input into the system. After that it is transferred to MSSRF server through file transfer protocol or email. At MSSRF, project staff gives the Meta data for each category including keywords using the OKN Management form available online. The content is stored in a server at MSSRF, and the metadata is transmitted to a server at the One World International *via* Internet. The knowledge workers use English for most of the meta tags. Most importantly this includes the description field, which offers a short description of the content of the item. For non-Tamil speakers this would be the point of entry for deciding whether this is an item worth translating into their own local language. The receiving end will connect Internet or World Space Radio receiver connected to a YAGI antenna, and a stored XSL style sheet, one obtains a daily news sheet in a pre-designed format. It is also possible to obtain the news sheet through the Internet without the use of World Space radio receiver.

Open Economics Centre, Harvard Law School, Accenture are other partners for this experiment.

South-South Exchange Traveling Workshops

Out of the positive experiences with these Village Knowledge Centres, the idea was born to share this good practice with other global organisations from other parts of the world, which also are involved in the field of ICTs and rural development. The objective is to learn from one another, be encouraged by a good example and be able to distinguish which experiences could be used straightaway and which ones need to be adapted to the local situation. The end result that the participants augment their capacity in ICT for rural development and thus deliver better quality services to their target groups. Based on this concept, NVA has been conducting south-south exchange traveling workshops regularly. These workshops are supported by HIVOS, The Hague, IICD, The Hauge, IDRC, Canada, Global Knowledge Partnership (GKP), Malaysia. Normally the time frame for these workshops is 10 days. During the workshop, the participants exchange their views with VRC and VKC village communities and Knowledge Workers, observe the different dissemination technologies, raise questions related to the content, connectivity, capacity building, management of VRC and VKC, *etc*. The details of the workshops so far conducted are as follows.

- Study tour of 25 Nenasala operators to the M S Swaminathan Research Foundation- Chennai & Telecentres in South India, on knowledge and experience sharing mission, 4 – 11 February 2008
- Traveling workshop on Pro-poor, Pro-women use of Information and Communication Technologies (ICT) for Social Development, 10-18 December 2006
- ICT enabled development: South-South exchange through traveling workshop, 28 November – 5 December, 2005
- ICT enabled development: South-South exchange through traveling workshop, 15-22 October, 2004
- ICT enabled development: South-South exchange through traveling workshop, 30 October-6 November, 2003
- ICT enabled development: South-South exchange through traveling workshop, 21-28 October, 2002

Knowledge on Wheels

On 7th August 2007, NVA launched a new programme called "Knowledge on Wheels". On 29 December 2006, NVA and Sankara

Nethralaya Medical Research Foundation signed a Memorandum of Understanding to provide eye care facilities and eye care awareness using a van for this purpose. The mobile unit has necessary ophthalmic equipments for diagnosis of eye related problems and is equipped with spectacle grinding and frame fitting facilities, to provide spectacles then and there at reasonable and affordable costs to the rural community. This mobile van visits VRCs regularly and provides the treatment for eye problems such as cataract, squint, glaucoma, *etc.* besides paediatric clinic, yag caps, retinal evaluation, field test, prescription of glass, *etc.* This is a joint initiative of Sankara Nethralaya Medical Research Foundation, Indian Space Research Organization (ISRO), World Diabetic Foundation (WDF), ESSILOR India Private Limited and M. S. Swaminathan Research Foundation (MSSRF).

With the help of HP and ISRO, NVA has got the soil and water testing van from HP-Kuppam project. Now the mobile van is providing service to poor farmers. The mobile laboratory has a pH meter, colorimeter, flame photometer, atomic absorption spectrometer, magnetic stirrer, electrical weighing machine, Oxygen cylinder and glasswares (supporting equipments). This laboratory is helping to test the different soil parameters such as pH of the soil and the irrigation water, electrical conductivity of the soil and the irrigation water, soil organic carbon, available phosphorus (Olsen method for neutral and alkali soils and brine method for acid soils), available potassium and available micro-nutrients (zinc, manganese, iron, copper, magnesium and calcium). This van has computers, LCD projector, solar backup and generator. With this facility, NVA plans to show a few documentaries on Soil Nutrient Management, Crop Cultivation Practices, Integrated Pest Management, Post Harvest Technologies, *etc.*

Virtual Congresses and Interactions

NVA is regularly organizing virtual congresses and virtual interactions between policy makers, experts and rural communities to deal with the different issues and bridge the prevailing gap between scientific "know-how" and field level "do-how". The details of the virtual congresses and interactions are as follows.

- 5th January 2006, First Virtual Congress of Farmers, Acharya N G Ranga Agricultural University, Rajendra Nagar, Hyderabad (Locations connected for discussion: ICRISAT, Hyderabad, Pudukkottai & Nagapattinam - Tamil Nadu, Moosapet - Andhra Pradesh, Pokran – Rajasthan, Koraput – Orissa, Anandwan & Waifad – Maharashtra, MSSRF - Tamil Nadu).

- 18 January 2007 & 14 April 2007, Virtual Conference on Medicinal Plants, Jeypore (Orissa) VRC (Locations connected for discussion: Kottakal, Kerala and Bangalore, Karnataka).
- 5th January 2008, First Virtual Congress of Mahila Kisan, Convocation Hall, Andhra University, Visakhapatnam, 95th Indian Science Congress (Locations connected for discussion: MSSRF, Chennai, Thiruvaiyaru - Tamil Nadu, Moosapet, Addakal - Andhra Pradesh, Pokran – Rajasthan, Koraput – Orissa and Waifad & Yavatmal – Maharashtra).
- 11th February 2008, Role of ICT in the management of Climate Change at the Grassroots Level, MSSRF, Chennai (Locations connected for discussion: MSSRF, Chennai, Thiruvaiyaru, Annavasal, Thangachi-madam and Nagapattinam – Tamilnadu, Waifad – Maharashtra and Jepore – Orissa).

Mobile Fisheries Application

During 2007, QUALCOMM (technology partner) supports this initiative in the form of financial and technical. Brew application is used for this technology. NVA (implementing agency) is responsible for updating the information required for the Fishery Information System as per the requirement of the community as well as the software. NVA interacts with the communities, consolidates the learning experiences. Tata Teleservices (service provider) disseminate the fisheries related information to their network. Astute Technology System, Indore (application provider) provides the software for updating the information, maintains the software in its server and integrates the software in the handsets. Based on the feedback from MSSRF, Astute Technology Systems will modify and fine tune the software to the requirements of the community. Through this system, fisher folk are getting the information on sea wave height, weather, potential fishing zone, rural yellow pages for fisher folk, government announcements, *etc.*

Monitoring and Evaluation of NVA activities

NVA is systematically monitoring its activities by developing monitoring and evaluation framework that comprises of both data collection and compilation plans. Data collection and compilation plan is prepared based on the indicators developed for different programmes of the NVA. Various methods and tools are developed to generate both qualitative and quantitative data at different intervals of time incorporating disaggregated information. The collected data are periodically analyzed using SPSS package and communicated to all the stakeholders at different levels to know the current status of the project activities and to take appropriate decisions and future plans. For example, NVA has documented the occupation of VRC/VKC users, frequency of their visits, purpose of the visit, type of users (one time, repeated users), number of times repeated users visiting VRC/VKC *etc.*, using the user registers.

In addition, NVA also conducts periodical internal reviews with the VRC staff to make necessary midcourse corrections wherever, necessary. Documenting the outcome of various activities of the NVA using case study and other qualitative and quantitative methods is a continuous process and helps to identify the changes among rural community in different dimensions as well as strengthen the project strategies and implementation process. The documented outcomes are shared among the partners in the form of publications.

Apart from the above, number of researchers, academic institutions and policy makers have studied the VRC and VKC activities in different aspects. The results are compared to other entrepreneurship models such as n-Logue, government initiatives (e-Seva) and corporate model (ITC e-Choupal).

Furthermore, the NVA has developed several methods to monitor the different activities of the VRCs and VKCs and has documented case studies of NVA Fellows, Microsoft Unlimited Potential Programme (MUPP) trainees, and users of VRCs and VKCs. NVA has also documented occupation of the users, frequency of the visitors, purpose of visit, repeated users, total number of persons who have been used VRCs/ VKCs, number of times of repeated visits to the VRCs/ VKCs by the users.

A few monitoring and evaluation methods for VRC and VKC activities are as follows.

- Developing LFA (Indicators for output, outcomes and impact, *etc.*).
- Data collection plan (methods of data collection, tools of data collection, frequency and responsible person).
- Data compilation and analysis.
- Reporting and decision-making.
- User register (conducting programmes, awareness programmes of VKC, how to convert non-users to users, *etc.*).
- Feed back from CNP readers through post card, phone, in person, meeting.
- Follow up of training programmes (through meetings, KWs, animators, NVA Fellows, *etc.*).
- Case studies.

- Attendance (MUPP, CALP and Intel) - Taking steps for bring back dropouts, *etc.*
- User meetings (identifying further needs, future programmes, *etc.*).
- VKC level interaction record.
- One-to-One interaction (staff and partner, users, *etc.*).
- Feed-back from partners meeting, VKC management meetings, government officials, *etc.*
- VKC records verification (registers of user, attendance, technical, MUPP, CALP, Intel, interaction, visitors, queries, daily news, public address system, asset, VKC MC meeting, staff visiting, *etc.*).
- Knowledge Workers interaction.
- Observation by staff during field visits.
- Follow up of visitors' remarks/ suggestions.
- Staff meeting (including action plan, unplanned activities, *etc.*) – Informal and monthly.

Research Findings and Lessons Learned

The following lessons were learned from the VRC and VKC activities.

- Connectivity, content and sustainability should receive concurrent attention.
- Constraints must be removed on the basis of a malady-remedy analysis; for example, wired and wireless technologies could be used where telephone connections are not adequate or satisfactory. Similarly, solar power can be harnessed where the regular supply of power is irregular. The approach should be based on the principle that there is an implementable solution for every problem.
- The information provided should be demand driven and should be relevant to the day-to-day life and work of rural women and men. Also, semi-literate women should be accorded priority in training to operate the centre, since this is an effective method of enhancing the self-esteem and social prestige of women living in poverty.
- Knowledge dissemination should be linked to access the inputs needed to apply the knowledge for economic activities.
- The Village Knowledge Centres should operate on the principle of social inclusion, thereby presenting a win-win situation for all.

- The programmes designed to empower rural families with new knowledge and skills should be designed on the *antyodaya* model, where the empowerment starts with the poorest and most underprivileged women and men.
- The local population should have a sense of ownership of the e-Community Centre. It should be client-managed and controlled, so that the information provided is demand-and user-driven.
- The local population should be willing to make contributions towards the expenses of the e-Community Centre, so that the long-term economic sustainability of the programme is ensured. Contributions in cash or kind generate a sense of ownership and pride and create an economic stake in the operation of the centre.

The VKCs create minor and major impacts such as, access to various services and information, access to government and private opportunities and entitlements, functional literacy, convenience and cost-saving services, functional literacy, protection of life, property and livelihoods, awareness raising, disaster mitigation, assertion of right to information, education of children, better health, new livelihood options, empowerment of women and marginalized, *etc*.

The project personnel as well as several outsiders often raise several questions related to "sustainability of the knowledge centre". Once a VKC is accepted by the community as a common property, the sustainability of the knowledge centre is assured. This is based on the "concept of fragmented returns". If the households have benefited significantly from the centre either in terms of an assured job or income or higher studies *etc*. then they should contribute in cash or kind to the VKCs. The members who benefit from VKCs must also make contributions to the sustainability of the VKCs.

However, it should be noted that although the VKC is linked to the knowledge system through its local hub (VRC) and beyond, it could also continue to act on its own in providing information. As such the facilitating agency (MSSRF) will need to play a supportive role for a very long time to come.

Community-based model requires ownership of knowledge centre by the community as a common property. This is an intensive process and could take anywhere from 3-8 years depending on the social context.

The knowledge services cover a very broad spectrum. Since the demands can be made in virtually any aspect of human endeavour and well being, meeting them all may not always be possible. Some demands may not be due to paucity of suitable knowledge and institutions.

Collection of success stories from the field is very important. Stories, whilst anecdotal, offer a rich picture of the impact of ICT interventions in local, complex and dynamic social settings.

Without developing strong linkages with knowledge institutions on the one hand, and farming communities on the other and establishing close contact with the poorest of the poor, it will be impossible to succeed. It is also very difficult to coordinate all the initiatives of large number of players through a small GGA secretariat. It needs region wise virtual secretariat to strengthen the partnership.

Likelihood of missing out the poorest people especially in the most inaccessible villages is very high unless conscious efforts are made to locate them. Since not all such villages are covered by the NGOs, the Mission or GGA may have to rely on the government machineries in such areas, which is ill-equipped to implement process-intensive, participatory projects.

There is need to develop the architecture, the structural as well as the physical design of a web based "Rural Knowledge Portal" which will allow the Alliance partners, VRCs, VKCs to interact together and also act as a common dynamic and evolving virtual embodiment of the Mission 2007, accessible from anywhere.

The community model is best suited for the poorest villages and most distressed districts in India. In contrast the commercial model is likely to work well in semi-urban and relatively prosperous districts where farmers have the purchasing power to pay for knowledge/ information services. However, there is scope for each model (community, government, corporate and entrepreneur) to learn from each other.

GGA secretariat would also need to play a critical role in ensuring that commercial and entrepreneurial models accommodate a few social and welfare services as part of their portfolios.

From the study of different models, GGA secretariat has arrived at the following findings.

- Community-based organizations must understand that user needs are the most important and should be provided.
- Resistance of the local policy makers to provide the e-Governance is a fundamental problem.
- Reaching the Unreached : Voicing the Voiceless is sometimes overlooked.
- Absence of location-specific demand-driven information in a ready usage format.

Fact sheet of M.S. Swaminathan Research Foundation's ICT-enabled development activities through Jamsetji Tata National Virtual Academy (NVA) in the form of Village Resource Centres (VRCs), Village Knowledge Centres (VKCs), Training School and Outreach, Extension and Partnership.

- Upto September, 2009, 18 VRCs and 101 VKCs have been set up by NVA, MSSRF, Chennai in the rural areas of Tamilnadu, Puducherry, Maharashtra (Vidarbha), Orissa (Koraput), Andhra Pradesh (Machillipattnam and Kakinada) and Kerala (Wayanad).
- Regarding the setting up and running of VKCs the MSSRF is working with 127 partners such as constitutional panchayat, traditional panchayat, grassroot NGOs, Village Development and Management Committee, Temple Trust, Parish Council, Community based organizations (Fisher and Farmers Associations, Youth Clubs, Milk Societies, Self-Help Groups, *etc.*), Government Institutions/ Financial Institutions (Fisheries Department, State Bank of India, *etc.*).
- For providing location-specific demand driven content and capacity building of rural communities, the VRCs and VKCs are working with 315 partners consisting of Government Departments - Agriculture, Animal Husbandry, Fisheries, Education, Rural Development, Health, *etc.*), NGOs, Academic and Research Institutions, Hospitals, *etc.*
- To spread the location-specific demand driven information/ knowledge to the community, the VRCs and VKCs are using number of new media tools such as ISRO uplink/ downlink satellite connectivity for conducing video conferencing and audio conferencing between experts and rural communities, rural communities and rural communities, WiFi based village level intranet to disseminate local news and early warning information, multiple Indian language intranet knowledge portal, Wireless Local Loop Audio Phones to connect experts and rural communities, CDs, SMS alerts to farmers and fisherfolk in their local language, NetMeetings, EduSat based children education programmes, *etc.*, audio-visual tools such as audio content dissemination through All India Radio, GSM based Wired/ Wireless Public Address System network to disseminate agriculture, fisheries, micro-enterprises, early warning, health, government department announcements and advisories from any where to the particular village and print-media/ traditional communication tools such as community newspaper in the local language, pamphlets, booklets, manuals, notice boards, tom tom, meetings at village level, *etc.*

- During 2006-2008, 1, 65,742 (Male: 58.5%; Female: 41.5%) rural women and men physically used these VRCs and VKCs (visited VRCs and VKCs personally and got the information). VRCs and VKCs are also set up in different dates and different years during this period.
- Upto September, 2010, 733 community newspaper issues were published in three local languages namely, Tamil, Marati and Malayalam. It reaches 552 villages. The community newspaper concept initiated by VRCs and VKCs since 2002. The newspaper also started different VRCs in different months during 2002-2008.
- 361 audio contents were developed for dissemination through All India Radio, Educational Institutions Campus Radios, IGNOUs Gyan Vani, *etc.*
- Under the mobile Tele-Ophthalmology services (Upto September, 2010), 14,685 patients screened in 170 villages through 194 camps. This programme was conducted with the help of Sankara Nethralaya, Chennai. During 2007-2008, 58 video conferencing consultations were conducted. 302 patients received the experts opinion. These video conferencing consultations covered 142 villages.
- Through Soil and Water Testing facilities, 394 farmers (2008) got the soil and health management advisories in Tamilnadu and 818 (2008) farmers got the soil and health management advisories in Maharashtra.
- Under the Microsoft Unlimited Potential Programme, VRCs and VKCs trained 12,041 rural communities in the topics of MS Office, Digital Media and Web Development. This programme covers 629 villages.
- In the Computer Aided Learning Programme, VRCs and VKCs trained 18,507 children in 643 schools. This programme provides several aspects such as how to improve the arithmetic operations, health aspects, flora and fauna of India, how fan works, how telephone works, how to improve the English vocabularies, *etc.*
- In the Intel Learn Programme, VRCs and VKCs trained 897 children in 181 schools. This is mainly focused on how to find out the solutions of different developmental issues through group of children as a project mode.
- During 2005-2008, VRCs and VKCs conducted 836 training and awareness programmes in different thematic areas. 42,637 rural communities participated in these events.
- Nine training modules (community needs assessment, general management of VRC and VKC, monitoring and evaluation, community

participation and facilitation skills, development and micro-planning, gender analysis, community based disaster preparedness, institution building and Microsoft Unlimited Potential Programme self-learning modules) were developed for VRC and VKC managers of India.

- 46 livelihood modules (agriculture, animal husbandry, health, fisheries, micro-enterprises, weather, technical, government schemes, *etc.*) were developed.
- 12 research students (both national and international) conducted research study in VRCs and VKCs programme.
- 31 publications published to explain the activities of VRCs and VKCs to different stakeholders.
- 985 grassroots academicians (NVA Fellows) were selected from 21 States. 25 grassroots academicians selected from 6 countries. 227 have been selected for Fellowships through 18 regional committees for the year 2009.
- In the GGA movement more than 350 partners are participating in the alliance.
- 18 policy advocacy publications were published based on the alliance partners suggestions.
- Nine rural applications (Village Centric Management System, Patient Logistics Management for Hospitals, Knowledge Kit for Goat Keepers, Application Tool for Bone Mineral Density Scan Analysis, Participatory Fisher-Folk e-commerce project, School Management Software, e-commerce web portal, primary eye care through rural vision centre and account maintenance of self-help groups) were developed under Rural Innovation Fund which was monitored by GGA secretariat.

Conclusion

Building partnership is important for establishing connectivity and technology, creation of time-and location-specific, demand-driven content and dissemination of the same for ensuring livelihood, food and health security of the rural communities.

6

SATCOM Application for Sustainable Rural Development

K. Narayana Gowda

ABSTRACT

Village Resource Centre (VRC) provides information services to the rural population to empower them to face day to day challenges. In this line, the Department of Space, ISRO & RRSSC is bringing its space technology to common man by setting up of Village Resource Centres, with a view to integrate the capabilities of communication and remote sensing satellites to address the changing and critical needs of rural communities. VRC is totally interactive VSAT based network for providing space enabled information related to land records, natural resources, drinking water wells, ground water recharge, soil types, alternate cropping pattern, waste lands, *etc.* VRCs also provide variety of services in its tele-education and tele-medicine programmes, online decision support, weather forecasting, water management, interactive farmers advisory services (video conferencing) *etc.* It is appropriate to establish one VRC in each hobli in the initial stage and thereafter for all the panchayats in the country in due course. It is a great challenge to the extension professionals to make use of this system for serving the farming community. Involvement of representatives of the group members of the local institutions such as SHGs, MPCSs and Gram Panchayats in the concerned programmes enhances the effectiveness of this information system.

Introduction

India is the second populace country in the world but occupying 7th place in terms of sharing geographical area. The total number of villages runs to over six lakh spread over in 2.2 lakh panchayats with a rural population of 71 per cent. The economy of the rural people is mainly agriculture and allied enterprises. The development efforts in the last five decades has brought about many useful changes particularly in attaining self sufficiency in food, milk, fish, poultry, fruits and vegetables. Although production and productivity levels have increased, the margin of profit has decreased continuously over the period resulting in loosing confidence by farmers in farming. But farmers are loosing confidence in farming in view of continuous increase in cost of production and decrease in profit margin. The production level of these produces need to be enhanced in view of growing population. The major issue that is coming in the way of sustaining production level and rebuilding confidence is the inadequate timely information support system catering to the needs of diverse entrepreneurs.

Many great personalities have also opined that the development of Rural India suffered due to inadequate, timely and dependable information support system. The situation further aggravates due to drying up of public extension system in the last one and a half decade in the country. The trend may not improve in the near future due to non-filling up of vacant posts. More demand for timely and reliable information by rural people necessitated in the recent years in view of liberalization, commercialization and globalization of agriculture.

The existing mass media namely; Radio, TV and Press will not be able to provide specific and reliable information to such a large number of rural people. It is a known fact that many of our qualified scientists, experienced personnel are stationed in urban areas, where as the majority of Indian population lives in rural areas, access to information and their education is major concern. It is the SATCOM CUM ICT which can provide the variety of reliable information timely at the easy reach of rural people even in remote areas. This is being addressed through the establishment of Village Resource Centres (VRCs) or Village Knowledge Centres (VKCs) in the recent years and it would be great boon to meet the unfulfilled needs of rural people.

The Indian Space Programme is driven by the developmental needs of the country and has endeavoured to reach out to the grassroots. Today advanced communication and remote sensing satellites address a variety of national needs including Agricultural, Health, Education, Communication and Natural Resource Management.

Specifically in the noble context of benefiting the grass root population, the Indian Space Research Organisation (ISRO) has successfully implemented a number of projects in the areas of Drinking water mission, watershed management, wasteland development, Tele-education, Tele-medicine or tele-health which is of great social relevance to the country for enabling specialty Healthcare to the remote rural and underserved population.

Village Resource Centre (VRC)

The Village Resource Centre (VRC) or Village Knowledge Centre (VKC) is a place to render variety of services from a single window to remote rural mass in a given geographical area through modern information and satellite communication technology *viz* Very Small Aperture Terminal (VSAT)/GRAMSAT/Leased line. The objective of this programme is focussed on reaching the benefits of satellite based information technology directly to the communities at the grass root level as a single window delivery of agriculture, health and variety of livelihood needs of rural mass. They are mainly information receiving centres partly creation centres as well as processing centres.

VRC provide information services to the rural population to empower them to face day to day challenges. In this line, the Department of space, ISRO & RRSSC is bringing its space technology to common man by setting up of Village Resource Centres, with a view to integrate the capabilities of communication and remote sensing satellites to address the changing and critical needs of rural communities. VRC is totally interactive VSAT based network for providing space enabled information related to land records, natural resources, drinking water wells, ground water recharge, soil types, alternate cropping pattern, waste lands, *etc.* VRCs also provide variety of services in its tele-education and tele-medicine programmes, online decision support, weather forecasting, water management, interactive farmers advisory services (video conferencing) *etc.*

Prof. Swaminathan states that this satellite based project aims for digital connectivity to remote villages for providing multiple services such as tele-medicine, tele-education and remote sensing applications through a single window. He said that unless we take the benefits of science and technology to the villages, we cannot get rid of poverty, ignorance and diseases. The information provided will be in the form of geo-referenced land record, natural resources, suitable sites for drinking water as well as sites for recharging to replenish ground water, water harvesting, waste lands that can be reclaimed, rural employment creation, watershed, environment, infrastructure, alternate cropping pattern, and so on. By suitably blending the information derived from earth observation satellites with ground derived and weather related information, local specific community advisory services can be provided.

VRC's, would be a great boon to meet these unfulfilled gaps. Establishing village resource centers to provide information on agriculture, health, a variety of livelihood related activities to empower rural mass is an urgent need. With the advent of communication technology especially satellite communication (SATCOM) combined with information technology we have means to extend the benefits of advanced agricultural and allied sectors to the remote and inaccessible areas.

Establishment of VRC

Indian Space Research Organization (ISRO) in its GRAMSAT (Developmental Network & VRCs) programme, started VRCs to provide various space enabled services such as Tele - medicine, Tele-education (non-formal), Geospatial information on natural resources, environment, in an integrated manner by establishing Village Resource Centres in the country in phase wise. The first VRC was inaugurated by Hon'ble Prime Minister of India which was established jointly by ISRO and the Prof. M.S. Swaminathan Research Foundation (MSSRF) on October 18th, 2004. Followed by this 85 VRCs were set up in Rajasthan, Gujarat, Maharastra, Orissa, Kerala, Karnataka, Tamil Nadu and Andhra Pradesh. Thereafter another 150 VRCs are being set up in the country in association with Universities, NGOs, Trusts, State and Central agencies.

Hon'ble President of India inaugurated the village resource network set up by ISRO in partnership with Amrita Vishwa Vidya Peetam Ettimadai, Coimbatore in Tamil Nadu and Kerala on July 6, 2005.

Rajiv Gandhi Institute of Information Technology (RGIIT), Tikarmafi, Amethi (A Campus of IIIT, Allahabad) has also launched Village Resource Centres (VRCs) in collaboration with ISRO which are located at: Raibarelly City, Lalganj, Unchahar, Jagdishpur, Gauriganj, Jayas, Amethi and RGIIT.

At present ISRO works in close liaison with the following agencies for the establishment of VRCs:

1. Amrtia Vishwa Vidya Peetam, Ettimadi, Coimbatore, Tamil Nadu.
2. Karuna Trust, Biligiri Ranga Hills, Chamarajanagar District, Karnataka.
3. MS Swaminathan Research Foundation (MSSRF), Chennai, Tamil Nadu.
4. Self-Employed Womens' Association (SEWA), Ahmedabad, Gujarat.
5. Kutch Nav Nirman Abhyan, Bhuj, Kutch District, Gujarat.
6. Mysore Resettlement and Development Agency (MYRADA), Bangalore, Karnataka.
7. Kerala State Planning Board, Government of Kerala.
8. University of Agricultural Sciences (UAS), GKVK, Bangalore.

Village Resource Centre in Bangalore Rural District

One VRC was inaugurated by V.L. Chopra, Honourable Member, Planning Commission GOI on June 9th, 2006 at Hadonahalli village, Doddaballapur Taluk of Bangalore Rural District which is established in collaboration with ISRO-DBT-UAS Bangalore. During the inaugural address Sri Chopra highlighted the importance of VRCs in providing information to rural masses for all-round development.

A Case Study

During 2005-06, the southern part of Karnataka has received the highest rain fall which was a record rainfall in the last five decades. The intensity of rain was more during the late Kharif coinciding with the harvest of the field crops particularly Ragi and Maize which are the major cereal crops of the area. Rural Bio-Resourse Complex (RBRC) Project operates in 72 villages covering 7500 families located in Tubagere hobli of Bangalore Rural District in Karnataka. One junior scientist works at the rate of 800-1000 families in the project area. The junior scientists provide timely information on various agro technologies including weather forecast. Timely weather forecast information helped many farmers in the project area to take crucial decisions such as whether to take up harvesting or postpone the harvest, heap the produce in case of onset of heavy rains.

Although quantification of economic benefits was not computed but it is informed by the farmers of the project area that majority of them were able to save the loss and obtained good grains due to the timely advice on the weather forecast. In fact the farmers of neighbouring hobli were envy to this and telling to us in some occasions what sin we have committed in not getting such vital information and we have lost heavily due to lack of timely information. Even the leftover grains are of poor quality unfit for human consumption. The information on weather forecast is also helping many farmers in the project area to schedule the plant protection measures particularly for high value crops like potato, tomato and grapes.

Expert Centre (EC)

The Expert Centres are also being established in different states to provide information support to the VRCs. In the ECs, opportunity is provided for two way audio video conferencing. One individual or group of individuals of VRCs can directly interact with the expert(s) in the Expert Centres. The expert can directly provide information to large number of rural people simultaneously through VRCs. Generally, the ECs are located in places where large number of experts are easily accessible.

One such centre is located at the University of Agricultural Sciences, Ghandhi Krishi Vignana Kendra, Bangalore. The two more Expert Centres are established in Karnataka one of them by Mysore Resettlement and Development Agency (MYRADA), Bangalore and other by KARUNA Trust, Mysore. There are 33 VRCs established in the State of Karnataka as on October 2006.

Requirement for Setting up of a Village Resource Centre

1. Establishing rapport with the local ISRO staff.
2. Identifying and suggesting suitable location for establishing VRC.
3. MOU regarding the infrastructure facilities to be provided by the host institution.
4. Trained manpower to monitor and use the system.
5. Minimum space and with electrical connections and a generator.
6. Minimum furniture for computer system and its accessories besides seating arrangements.
7. Space to install satellite signal receiving antenna or communication equipment.

Benefits of Village Resource Centre

Provide information on alternate crops suitable to different clientele directly from the experts.

1. Provide information on latest cultivation practices of different crops and enterprises including, Integrated Nutrient and Pest Management.
2. Provide information on weather forecast such as rainfall, temperature, RH, wind velocity, wind direction, cyclone, earthquake, *etc.,* for taking suitable decisions and action.
3. Provide information on marketing opportunities, warehouse facility, value addition and processing for taking right decisions.
4. Availability of various subsidies, services from government and corporate bodies and NGO's for utilizing such benefits.
5. Availability of loan provisions for different Vocations, rate of interest for different enterprises and clientele.
6. Provisions of crop, health and market insurance.
7. Enhancing the knowledge base on water, plant and other natural resources of the area including Indigenous Technical Knowledge for taking rational decisions.
8. Source of availability of various inputs, nursery seedlings, rates, quality, quantity including availability of transport facility and fares.

9. Information on health related knowledge, including facility of super speciality hospitals.
10. Farmers, SHG's, students, patients and artisans can interact directly with experts on specific problems with out loss of time through video conferencing.
11. On line degree programmes can also be taken up.
12. In the context of cost benefit ratio this method is less cost and more effective.
13. Provide information on legal aspects.
14. Information on remedies for simple aliments.
15. Village stories with moral values, village history, village folk songs, village culture and traditions.
16. Employment information to needy people.
17. Availability of local artisans and skilled persons such as black smithies, Basket weavers, toys maker, Electrician, pump set repairer, tractor power tiller repairer *etc.*

Ways to Enhance the Performance of VRCs

1. Providing handouts to all the stakeholders in the area on the availability of various information at VRC.
2. Office bearers of SHGs and other local institutions need to be involved continuously for participation on the topic of concern.
3. Inviting the stakeholders on the important occasions and briefing them about usefulness of VRC.
4. Recognizing and encouraging the users of VRC on the important occasions.
5. Establishing constant linkage with all the relevant institutions and organizations for information support on up to date basis, particularly with ECs.
6. Fixing minimum service charges for printouts of vital or specialized information.
7. The personnel in-charge of VRC should be trained constantly on the skill of updating, interpreting and maintenance of the centre.
8. The personnel working at VRC need to be made accountable with suitable recognition and encouragement.
9. The system can also provide information support to local officials, NGO's and others working in the area.
10. Relay of important recreation based programmes.
11. Success story on the users of such information.

Conclusion

The Indian farmers have suffered over the years partly due to lack of timely comprehensive information support for taking rational decisions. In this context, VRCs have great role to play in providing timely reliable information at a reasonable cost to the farmers in particular and rural people in general for their development. Therefore, it is appropriate to establish one VRC in each hobli in the initial stage and thereafter for all the panchayats in the country in due course. It is a great challenge to extension professionals to make use of this system for serving the farming community. Involvement of representatives of the group members of the local institutions such as SHGs, MPCSs and Gram Panchayats in the concerned programmes enhances the effectiveness of this information system. Similarly, it is an opportunity for the extension researchers to explore how to enhance the system in terms of reaching every clientele with required effectiveness, as well as computing Cost benefit ratio over other methods.

7

e-Village for Agriculture and Rural Development: Integrating ICTs with Traditional Agricultural Extension Methods in North-East India

R. Saravanan, C. Kathiresan, Narmi Darang, T. Indra Devi, Supriya Devi and Lizamoni Chungkurang

ABSTRACT

India's North-East is predominantly hill agriculture based and endowed with rich natural resources for the development. The average rainfall in the region is the highest in the country. However, low agricultural productivity and food in-security are largely prevalent in the North- East India. Lack of adequate information on advanced farm technologies, market intelligence and rural development schemes led to the income poverty in the region. The limited technical manpower, lack of transport and communication facilities, inadequate financial support to the technology transfer and less infrastructure facility creates huge technological gap among rural tribal farming community. Considering the disappointing agricultural and rural development scenario and realising the potential of ICTs, the Central Agricultural University (CAU), College of Horticulture and Forestry, Pasighat, Arunachal Pradesh and Centre for Development of Advanced Computing (C-DAC), Hyderabad have initiated a pilot project named "Model e-Villages in North-East India", an innovative e-Governance initiative committed to the agricultural and rural development. The Model e-Village project implemented in the selected ten remote tribal villages of East Siang District of Arunachal Pradesh State and provides regular e-awareness and e-literacy programmes and also experimenting local need based IT applications in the area of agriculture and allied sectors. The e-Village project activities includes; Village Awareness Meetings and Identification of e-Villages, Villagers' Information Needs Assessment, ICT Capacity Building Workshop for Facilitators, facilitating Online and Offline Access to the Agricultural Resources by the tribal farmers, Convergence of University Education, Research and Extension Programmes with e-Village Centre, Advanced Farm Technology Demonstrations by the e-Village Project Fellows, KVK Extension Activities at the e-Village Centres, Village Adoption Scheme of the Central Agricultural University, Rural Horticultural Work Experience Programme (RHWE), Environmental Awareness Programmes, Computer Education and e-Literacy, Multimedia CDs, National Seminar & Exposure Visits and TV-DTH: Awareness, Education & Entertainment. Field experiences shows that farmers are very keen to get agricultural related information from the e-Village centre. However, they feel that along with e-Resources, field level demonstrations are must to understand advanced farm technologies. Even though, they value the importance of agriculture information for livelihood generation, they are unwilling to pay even a nominal fee. Because, the feel that providing agricultural information is a public good and welfare activity of government institutions. But, interestingly, nine out of ten e-Village centres are hosted at the infrastructure (house & furniture, electricity connection) provided by the community at free of cost. Even, some of the e-Village centres farmers are willing to share expenditure on consumables (printer cartridge, paper *etc.*). Computer training to the village children are very much appreciated by the villagers. The community members voluntarily participated in large numbers in all the e-Village programmes. Considering, inadequate local need based content, language problem, inadequate ICT infrastructure (telephone, internet), road and other communication facilities in Arunachal Pradesh, integrating conventional extension approaches with ICTs (e-Village centres) emerged as a appropriate mechanism for agriculture and rural development in remote and backward rural areas.

Introduction

India's North-East is endowed with rich natural resources for the development. The average rainfall in the region is the highest in the country. However, the region has exhibited most backwardness in most important indicators of human development. The non-income poverty in terms of inadequate information on advanced farm technologies, market intelligence and rural development schemes produces the income poverty in the region. The limited technical manpower, lack of transport and communication facilities, limited financial support to the technology transfer and less infrastructure facility creates huge technological gap among rural tribal farming community. Further, difficult terrain, mountainous periphery and frequent natural disasters hinder the development of the region. Due to non-availability of improved technological information to the tribal farmers, agriculture exhibits low unstable productivity, which creates food insecurity problem and also poses serious developmental question to the policy makers. Among eight north-east states, Arunachal Pradesh State is having the lowest agricultural productivity and exhibited least performance in almost all human development indicators. The "Connectivity" remains a major obstacle for the development. It is expected that application of new ICT tools play a major role in the development of the region. However, the e-readiness assessment report of Government of India indicated that all the eight states of the North-East India categorized under below average and least achievers category. The report also recommends increasing the awareness of potential benefits of ICTs in rural development. Considering the existing scenario and potential of ICTs, the Centre for Development of Advanced Computing (C-DAC), Hyderabad and the Central Agricultural University (CAU), College of Horticulture and Forestry, Pasighat, Arunachal Pradesh jointly implemented a research project entitled "Creating Model e-Villages in North-East India", an innovative e-Governance initiative for the agricultural and rural development, since 2008. The Model e-Village project implemented in the selected ten remote tribal villages with the main objective of creating the computer infrastructure in the remote tribal villages and to provide regular e-awareness and e-literacy programmes and also to experiment local need based IT applications in the area of agriculture, health, education, governance and other areas (www.modelevillage.in).

Project Methodology

The e-Village project implemented in the selected 10 remote tribal villages in East Siang district of Arunachal Pradesh State, North- East India. The selected villages are located at the center of rural area, so that people from surrounding villages will have an access to the IT infrastructure. One project facilitator was selected from the same or near by village to act as a facilitator and trainer in the village IT centre. The project facilitator and group of selected village council members are responsible for conducting regular e-awareness and e-literacy programmes. Regular computer training classes were conducted to the children, village youth and others. The project research fellows and facilitators are experimenting local need based IT applications in the area of agriculture, health, education, governance and other areas.

Project Activities : Part-I

a. Village Awareness Meetings and Identification of e-Villages

During July 2008 to September 2008, village level ICT awareness meetings and focussed group discussions (FGDs) were conducted to identify suitable villages to host ten e-Village centres in the East Siang district of Arunachal Pradesh.

b. Villagers' Information Needs Assessment

After the identification of the villages, villagers' information needs on agriculture, education, health, governance *etc.*, were identified through Participatory Rural Appraisal (PRA) methods, group discussions and semi-structured interview schedules.

c. ICT Capacity Building Workshop for Facilitators

Twenty number of e-Village and e-Arik project staff participated in the six day workshop on "ICTs for e-Village Knowledge Facilitators". Technical sessions on e-Village project-backround, objectives, working mechanism, basics and advances of IT, internet and its tools, InDG web portal and national level web portal initiatives for the agricultural and rural development, use of Television and radio in the farm technology transfer, application and entrepreneurial aspects of IT and they also visited the e-Arik village knowledge centre, Yagrung village of East Siang District of the Arunachal Pradesh State. During the workshop, "Training Manual on Fundamentals of IT for Knowledge Centre Operators", a practical manual published by the India development gateway (InDG) initiative of the C-DAC, Hyderabad has been distributed to the participants.

Project-Activities : Part-II (e-Agriculture)

1. Online and Offline Access to the Agricultural Resources

The e-Village centres are equipped with wireless internet connectivity for online access of agricultural and rural development related resources. However, considering frequent power failure and weak signal for internet connectivity, number of offline resourses is made available at the centres. Especially, InDG webportal (www.indg.in) contents are made available to the villagers through e-Village centres.

2. Convergence of University Education, Research and Extension Programmes with e-Village Centre

All the research and extension activities of the College of Horticulture and Forestry, CAU and Krishi Vigyan Kendra, East Siang district, Pasighat are carried out at the ten e-Village centres. Further, final year B.Sc (Horticulture) students internship (Rural Horticultural Work Experience) Programme also conducted at the e-Village centres. Farm demonstrations, Scientist-farmers interactions, Kissan mela and awareness programmes were conducted at the e-Village centres.

3. Advanced Farm Technology Demonstrations by the e-Village Project Fellows

Advanced farm technology demonstration on rice and khasi mandarin cultivation, organic farming, mushroom cultivation and processing & post harvest aspects were demonstrated to the tribal farmers using e-Village ICT facilities.

4. KVK Extension Activities at the e-Village Centres

The Krishi Vigyan Kendra, CHF, CAU, Pasighat also adopted all ten villages having e-Village centres for conducting vocational training, front line demonstrations and on-farm research. The KVK uses resources and infrastructure of e-Village centres to impart effective training to the rural tribal farmers.

5. Research Projects

Field based research activities also carried out by the faculty and research team members of the College of Horticulture and Forestry, Pasighat in the ten villages which is hosting e-Village centres. The farmer participatory research approaches are proved successful in ensuring farmers' participation.

6. Village Adoption Scheme of the Central Agricultural University

The College of Horticulture and Forestry, CAU, Pasighat and e-Village team to adopted Berung village for popularizing the "Giriraja & Girirani", a high yielding poultry breed. One thousand Giriraja chicks were distributed to more than 20 farmers at the e-Village centres. Multimedia based training programme on "Giriraja- For Profitable Poultry Farming" was carried out under the Village Adoption Scheme of the Central Agricultural University, College of Horticulture and Forestry, Pasighat.

7. Rural Horticultural Work Experience Programme (RHWE)

Every year, the final year B.Sc (Horticulture) students of College of Horticulture and Forestry, Central Agricultural University (CAU), Pasighat along with e-Village research team engaged in the basic socio-economic data collection and conducting awareness programmes, tree plantation campaign, mushroom cultivation, vermi-compost, soil sampling and testing, horticultural products processing, landscaping and flower arrangement trainings were conducted at the e-Village centre during their month long village stay internship programme called Rural Horticultural Work Experience Programme. Students also conducted kissan mela by demonstrating need based low cost and green technologies by conducting village survey and farmers technological need assessment. The technologies such as; mushroom production, vermi-compost, flower arrangement and bouquet preparation, low cost zero energy chamber for storage of vegetables and fruits, processing of vegetables and fruits, multimedia awareness campaign on potential on rubber cultivation, management and processing and marketing in Arunachal Pradesh was demonstrated practically and also exhibited in the kisan mela stalls. The multimedia presentation was organized at village community hall.

8. Environmental Awareness Programmes

The e-Village team periodically undertakes environmental awareness programmes by showing multimedia based awareness programmes, tree planting, awareness lectures, discussion, quiz competition and cultural programme *etc.* On 5th June, 2009 e-village team joined with villagers to celebrate the World Environment Day with the theme of *"Your Planet Needs You! UNite to Combat Climate Change"* and joined the United Nations Environment Programme (UNEP)'s "Billion Tree Campaign". Final year B.Sc (Horticulture) students of College of Horticulture and Forestry, Central Agricultural University (CAU), Pasighat adopted Berung e-village to celebrate world environment day and Berung joined "UNEP's Billion Tree Campaign", each family of Berung village, e-Village team members and College of Horticulture and forestry students planted nearly one hundred tree saplings.

9. Computer Education and e-Literacy

Regular computer education and e-literacy programmes were conducted for the benefit of tribal children and youth and other villagers in the ten e-village centres. Three month training programme on basics of computer and internet were conducted for the school going tribal students. So far more than 500 students have completed computer training at the ten e-Village centres.

10. Multimedia CDs

Advanced crop production technologies, Rubber cultivation, Integrated Pest and Diseases Management (IPM), Citrus e-Clinic CDs, Medicinal & Aromatic plants and Nutrition CDs by the C-DAC, Hyderabad, Cure@home CD by C-DAC, Pune *etc.*, were distributed to the e-Village centres and shown to the tribal farmers regularly.

11. National Seminar & Exposure Visits

The national seminar on "ICT4ARD" was conducted at Pasighat, Arunachal Pradesh on 9th to 11th September, 2009 and 80 participants from India, one from USA, members of the e-Village research team and large number of farmers were participated and seminar delegates had exposure visits to the e-Village centres and had an interaction with the farmers of the e-Villages.

12. TV-DTH: Awareness, Education & Entertainment

The e-Village centres also having TV and direct to home connection and villagers are shown awareness, educational and also entertainment programmes at the e-Village centres.

Conclusion

Field experiences shows that farmers are very keen to get agricultural related information from the e-Village centre. However, they feel that along with e-Resources, field level demonstrations are must to understand advanced farm technologies. Even though, they understand the importance of agriculture information for livelihood generation, they are unwilling to pay even a nominal fee. Because, they feel that providing agricultural information is a public good and welfare activity of government institutions. But, interestingly, nine out of ten e-Village centres are hosted at the infrastructure (house & furniture, electricity connection) provided by the community at free of cost. Even, some of the e-Village centres farmers are willing to share expenditure on consumables (printer cartridge, paper *etc.*). Computer training to the village children are very much appreciated by the villagers. The community members voluntarily participated in large numbers in all the e-Village programmes. Considering, inadequate local need based content, language problem, inadequate ICT infrastructure (telephone, internet), road and other communication facilities in Arunachal Pradesh, integrating conventional extension approaches with ICTs (e-Village centres) emerged as a appropriate mechanism for agriculture and rural development in remote and backward rural areas.

8

India Development Gateway Initiative - A National Level ICT Platform to Reach the Unreached

C. Kathiresan, B. Vijayalakshmi and N. Sarat Chandra Babu

ABSTRACT

ICT has seen its vertical as well as horizontal growth in various domains, be it rural livelihood sectors and governance. India is well poised to take the benefits of ICT to the common man. There has been consistent progress in addressing the challenges that exists - be it infrastructure or social barriers. The India Development Gateway (InDG) initiative is in response to limitations that exist at the social front. InDG facilitates rural empowerment through provision of relevant information products and services that respond to the real and strategic needs of the rural communities, in their local languages. In the process, it catalyzes the use of ICT for collaboration and knowledge sharing among development stakeholders. The multilingual portal www.indg.in developed as part of this initiative hosts information in six livelihood oriented sectors (Agriculture, Health, Rural Energy, Primary Education, Social Welfare and e-Governance) in eight languages (Hindi, Bengali, Assamese, Marathi, Kannada, Tamil, Telugu and English). Currently, over 300 organisations associate with InDG for generation of content on about 2400 topics and about 80 experts offer their services for various content management activities. Value added services like Ask An Expert, Buyer Seller Platform, Dynamic Market Information, Weather Forecast and multimedia CDs introduced as part of this initiative had been pilot tested in selected locations and are well received by the rural community. As the Rural landscape is set to take advantage of the flourishing ICT initiatives, by various institutions, and more specifically the Common Service Centers (CSCs), InDG offers the much required 'Content and Services' that makes the difference in the lives of rural people. A synergy between these two initiatives will lead to a giant leap in utilizing ICT for Rural empowerment in India. This nation-wide initiative is supported by the Department of IT, Ministry of Communications and IT, Government of India and executed by the Hyderabad centre of Centre for Development of Advanced Computing (C-DAC).

Introduction

India today is on the fast track in terms of technology use for development. The key elements for an inclusive growth in the use of technology for development are those of access, affordability and utility. Information and Communication Technologies (ICT), in the recent past, have emerged as powerful catalysts that can accelerate inclusive growth and development. Appreciable polices and initiatives that aim at improving the access and affordability of ICT to common man, especially those of rural communities, are underway. However, the strength of the chain is in its weakest link - utility. Social barriers such as illiteracy, language and cultural backgrounds often limit the utility and scalability of ICT for development. It is an established fact that Information, when properly provided, is a positive power of unequalled strength. Powerful ICT tools like internet and mobiles have shrunk the barriers of time and distance for rural communities to access crucial information. However, proliferation and utility of information can best happen, if it is in regional languages, understandable by the rural community and is the collective wisdom and experience of various development stakeholders. The India Development Gateway (InDG) initiative, hence, is driven by an inclusive vision to harness the use of technology for the collective sharing of knowledge and wisdom that yields benefit to the unreached.

This nation-wide initiative is supported by the Department of IT, Ministry of Communications and IT, Government of India and executed by the Hyderabad centre of Centre for Development of Advanced Computing (C-DAC).

Need for an Multilingual ICT Platform

India is well poised to take the benefits of ICT to the common man, particularly in terms of the ICT infrastructure build up. However, various challenges still exist in bridging digital divide. Absence of relevant, authentic local language content is often viewed as a key challenge. A step further is the availability of local language information, products and services related to all the livelihood areas of interest to rural communities, on a single platform. On the other hand, rich experiences and well thought of schemes and plans for development that have, or likely to create an impact on the society, are often shared within a limited geographical boundary and communities.

India Development Gateway (InDG) is thus an attempt to meet the local language content requirement of the rural people related to the sectors that affect their livelihoods through a platform that enables knowledge networking among development stakeholders. Use of ICT, in the process, enables anytime anywhere sharing and access of relevant local language content for rural development.

India Development Gateway (InDG) Initiative - Concept

India Development Gateway (InDG) is a national level initiative dedicated to rural empowerment, provides credible information, products and services in local languages that respond to the real and strategic needs of the rural people. As the rural landscape in India is set to take the advantage of the flourishing ICT initiatives pioneered by various institutions, more specifically the Common Service Centers (CSCs), InDG initiative offers the much required content and services in local languages that makes the difference in the lives of the rural people.

InDG catalyzes the use of Information and Communications Technology (ICT) for collaboration and knowledge sharing among development stakeholders representing from Government, NGOs, community based organisations, academic, research institutions and private sectors, leading to development.

Objectives and Strategies of InDG Initiative

InDG initiative's main objective is 'Reaching the Unreached' by facilitating rural empowerment through provision of responsive information, products and services in local languages.

To meet the above objectives, the following strategies are being followed.

- Leverage the unique partnership structure of the Gateway to build strong credibility and brand equity
- Deliver value added content and services through the identified verticals on the Gateway, through online and offline presence
- Use local languages to the extent possible for content delivery and services
- Ensure strong pro-poor and pro-women orientation in all programmes

- Utilising existing network of ICT initiatives in India for wider dissemination of InDG products and services
- Formation and strengthening of region wise domain specific content consortiums to enable relevant and regional content flow in the portal
- Evolving various models of sustainability for the initiative

Salient Features of this Initiative

- Content and services in local languages which can be accessed through the widespread infrastructure available in India, like Village Information Centres, Common Service Centres (CSCs), Corporate Social Responsibility initiatives and similar ICT initiatives by different organisations.
- The portal is to serve as a Google for rural development workers. The six focus areas of the portal, i.e., Agriculture, Health, Primary Education, Social Welfare, Rural energy and e-Governance provides focused information in local languages
- Partnership with various developmental stakeholders working at grassroots level for the development of the rural communities.
- Synergy of Online (www.indg.in portal) and Offline initiatives for better outreach.
- A potential self sustaining initiative with a social perspective.

MULTILINGUAL PORTAL – www.indg.in

The multilingual portal (www.indg.in) developed as part of this initiative, hosts information related to 6 important livelihood oriented sectors namely Agriculture, Primary Education, Health, Rural Energy, Social Welfare and E-governance. The portal currently hosts information in 8 languages (Hindi, Bengali, Marathi, Assamese, Tamil, Telugu, Kannada and English) and two more languages (Gujarati and Malayalam) are being added. This multilingual portal (www.indg.in) was formally launched by the Hon. President of India in July 2008.

www.indg.in has been developed using 'Plone', a user-friendly and secured open source Content Management System (CMS), which supports multilingual content (Fig 1). The use of CMS enables shared ownership of the portal with various stakeholders. Any authenticated user can upload content to a specified topic in the portal. CMS helps the user to directly upload the content into the portal by following simple steps. It saves the time in content flow from content providers and user end. All the registered members will be able to access this facility. Special privileges like 'edit/modify the content' will be given to such partners who validate the content online.

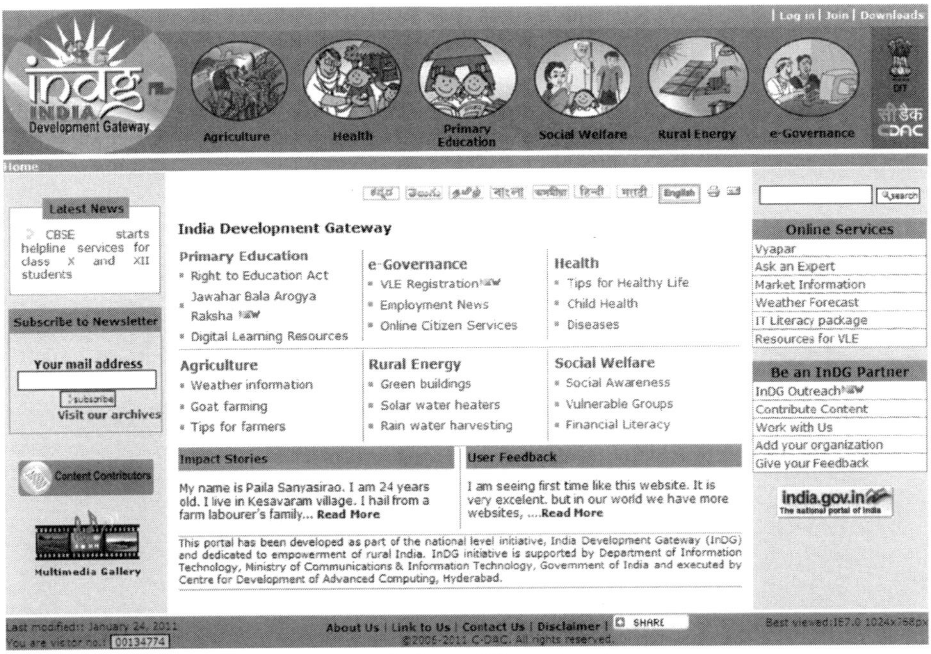

Figure 1. Home page of the www.indg.in multilingual portal

Content Generation Process in InDG

The content available in this portal is supported by various organizations working in these 6 sectors across India. Series of multi-stakeholder workshops are being conducted in selected states to identify information needs of that region, select right content partners in that state and associate with them to collate region specific content required for the portal. InDG initiative is driven by unique partnership model, wherein many like minded organisations representing NGOs, Government and Private are supporting content generation and outreach activities. Grassroots level piloting of services are being undertaken in association with partners in identified states.

InDG follows a decentralized Content management system for hosting content on the multilingual portal. An automated content flow process (Fig 2) that ensures free flow of relevant, unbiased, value added content is the key feature of the portal.

Various interesting features that aim at providing interactivity and comfortable use by the rural audience are the focus. They include discussion fora, mechanisms for user's feedback, Ask an expert, online partnership form and content sharing processes.

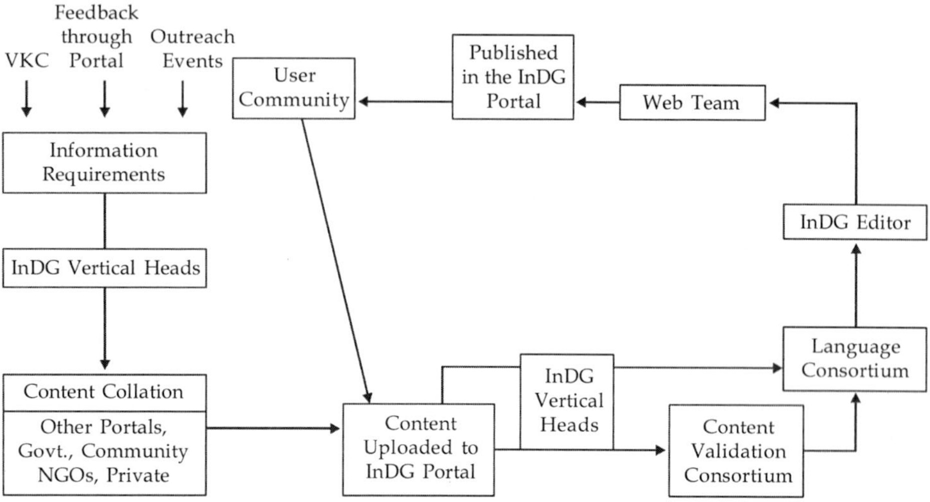

Figure 2: Content Generation process followed by InDG

Focus Areas of the Initiative

InDG initiative mainly focuses on providing information and services in the following six verticals (domains) related to rural development.

Agriculture: Sustainable use of resources and gainful employment

The Agriculture vertical of www.indg.in focuses on technologies and processes for sustainable use of resources, besides creating avenues for employment and supplementary income for the rural communities. The portal host information related to Agriculture and its allied sectors like Animal husbandry, horticulture, fisheries *etc.* To mention a few areas - Agricultural credit and insurance, crop production techniques, animal husbandry, fisheries, market and weather, policies and schemes, farm-based enterprises and best practices. Several products and services that address the needs of farming communities along the supply chain are also available for use. Log on to www.indg.in/agricutlure for more details.

Health - Improving the quality of health in rural India

The Health vertical of www.indg.in is a platform for providing health related information to rural people, with a special focus on women and child health. This section provides useful information related to diseases, nutrition, women and child health, first aid, hygiene, lifestyle disorders and important database related to health in India. Value added services like multimedia resource materials and assessment tools are available for use by the health workers to deliver their services in best way possible. Visit http://www.indg.in/health/ for more information.

Primary Education - Establishing the right of a child to quality education

Primary education vertical of www.indg.in has a combination of both and focuses on Primary education as a Fundamental Right and improving quality of Primary education. It holds the variety of Digital learning resource material on Mathematics, Social Science, English and Languages. Primary Education vertical also focuses providing the information and multimedia awareness material on the Child Rights, latest policies & schemes, teaching methods & best practices and discussions on the latest happenings in the Education system in Indian regional languages.

Social Welfare - Empowering vulnerable groups through ICT initiatives

To provide authenticated multilingual information, content and ICT products and services to the Women, Children, Senior Citizens, SC/ST/BC, Persons with Disabilities & other vulnerable groups on social, political, legal, educational, economical, financial, employment and other opportunities (Government Schemes, Food, Housing, Education, Constitutional Rights, Legal awareness, Financial Literacy, Microfinance, Livelihood opportunities *etc.*)

Rural Energy - Making Rural India Self Sustained Energy Communities

Rural energy vertical of www.indg.in focuses on three major areas, by way of providing information. These are energy conservation, energy efficiency and energy production from renewable sources. Basic concepts, Technologies, processes, best practices and policy support related to energy and environment find place in the portal. Visit http://indg.in/rural-energy for more information.

e-Governance - Compliment E-Gov Movement in the Country

e-Governance vertical of InDG is dedicated to promote the usage of online citizen and business services. India Development Gateway (InDG) has hosted information related to: Citizen Services, legal services, Right to Information Act, e-Governance initiatives in states, Mobile Governance, Employment News, VLE Corner *etc.* All these services are powered by Ask an Expert and Discussion Forum tools which provide citizens to communicate with InDG's expert team and discuss issues with InDG's user groups.

Value Added Products and Services

'Vyapar' – Buyer Seller Platform: This online service provides the community to post the details of any items (Agricultural produce, livestock, electronic goods, machineries *etc.*.), they are interested to sell or buy. They can also avail

the services like rent and consultancy. Also, the village level entrepreneurs managing the Common Service Centres can utilize this platform for generating regular income by providing marketing service to the community.

Ask An Expert: This web-based application available in the www.indg.in portal, provides an opportunity to the community to get expert solutions online in their local languages.

Dynamic Market Information: This service is being provided by InDG in association with Tamil Nadu Agricultural University. The wholesale and retail prices of 161 perishable commodities (vegetables, fruits, flowers and spices) from 13 major markets in South India (Tamil Nadu, Kerala and Karnataka) are provided on daily basis through mobile (via SMS) and website (English & Tamil). About 24000 farmers are accessing the market information through mobile on daily basis and reaping the benefit. This platform also provides market profiles, market trend, database of traders and daily tips to farmers

Weather Forecast: Presently InDG provides detailed 72 hour weather forecast up to Panchayat level, covering nearly 32,000 stations in six states. Major weather parameters covered are Temperature, Relative Humidity, Wind speed and Chance of rainfall. This application also provides location-wise search option.

Multilingual Multimedia CDs: Considering the need of the community series of offline multimedia CDs are being produced in association with partnering organisations. 'Nutrition and Health' is one such multilingual CD developed in association with National Institute of Nutrition. This CD is presently available in 6 languages and is a useful one for children, housewives, paramedics and students. Similarly, Production and Marketing of Medicinal, Aromatic & Dye Yielding Crops, is an another attempt by InDG team to provide information to farmers on how to produce and market 54 commercially important medicinal, aromatic and dye crops supported by National Medicinal Plants Board, Government of India.

Outreach Activities – *Connecting to Grassroots*

Participation of rural communities and the people working closely with them, at every stage of the portal development and piloting of services, has been one of the unique characteristics of India Development Gateway. This was made possible through the outreach activities of InDG, organised at state/ District/Block level in identified states, with following objectives.

- Identification of information needs of the rural people and the first level service providers in the five identified verticals

- Capacity building of the first level service providers and Kiosk Operators to enhance the effective utilization of the InDG portal content
- Field testing and demonstrating the utility of www.indg.in portal to the rural communities

The following training manuals are developed and being distributed to the kiosk operators as part of capacity building training programmes.

1. Manual on Fundamentals of IT
2. Manual on Basic Hardware troubleshooting
3. Resource guide on Basic Documentation skills

Impact and Utility of InDG at Grassroots Level

Initial analysis of the users' feedback indicates the utility of the portal in the following areas:

1. *Access to relevant information that address the latent needs of rural people:* Health information especially that of reproductive health have helped women who generally are shy to openly talk about such issues in rural areas.
2. *Access to Value added Services:* the available online services like market information, weather forecast, Ask An Expert and Vyapar provides the users to enhance their income generation
3. *Access to information that saves time and reduces the drudgery:* the availability of online forms, links to e-governance applications such as "check the status of your passport application", case status in high court, *etc.*
4. *Livelihood opportunities* - Simple tips for managing pests in crops, on and off farm enterprises, credit availability, access to market information which helps farmers and farmer federations take informed decision on selling farm produce *etc.*
5. *Capacity building in terms of IT skills :* Village knowledge operators and first level service providers are now more confident in using local languages in computers (be it typing or searching the internet)
6. *Linkages among stakeholders* - The development stakeholders get access to experiences and materials generated by various agencies in a single platform. It helps prevent reinvention of the same in most cases.

Uniqueness of the Initiative

InDG is the unique ICT initiative which is providing region specific multilingual information, products and services related to six important livelihoods oriented sectors and dedicated to rural empowerment in India. The following strategies make InDG project innovative and unique one.

a) *Multilingual Content in Livelihood Domains*: The multilingual portal (www.indg.in) developed as part of this initiative hosts information, products and services in 6 Indian languages. Over 2000+ topics in eight languages (Hindi, Tamil, Telugu, Marathi, Bengali, Assamese, Kannada and English) are made available for public use.

b) *Content Generation Process*: Assures demand driven, authentic information provision for the benefit of the user communities. Rural communities are looked at as generators of information than just users of information

c) *Unique Partnership Models (Content & Outreach):* Wide partnership base of over 225 institutions from varied background including CBOs, national and international organisations have shared content for the portal. Apart from that selective partners have been identified for outreach activities in specific states, including the SCAs in 7 states.

d) *Online & Offline Presence*: The www.indg.in portal assures anytime, anywhere access to the information and services. The offline mode in the form of products complements the online mode, wherever connectivity and other delivery issues are present.

e) *Connecting to Grassroots*: InDG derives its strength in connecting itself to the grassroots through the outreach activities in association with its outreach partners. Series of ICT awareness programmes, capacity building trainings to kiosk operators, state level introductory workshops, interaction with community at regular intervals, piloting of services in selective locations help the InDG team to understand and develop products and services as demanded by the community.

f) *Delivery of Value Added Products & Services*: All the products and services provided by InDG have been formulated based on the demand and support by the partnering organizations (For eg. 'Dynamic Market Information for perishables' with Tamil Nadu Agricultural University; 'Nutrition & Health' multilingual product with National Institute of Nutrition)

Sustainability Model

InDG's target group, the unreached and backward, primarily women and the poor cannot be expected to have any purchasing capacity at the inception of the program. In keeping with the project objectives, information products and services delivered in first phase will be directed towards enhancing the target groups productive capacities which will ensure that InDG would have a steady grassroots clientele when it moves from information to products and services transaction in second phase. However various other options like consultancies, capacity building trainings, membership fees and joint projects with potential partner organisations are being explored to sustain this initiative.

Conclusion

The existing ICT initiatives that we have reviewed for this project clearly indicates that despite high user interest, the availability and affordability of relevant content in local languages and the information network remain major problems. As a result, most projects are not financially viable and have not met user or provider expectations on content or usage. The hope that ICT can surmount at least some of these social, political, and administrative challenges and become a viable technology for the provision of health, education, and other social services is thus ICT's strongest calling card for Socio-Economic-Development. India Development Gateway initiative will answer the above problems in the right way, reaching the rural India's 700 million people. Various options for revenue generation that would enable the sustainability of the initiative are being worked out. As the Rural landscape is set to take advantage of the flourishing ICT initiatives, by various institutions, and more specifically the Common Service Centres, InDG offers the much required Content that makes the difference in the lives of rural people. A synergy between these two initiatives will lead to a giant leap in utilizing ICT for Rural empowerment.

Acknowledgement

We thankfully acknowledge the contribution of various development stakeholders, who have been the key resources in shaping the initiative to its present form. The financial support and guidance of the Department of Information Technology, Ministry of Communications and Information Technology, GOI is also thankfully acknowledged.

References

Anandaraja, N., Sriram, N., Kathiresan, C., Sebastian, S. and Vadivel, E., 2009. Web cum Mobile based Dynamic Market Information Support for Agri-Horti Produces: Linking the Farmers with Market. Paper Presented at 4th International Conference for Internet Technology and Secured Transactions (ICITST-2009), November 2009, London, UK.

Jagadish Babu, M., 2010, InDG Initiative- Empowering Rural India, Paper Presented at eIndia2010 conference, 5th August 2010, Hyderabad, India

Vijayalakshmi, B, Kathiresan, C and Sarat Chandra Babu, N., 2009, India Development Gateway Initiative. Paper Presented at IEEE Conference on Technologies for Humanitarian Challenges, 28th August 2009, Bangalore, India

9

Linking the Farmers with Market through Web and Mobile

N. Anandaraja, N. Sriram, C. Kathiresan, Shibi Sebastian and E. Vadivel

ABSTRACT

Farmers in India are at the mercy of traders especially the middlemen who fix lower rates at the farmers end. The farmers are desperate as they are not aware of the prices elsewhere and they have meager produce at their hand. Market led agriculture is need of the hour which needs market intelligence. Tamil Nadu Agricultural University, Coimbatore, Tamil Nadu in collaboration with Centre for Development of Advanced Computing, Hyderabad centre initiated the project for the benefit of farmers and traders covering six major markets of South India as Pilot project. Main emphasis of the project is covering Perishable items of 161 commodities, updating the daily wholesale and retail price through www.tnau.ac.in/ www.indg.in @ 1.00 PM with the help of Market Analysts recruited for this purpose. Ensuring the last mile connectivity and reaching the large number of users, the price information is being delivered through Mobile phone. This paper also explains about its additional features.

Introduction

India has made tremendous strides in its path towards development. The farm sector shoulders much of the responsibility in this endeavor. Agriculture plays a vital role, both as an essential infrastructure and as a development component in generating and sustaining a higher national income. India has high population pressure on land and other resources to meet its food and development needs. The natural resource base of land, water and bio-diversity is under severe pressure. The massive increase in population (despite the slowing down of the rate of growth) and substantial income growth, demand an extra of about 2.5 mt of foodgrains annually, besides significant increases needed in the supply of livestock, fish and horticultural products. Under the assumption of 3.5 per cent growth in per capita GDP (low income growth scenario), demand for foodgrains (including feed, seed and export) is projected in the year 2020 at the level of 256 mt comprising 112 mt of rice, 82 mt of wheat, 39 mt of coarse grains and 22 mt of pulses. The demand for sugar, fruits, vegetables and milk is estimated to grow to a level of 33 mt, 77 mt, 136 mt and 116 mt respectively. The demand for meat is projected at 9 mt, fish 11 mt and eggs 77.5 billion (Vision 2020, GoI).

The agriculture sector, with more than 60 per cent population dependence, contributes only about 18 per cent of the total Gross Domestic Product (GDP), resulting in low per capita income in the farm sector (Datt and Sundharam, 2009). Consequently, there is a large disparity between the per capita income in the farm sector and the non-farm sector. The income levels are determined by the overall production, supported by reasonable levels of yield and prices realised by the farmers. Several constraints such as preponderance of small and marginal holdings accounting for about 82 per cent of total holdings, imperfect market conditions and lack of backward and forward linkages affect the income levels of farmers adversely.

Current Farm Market Structure in India

India produces 150 million tonnes of fruits and vegetables. It is estimated that about 50 per cent of the agricultural produce is available as marketable surplus. Post-harvest losses are estimated to be of the order of 5-7 per cent in food grains and 25-30 per cent in the case of fruits and vegetables. More than 72 per cent of the vegetable and fruits are wasted in the absence of proper retailing (The Economic Times, 2008).

Most agricultural produce in India is sold by farmers in the private sector to moneylenders (to whom the farmer may be indebted) or to village traders. Produce is sold in various ways. It might be sold at a weekly village market in the farmers' own village or in a neighboring village. If these outlets are not available, then produce might be sold at irregularly held markets in a nearby village or town or in the *mandi*. Farmers also can sell to traders who come to the work site.

Assured and remunerative marketing opportunities hold the key to continued progress in enhancing farm productivity and profitability. Several significant market reforms have already been initiated by the Central and the State governments. The various central government organizations involved in agricultural marketing, include the Commission for Agricultural Costs and Prices, the Food Corporation of India, the Cotton Corporation of India and the Jute Corporation of India. There are also specialized marketing boards for rubber, coffee, tea, tobacco, spices, coconut, oilseeds, vegetable oil and horticulture. A network of cooperatives at the local, state and national levels assist in agricultural marketing in India. The major commodities handled are food grains, jute, cotton, sugar, milk and arecanuts. Established in 1958 as the apex of the state marketing federations, the National Agricultural Cooperative Marketing Federation of India handles much of the domestic and most of the export marketing for its member organizations.

Large enterprises, such as cooperative Indian sugar factories, spinning mills and solvent-extraction plants mostly handle their own marketing operations independently. Medium and small-sized enterprises, such as rice mills, oil mills, cotton ginning and pressing units and jute baling units, mostly are affiliated with cooperative marketing societies.

The Indian government has adopted various measures to improve agricultural marketing. These steps include establishing regulated markets, constructing warehouses, grading and standardizing produce, standardizing weights and measures and providing information on agricultural prices over All India Radio (Akashvani), the national radio network (www.indianchild.com). These reforms provide more options to farmers for selling their produce, allowing the private sector, including cooperatives, to develop markets, promote direct sales to consumers, processors and retail chain suppliers/ exporters and remove scope for corruption and harassment.

Issues in Indian Farm Markets

However, over a period of time, these markets have, acquired the status of restrictive and monopolistic markets, providing no help in direct and free marketing, organised retailing and smooth raw material supplies to agro-industries. Exporters, processors and retail chain operators cannot procure directly from the farmers as the produce is required to be channelised through regulated markets and licensed traders. There is in the process, an enormous increase in the cost of marketing and farmers end up getting a low price for their produce. Monopolistic practices and modalities of the state-controlled markets have prevented private investment in the sector.

The process involved in the disposal of a substantial produce of great economic importance are significant not only for the farmers but also for the country as a whole. The unreasonably low return that the farmer gets for his produce and the excessive margin of profit retained by the intermediaries is a great concern to all policy makers (www.krishiworld.com). New opportunities (and threats) for trade in international markets have also added a new challenge for Indian farmers (www.india.gov.in). Contributions of small holders in securing food for growing population have increased considerably even though they are most insecure and vulnerable group in the society. The sector is constrained by widespread fragmentation in the supply chain, low productivity levels and huge post harvest losses arising out of inadequate storage, cold chain and transport infrastructure, logistics and supply chain management. Farmers require authentic advice based on meteorological, marketing and management information for land-use decisions and investments. Infrastructure support would be put in place to minimise post-harvest losses and enable agro-processing and value-addition at the village level itself to increase employment and income. Farmers' organisations and other entities like cooperatives and small farmers' estates would be encouraged so that farmers can get a fair deal (agricoop.nic.in).

Market Information Dissemination Channels

In India a number of projects have been initiated to bring market information to the farmers. A few of the initiatives giving market information is given below :

Marketing Research and Information Network (AGMARKNET) : It is a central sector scheme which was launched by the Department of Agriculture and Cooperation in March 2000. The scheme aims at progressively linking important agricultural produce markets spread all over India and the State Agriculture Marketing Boards/ Directorates and the Directorate of Marketing and Inspection (DMI) for effective exchange of market information. The market information network, AGMARKNET, is being implemented jointly by DMI and National Informatics Centre (NIC), using NICNET facilities available throughout the country. The objective of the scheme is to facilitate collection and dissemination of information for better price realization. The portal covers market price, infrastructure and promotion related information for efficient marketing. The markets are reporting daily prices and arrivals data using a comprehensive national-level database at Agmarknet Portal (www. agmarknet.nic.in). Information on wholesale prices and arrivals in respect of 300-plus commodities and 2000 varieties are being disseminated through the portal on a daily basis. More than 1900 markets have been linked to the Central Agmarknet Portal. An arrangement has been worked out with Indian Farmers Fertilizer Cooperative Limited (IFFCO) for regular transmission of prices and arrivals data from Agmarknet to the touch screen multimedia kiosks being installed by them at the rural cooperative societies (india.gov.in).

State Marketing Boards

In India, Agriculture is being considered as State Subject. In this context, each state government took an initiative to establish a new development department (line department) called Directorate of Agricultural Marketing and Agri-Business. Almost all the state governments in India has launched a webpage for the directorate and is providing details on schemes, welfare activities, policy notes and in fact Marketing Information too. Such initiatives have been made by Maharashtra State Marketing Board (http://www.msamb.com/), Andhra Pradesh Agricultural Marketing Board (www.ap.nic.in), Delhi Agricultural Marketing Board (http://www.delagrimarket.org/), Karnataka State Agricultural Marketing Board (http://www.ksamb.gov.in/), http://maratavahini.kar.nic.in/), Gujarat State Agricultural Marketing Board (http://agri.gujarat.gov.in/boards_corporations/gs-agri-mark-board/index.htm), Rajasthan Agricultural Marketing Board (http://www.rajamb.com/) and Tamil Nadu State Agricultural Marketing Board (http://www.tnsamb.com/othractivities.html). These websites provide daily information on Non-Perishable commodities such as Cereals, Pulses, Oil Seeds and Cash crops like Cotton and Sugarcane. Further, they have networked with Regulated Markets to inform the market arrivals, minimum, maximum and model prices.

Market Information Related ICT Initiatives

e-Arik: e-Arik ("Arik" means "Agriculture" in the *Adi* tribal dialect of Arunachal Pradesh State), a research project to experiment the application of ICTs in agricultural extension services provision and also to measure its impact on the tribal farmers has been implemented in "Yagrung" village of East Siang District of Arunachal Pradesh State. Funded by the Department of Scientific and Industrial Research (DSIR), Ministry of Science and Technology, Government of India the project proposes single window system for the improved agricultural information and technology delivery to the tribal farmers by using computer, internet, phone, radio and television. The project provides local market information for vegetables, fruits, meat and fish on daily basis (www.earik.in).

TNAGMARK: Tamil Nadu Agricultural University and the Department of Agricultural Marketing and Agri-business, Government of Tamil Nadu initiated the Domestic and Export Market Intelligence Cell (DEMIC) which provides market data as well as price behaviour at regular intervals in Tamil Nadu. Further, the Market Intelligence Cell displays data on arrival and transaction of important agricultural commodities in the Regulated Markets of Tamil Nadu which is available in AGMARKNET. Using this data, the Market Intelligence Cell forecasts the prices of these commodities in the forthcoming months and the same is transmitted to the Regulated Markets and the farmers through web, All India Radio, television and newspapers. The prices prevailing in nearby states are informed to the farmers so that they get better prices for their produce which will help them in planning cropping pattern and the right time to sell their crop output (*http://www.tnagmark.tn.nic.in*).

Indiaagronet: This commercial website, combines the market forecast details and market news available from various sources on various commodities like cotton, fibre, flowers, fruits, meat, oilseeds, sugarcane, eggs, fish, food grains, jute, milk products, vegetables and the like. Further, this site provides farm technological information, exhibition details and job avenues in the farm sector (*http://www.indiaagronet.com*).

Snxindia: The Safal National Exchange of India Limited (SNX) is a seamless delivery based electronic spot market for horticulture commodities, aimed at building a National Integrated Produce Market in an Exchange format. It is transparent and guarantees payment and delivery with quality for the benefit of sellers in e-Enabled Rural India and the large number of buyers across the country (*http://www.snxindia.com*).

Mahindrakisanmitra: It is a place for the Indian farmers to access wealth of information which is updated on a daily basis. Farmers can check daily mandi (Market) prices, weather updates, latest crop advisories and agri related news. Farmers can also read and benefit from the success stories of other fellow farmers. The site also provides information under various other sections such as crop information, loans, insurance, mandi database, cold storages/ warehouses and agricultural related events. It also gives the futures and spot price details from the NCDEX (National Commodity and Derivative Exchange Ltd) and NMCE (National Multi Commodity Exchange of India Ltd) (*http://www.mahindrakisanmitra.com*).

Agriwatch: It is another commercial website that deals with spot and futures market prices of commodities across India (http://www.agriwatch.com).

aAQUA: aAQUA(almost All QUestions Answered) - The aAQUA, e-Agri Service answers agri-related queries in 24 to 72 hours depending on your question type. Anyone can answer aAQUA questions since the forum is open to all on the Internet.

e-Bay : The e-Bay project is offering space for online marketing of various goods. Such innovations in agricultural marketing will add revenue to the farmers due to direct trading.

World over the experimentation on disseminating market information is taking momentum. Green (Diane Green, 2009) has created a model for setting up farmers markets and continue to update and extend information for new and seasoned gardeners and farmer marketers. The impact of globalisation and trade liberalisation has created the need for information on local markets and prevailing prices (Chisenga, 2007).

The above said Web Portals are basically concentrating information on non perishable farm commodities. These agencies source the data from Govt. owned or sponsored societies/organizations. But in reality, the actual information available in the database does not match with the farmers' actual need and interest.

Efforts of Tamil Nadu Agricultural University (TNAU)

The Directorate of Extension Education, Coimbatore, Tamil Nadu Agricultural University (TNAU) in collaboration with Centre for Development of Advanced Computing (CDAC), Hyderabad, Ministry of Communication and Information Technology, Govt. of India attempted to provide real time dynamic market information which will be useful and it helps the farmers in South India to make a meaningful decision on the harvest of produces. The DMI was started on 02, June 2008 with the following objective

- To collect dynamic market data for Agri-Horti produces from major markets in South India on experimental basis and delivery to the farming community by using Information and Communication (ICT) tools and devices.

Selection of Markets in South India

Thirteen markets chosen for this initiative are from three South Indian states namely Cochin in Kerala; Coimbatore, Oddanchatram, Trichy, Madurai, Panrutti, Thalaivasal, Mettupalayam, Tirunelveli, Hosur, Kumbakonam and Chennai in Tamil Nadu and Bangalore in Karnataka. The reasons for selecting these markets are as follows

- These are the major markets for perishable commodities in South India and handle more than hundred tonnes of produce on daily basis.
- It has space for both wholesalers and retailers.
- These markets influence the flow of perishable commodities in the south Indian states.

- Further, the Market Committee functioning in each market are ready to provide the information for the benefit of farming community.

Placement of Market Analyst

Six market analysts (Graduates in Agricultural/ Allied Sciences) are placed at the different markets to collect the market price from both wholesalers and retailers. The entire market analysts are instructed to stay in market town which makes them easy to do the service. In the initial period, market analyst had spent one week time in the market to obtain a high end rapport with the traders of each market and studied the market, its nature and functioning. In order to collect the exact price, the market analysts make it a point to be at the markets during the time of trade, be it early in the morning or later in the day. A data scouting sheet has been formulated and handed over to each market analysts for daily collection of information in-person. The retail prices of the respective commodity are being obtained from the nearby retail vendors like Reliance fresh, More Markets, Pazhamuthir Nilayam *etc.*

Daily Scouting on Wholesale and Retail Price Information

The Wholesalers buy the produce from the farmers, while the retailers buy the produce from the wholesalers and sell it to the consumers. As on date, 161 commodities have been added to the DMI list. The data base is created in PHP (Personal Home Page) format.

Sl. No	Name of Commodity	No. of Produces
1	Vegetable	70
2	Fruits	36
3	Flowers	33
4	Spices	14
5	Plantation Crops	08
	Total	**161**

Content Uploading and Publication

The market analysts compile the data collected from the market in a previously formatted excel sheet. This is done to maintain uniformity among the markets during display. The data is collected and sent by 11 am every day to TNAU which acts as the hub centre. The data received is uploaded in the website server www.indg.in in TNAU. The server is maintained by CDAC, Hyderabad. In order to ensure that only the correct information is uploaded, the data is cross verified with the previous days' data. In case of any discrepancies the data is verified and triangulated by calling over to the market.

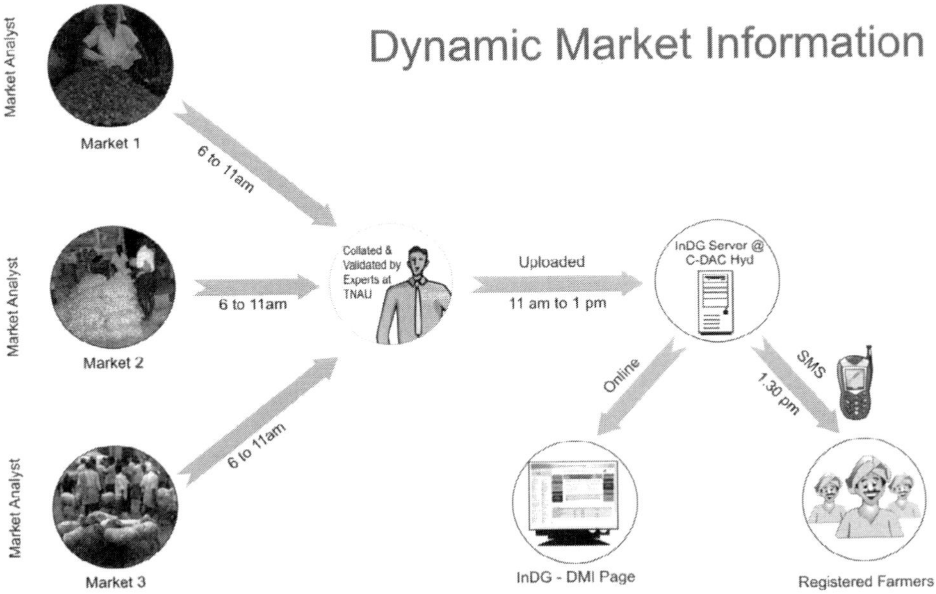

Figure 1 : Schematic diagram on Dynamic Market Information

Information Details

The rates are ready to be viewed by 2.00 pm daily. The market data is presented both in English and Tamil.

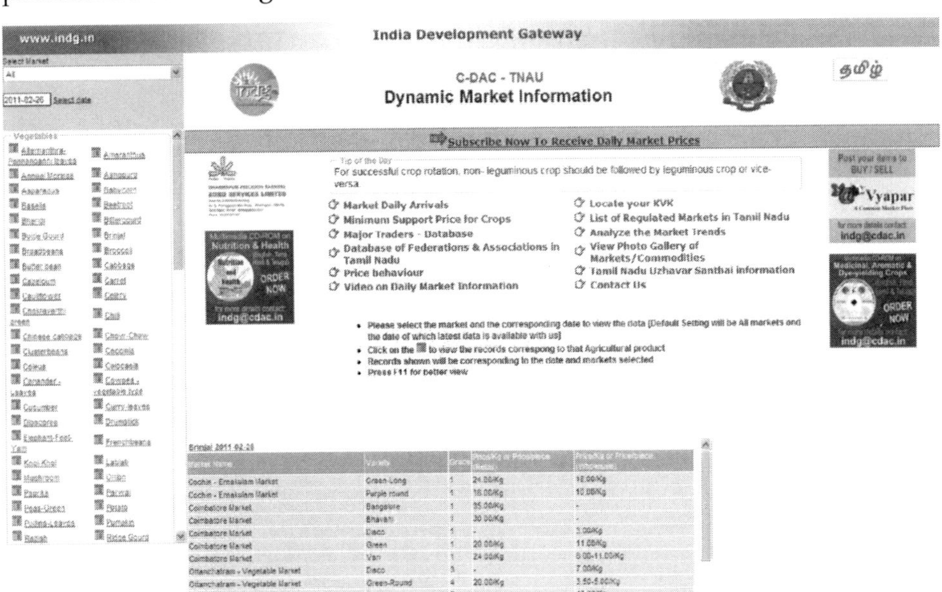

Figure 2 : Market Information on Display Sheet in English language

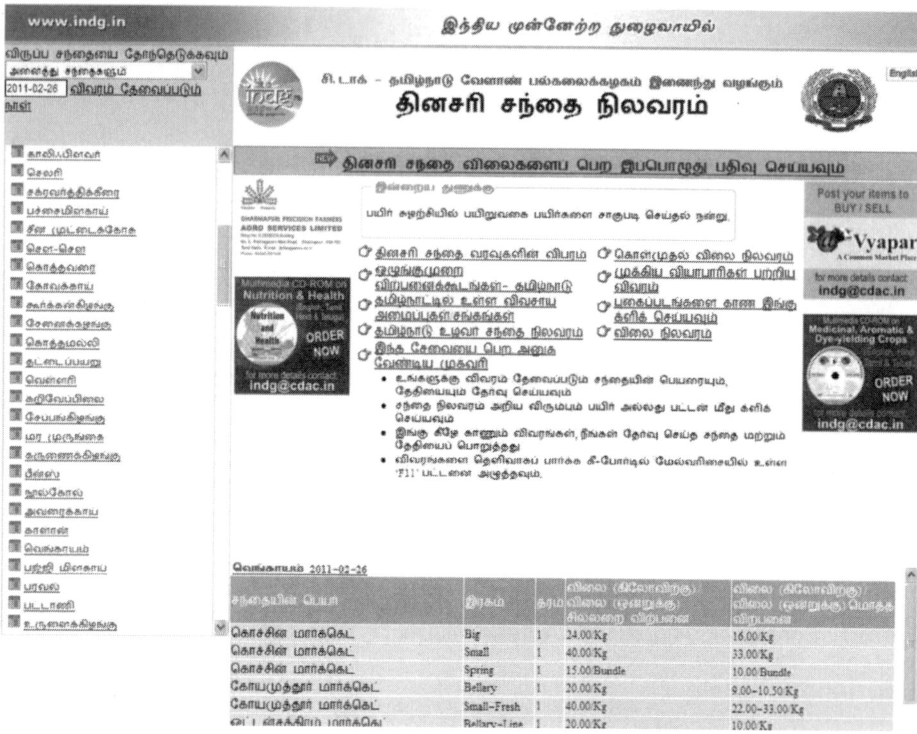

Figure 2a : Market Information in Display Sheet in Tamil language

Additional Features in the Website

The additional features in the website include:

- *Back-up database facility*: The present information delivery system is not advocating the price forecasting information. Instead of that it provides back up data facility from the date of inception.
- Market arrivals on daily basis.
- Market profiles of the six markets with photographs explains auction time, preferred package details, toll gate charges *etc.*
- Trader's database from each market includes name, contact number and the commodity he/she is dealing.
- Farmers association details.
- Best practices adopted by other successful farmers.
- Minimum Support Price of major commodities.
- Regulated Market details in Tamil Nadu.
- Krishi Vigyan Kendras (Farm Science Centers) in Tamil Nadu.

For wide coverage, Tamil Nadu Agricultural University had given a link of DMI through www.tnau.ac.in. It is proposed to include a cover story of each market and also success stories of traders in the market.

Exploring Mobile based Price Information Delivery

India is fast booming in mobile technologies. Its subscriber bases are increasing rapidly. The total number of telephone connections reached 784.16 million at the end of December, 2010. Total Broadband, subscriber base has reached 5.28 million by the end of November 2008. The total wireless connections in Tamil Nadu was 23,607,338, while the wireline subscriber base was 2,315,785 during November 2008 (TRAI, 2008).

Although tele-density in the country has reached over 32.34 per cent, the spread isn't even. The Rural tele-density is still hovering around 8 per cent. The government's target of 25 per cent rural density will require collective efforts from both the government and the industry.

Rural Wireline Subscriber base stood at 11.33 million in quarter ending June 2008 as compared to 11.64 million in quarter ending March, 2008. The major market being held by BSNL Number of Village Public Telephones (VPT) have increased from 5.60 lakhs in quarter ending March 2008 to 5.63 lakhs in quarter ending June 2008.

The use of internet is also quite impressive. There are 11.66 million Internet subscribers at the end of June 2008 in India. Besides above, there are 75.97 million wireless data subscribers at the end of June 2008 (capable of accessing data services including internet through mobile handsets) (Parameswaran, 2008).

The farmers may not have adequate knowledge and exposure to access the internet for market information. By understanding the existing nitty-gritty, the on-line market information data base has been linked with each farmers association for easy reach. At the same time, mobile based delivery of information was also explored as part of the initiative.

Short Message Service (SMS) : The price details of selected commodities are sent to the farmers on trial basis through SMS. Right now 200 farmers and fifty KVK scientists are linked with the database to generate automatic mobile based information at 2.30 pm on all working days which gives information on one commodity or produce with two market wholesale rates. To enhance the mobile delivery, TNAU has organized buyers-seller meet inviting 200 progressive farmers across Tamil Nadu and important traders from these six markets to make an interface mechanism and linkage.

The future plans are:

1. On-line trading.
2. Linkage with the remaining farmers federations to promote use of available information.
3. Exploring the possibility to link with GIS and GPS to provide spatial assessment for market forecast.
4. Popularization with partnership like inter and intra institutional and service organization for reaching the unreached.
5. The holistic marketing related information namely trends, quality preference, daily market prices, demand and supply analysis for the different produces, composition of suppliers, packaging and transport mode for different produces will be analyzed and informed to the producers, buyers and consumers through online portals (www.tnau.ac.in & www.indg.in).
6. Periodical publishing of market support information with regional specific inferences in different produces for the benefit of all the stakeholders.

Conclusion

The project is all set to reach the following potential benefits:

Faster Reach

The market data is not available to the farmers by any means. The use of internet is also meager. However, the proliferation of mobile has been in leaps and bounds. Information by mobile can reach the farmers at hand. In this venture, the farmers are able to know the market rate of major South Indian markets. Market data reaches the farmers in time over mobile phone for decision making and maximizing profits.

Better Price Realization

Often farmers are exploited by the middle men and unable to get actual benefit for their end produces. This joint initiative will facilitate the farmers and field extension officials to understand the daily prices of important markets in south India. Ultimately this would help the growers to mobilize the commodities for getting the actual market prices. By calculating the costs involved and the likely profits, the farmers can take their produce to the market providing the best prices.

Organizing Farmer Groups

The unorganized and resource poor farmers are most likely to sell their produce in the local markets. This will cause loss to the farmers and more profits to the traders. If the poor and marginal farmers are unable to mobilize their produce individually, they can organize into groups and take their commodities to the market that gives them better economic returns. This leads to collective bargaining by the farmers.

Huge Market Data

The current database will also serve as a document for further analysis and forecasts. This will be useful for developing strategic plans in production cycles to enhance profits.

This initiative adds value to the information and facilitates decision making at various levels. Farmers will be able to reap better prices for what they produce by making meager modifications in their supply chain system.

References

Chisenga, Justin, 2007. Impact of Globalization on the Information Needs of Farmers in Ghana: A Case Study of Small Scale Poultry Farmers. *Proceedings of World Library and Information Congress.* http://www.ifla.org/iv/ifla73/index.htm

Datt, Ruddar and Sundharam, K.P.M., 2009. *Indian Economy.* 28. pp. 485–491

Diane Green, 2009. Developing a Sustainable Market for Small Farms in a Rural Community. http:://wsare.usu.edu

Parameswaran, N., 2008. TRAI's Principal Advisor, Increasing Rural Teledensity 25 Mar, 2008, http://www.vnl.in/blog/2008/increasing-rural-teledensity/

The Economic Times, 2008, 72% of India's Fruit, Vegetable Produce Goes Waste. http://economictimes.indiatimes.com/Agriculture/72_of_Indias_fruit_vegetable_produce_goes_waste/articl eshow/3033498.cms (Accessed on May 12th, 2008.)

TRAI, 2008. Telecom Regulatory Authority of India- Press Release No. 110 of 2008 TRAI. http://www.trai.gov.in/WriteReadData/trai/upload/PressReleases/631/pr29dec08no110.pdf

Vision, 2020, *Policy Note,* Government of India.

http://www.indianchild.com/agricultural_marketing_in_india.htm

http://www.krishiworld.com/html/agri_storage1.html

http://india.gov.in/outerwin.php?id=http://agricoop.nic.in/annualReport06- 07/Content.htm

http://agricoop.nic.in/NPF/npff2007.pdf

http://www.earik.in/about.html

10

ICT based Kisan Mobile Sandesh – An Innovative Approach

Abhay Wankhade, Nalin Khare, Surendra Pannase and Druv Shrivastava

ABSTRACT

The Cellular technology (mobile phone) is becoming popular day by day with companies up with low price handsets and nearly one-fourth of the India's population possess a mobile phone even in the rural areas. The Indian government has set a target of 500 million users by 2010; half of this, 250 million, has reached till 2007 (IANS). Keeping in view the scenario, messaging agricultural information through computers to the end users on mobile phone is a means, which will go a long way for the farmers to receive timely information. There are different media platforms through which farmer's access information in India. This includes radios, television, print, person to person and so many others. However, with the rapid growth in technology, there is a strong belief that there is a need to extend the mobile phones' use to agriculture sector. Currently, mobile phones are widely used in day to day activities of social, economic and political spheres of life evidenced worldwide. Many ICTs are used but JNKVV and ZPD unit –VII, has realized that the mobile phones are one of the most accessible technologies. The KVK Chhindwara has pioneered for the first time in IT enabled service *i.e.* "Kisan

Mobile Sandesh" aiding instant bulk messaging from KVK to individual farmers, extension personnel and agro-input provider for extending Agricultural information through bulk SMS. Bulk SMS is a special type of system by which an institution or an organization becomes able to send or receive a large volume of SMS. The most obvious and useful feature of Bulk SMS could be found from its name. A single SMS could be sent to number of users using the concept of Bulk SMS. The Bulk SMS sender could set the number of recipients. A unique sender Identification tag or ID Brand is generated for each and every user. Every time the sender has to use that ID brand while sending Bulk SMS. The present study was conducted in Chhindwara and Seoni district of Madhya Pradesh where Krishi Vigyan Kendra was engaged in operating Kisan Mobile Sandesh. Five development blocks from each district were selected for study. In total 400 KMS users having their own mobile phone were identified for the study, out of which 250 farmers, 100 extension personnel and 50 agro-input providers were selected from 217 villages. Selection of villages was based on availability of accessing mobile network. Two SMS in a week *i.e.* Tuesday and Friday are issued on various agricultural developments like weather forecast, disease forecast, availability of input, critical stages, Plant protection measure, Livestock production and management and market information. Kisan Mobile Sandesh is also being used as a medium to send information on important trainings and other programmes to the contact farmers of KVK. All selected KMS users who are having their own cellular mobile handsets supporting all major GSM/CDMA networks within the study area can receive the SMSs from KVK Chhindwara and Seoni separetly without incurring any cost. The KVK bears all costs of sending the SMS. KMS users say that you feel the man of the day once you receive many SMS's at a go, even you are at a field. People come to you to ask what the SMS is about and you feel proud explaining the content to them. The KMS service has been well appreciated by the farmers, extension personnel and agro-input provider and it also has found extremely cost-effective, reachability and storage facility for the rural farmers. This article therefore, attempts to discuss the potential role of mobile phones for dissemination of information to the farmers during the crop period as well as awareness campaign of the new technology for consequent adoption. It also discuss the relevance of the method in the present day context since it is a cheap, yet personal tool to keep in touch with a large number of clientele with need based, location specific and timely technological information dissemination. The paper further discusses the relevance of mobile phones as means for receiving information related to pest and disease forecast; their management; market intelligence and weather forecast.

Introduction

The Cellular technology (mobile phone) is becoming popular day by day with companies up with low price handsets and nearly one-fourth of the countries population possess a mobile phone even in the rural areas. The

Indian government has set a target of 500 million users by 2010; half of this, 250 million, has reached till 2007. (IANS). Keeping in view the scenario, messaging agricultural information through computers to the end users on mobile phone is a means, which will go a long way for the farmers to receive timely information.

There are different media platforms through which farmer's access information in India. This includes radio, television, print, person to person and so many others. However, with the rapid growth in technology, extension professionals should extend the mobile phones' use to agriculture sector. Currently, mobile phones are widely used in day to day activities of social, economic and political spheres of life evidenced worldwide.

Many ICTs are used but JNKVV and Zonal Project Directorate (ZPD) - VII has realized that one of the most accessible technologies is the mobile phone. The KVK Chhindwara has pioneered for the first time IT enabled service *i.e.* "Kisan Mobile Sandesh" aiding instant bulk messaging from KVK to individual farmers, extension personnel and agro-input providers for extending Agricultural information through bulk SMS. Bulk SMS' is a special type of system by which an institution or an organization becomes able to send or receive a large volume of SMS *i.e.* SMS in a bulk. The most obvious and useful feature of Bulk SMS could be found from its name. A single SMS could be sent to number of users using the concept of Bulk SMS. The Bulk SMS sender could set the number of recipients. A unique sender Identification tag or ID Brand is generated for each and every user. Every time the sender has to use that ID brand while sending Bulk SMS.

The present study was conducted in Chhindwara and Seoni district of Madhya Pradesh where Krishi Vigyan Kendra were engaged in operating Kisan Mobile Sandesh. Five development blocks from each district were selected for study. In total 400 KMS users having their own mobile phone were identified for the study, out of which 250 farmers, 100 extension personnel and 50 agro-input providers were selected from 217 villages (Table - 1). Selection of villages was based on availability of accessing mobile network. Two SMS in a week *i.e.* Tuesday and Friday are issued on various agricultural developments like weather forecast, disease forecast, availability of input, critical stages, plant protection measure, Livestock production and management, market information. Kisan Mobile Sandesh is also being used as a medium to send information on important trainings and other programmes to the contact farmers of KVK. All selected KMS users who are having their own cellular mobile handsets supporting all major GSM/CDMA networks within the study area can receive the SMSs from KVK Chhindwara and Seoni separately without incurring any cost. The KVK bears all costs of sending the SMS. KMS users say that you feel the man of the day once you receive many SMS's at a go, even you are at a field. People come to you to ask what the SMS is about and you feel proud explaining the content to them. The KMS service has been well appreciated by the farmers, extension personnel and agro-input providers and it also has found extremely cost-effective, reachability and strong storage facility for the rural farmers.

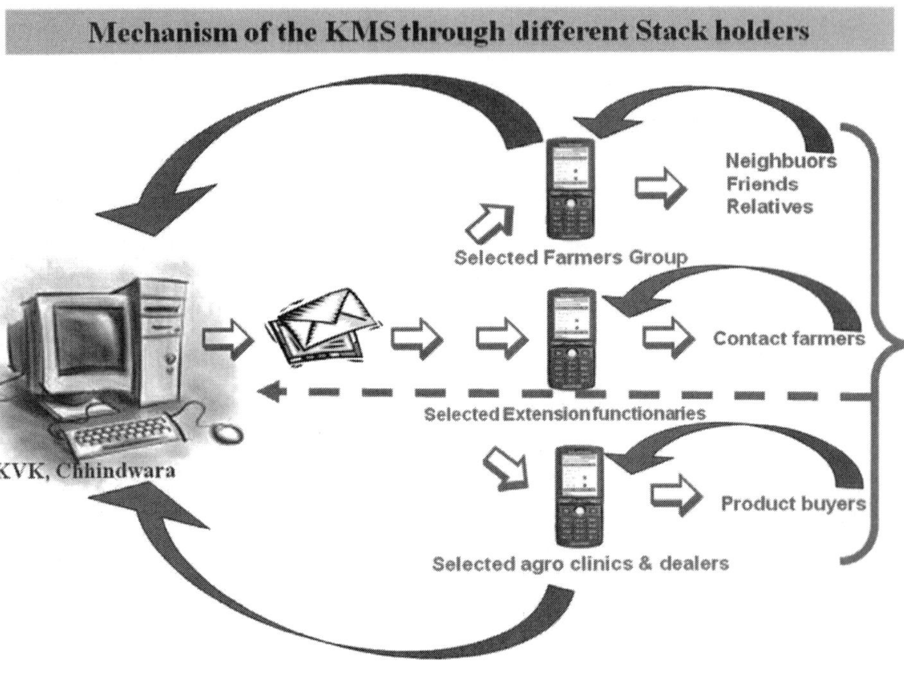

Figure 1 : Mechanism of the KMS

Table 1 : KMS User's Category

Users category	No. of Users	Percentage
1. Farmers a. Vegetable growers group b. Fruits growers group c. Practicing farmers group d. Dairy farmers group e. Cotton growers group	250	62.50
2. Extension Personnel (Selected from the Agril. Department, Horticulture Department and NGOs)	100	25.00
3. Agro- input poviders	50	12.50
Total	**400**	**100**

Table 2 : Area Covered under KMS through Farmers Group

Name of District/ Block	No. of Villages	No. of Farmers
District - Chhindwara		
1. Block - Chhindwara	10	25
2. Block - Mohkhed	15	25
3. Block - Pndhurna	19	25
4. Block - Sausar	15	25
5. Block - Parasia	19	25
District - Seoni		
1. Block - Seoni	16	25
2. Block - Kurai	21	25
3. Block - Barghat	14	25
4. Block - Chapara	16	25
5. Block - Lakhanadon	15	25
Total	**160**	**250**

Table 2 shows that the coverage area under KMS through farmers group. It is clear from the table that selection of farmers representative from different enterprise based group are equal *i.e.* 25 selected from five developmental block of Chhindwara and five developmental blocks of Seoni district. 250 farmers were selected from 160 villages.

Table 3 : Area Covered under KMS through Extension Personnel

Name of District / Block	No. of Villages	No. of Extension personnel
District - Chhindwara		
1. Block - Chhindwara	58	10
2. Block - Mohkhed	62	10
3. Block - Pndhurna	73	10
4. Block - Sausar	67	10
5. Block - Parasia	56	10
District - Seoni		
1. Block - Seoni	76	10
2. Block - Kurai	83	10
3. Block - Barghat	54	10
4. Block - Chapara	81	10
5. Block - Lakhanadon	77	10
Total	**687**	**100**

Working area of each extension personnel is 16 to 27 villages.

The table 3 depicts that the coverage area under KMS through extension personnel. The extension personnel covered more area (villages) because their working area was very large and regularly contacted a large number of farmers. KMS message was also helpful for their effective job performance.

Table 4 : Area covered under KMS through Agro-Input Providers

Name of District/Block	No. of Villages	Agro-Input Providers
District – Chhindwara		
1. Block - Chhindwara	70	5
2. Block - Mohkhed	30	5
3. Block - Pndhurna	50	5
4. Block - Sausar	60	5
5. Block - Parasia	60	5
District – Seoni		
1. Block – Seoni	60	5
2. Block – Kurai	40	5
3. Block – Barghat	45	5
4. Block – Chapara	50	5
5. Block – Lakhanadon	45	5
Total	**510**	**50**

Approach area of one agro-input provider is 5 to 25 villages.

The table 4 depicts that the coverage area under KMS through agro-input providers was also large *i.e.* 510 villages of study area. In present condition, Agro-input providers play very effective and major role for transfer of agriculture information because they are closely related to the farmers group and provide need based agro inputs.

Table 5 : Total Area Covered under KMS

Name of District/Block	No. of KMS Users	Villages Covered
1. Through Farmers groups	250	160
2. Through Extension Personnel	100	687
3. Through NGO's and Agro Input Providers	50	510
Total	**400**	**1357**

No. of villages in Chhindwara ditrict - 1958
No. of villages in Seoni ditrict - 1738
Total - 3696

Table 6 : Impact of Kisan Mobile Sandesh

Impact Statements	Source of information - KMS Farmers (92)	Extension Personnel (50)	Agro-input providers (25)	Mean Score	Other Source of information (167)
Received need based information	92	47	24	163	62
Spend few seconds for receing information	92	50	25	167	22
Easy to understand	71	50	25	146	84
Appropriate time of information	92	50	25	167	77
Develop information Bank	64	50	24	138	30
Increase social contacts & importance	68	42	22	132	23
Saves time and money	64	44	16	124	41
KMS also work as a reminder	78	40	18	136	28
Possible for giving feedback	39	27	15	81	22
Strong linkage with KVK	92	50	24	166	08

To know the impact of Kisan Mobile Sandesh the content of evaluation was studied in ten dimension *viz.*, received need based information, spend few seconds for receiving information, easy to understand, appropriate time of information, develop information bank, increase social contacts and importance, save time and money, KMS also work as a reminder, possible for giving feedback and strong linkage with KVK. The table-6 depict that the spend few seconds for receive information and appropriate time of information was ranked first (167), strong linkage with KVK ranked second (166), received need based information ranked third (163), easy to understand ranked fourth (146) followed by develop information bank (138), KMS also work as a reminder (136), increase social contacts & importance ranked (132), save time and money (124) and possible for giving feedback (81). It is clear that the table that Kisan Mobile Sandesh are playing effective and positive role for technology transfer in a very low cost with area specific and farmers need based issues.

The KMS has the potential role of mobile phones for dissemination of location and farmer specific information to the farmers during the crop period as well as awareness campaign of the new technology for consequent adoption. It is also cheap, yet personal tool to keep in touch with a large number of clientele with need based, location specific and timely technological information dissemination.

11

ICT Initiatives for Agricultural Development in Kerala State

S. Nazreen Hassan and N. Kishore Kumar

ABSTRACT

Information and Communication Technology (ICT) is the driving force for the fundamental economic and social revolution. It accelerates globalization, makes access to knowledge and information much easier for the people in respect of creation of technologies, diffusion of old innovation and diffusion of human skills for rural prosperity. The National Conference on Informatics for Sustainable Agriculture Development (ISDA) organized by the Ministry of Agriculture and National Informatics Centre (NIC), in May 1995, has given much needed "roadmap" to usher in "ICT led agriculture development" in the country. The National Agricultural Policy (2000) lays emphasis on the use of ICT for achieving a more rapid development of Agriculture. ICTs diffusion have many potential applications spanning the breadth of the agricultural industry, at all scales of organization; farmers, cooperatives, professional bodies, farm machinery vendors, fertilizer and chemical companies, insurance, regulators and commodities, agronomists, consultants, and farm advisors. The KISSAN KERALA, co-ordinated by IIITM-K, State Department of Agriculture and Kerala Agricultural University. The Akshaya project is the idea to bring ICT opportunities to the common man in Malappuram came from the local self-governments (local Panchayats). The State Government in Kerala initiated an e-governance transaction enabling mechanism called FRIENDS (Fast, Reliable, Instant, Efficient Network for Disbursement of Services) which offered integrated computerized payment counters for citizens at each of the district headquarters, for over 1000 combinations of services.

Introduction

The ICT revolution is having an impact on economic and social conditions around the world, including the developing countries. The experience of information users around the world provides ample evidence that ICT could play an important role in agriculture and poverty alleviation. Modern ICT has much to offer in meeting the information needs of rural communities, other than agriculture like; information related to health, micro credit and government services, create direct employment opportunities, provide training and education to people, and support the poor in the production, storage and marketing of farm and non-farm products. Hence, access to information is a key to building human capabilities. The most critical challenge is how to effectively and efficiently make use of the facilities available to the farmers in rural and remote areas and develop appropriate content so that disadvantaged sections of the population will be able to benefit from it.

Need for ICT in Extension in India

Knowledge and information are essential for empowering rural communities. Communication is central to this process. Without effective communication there can be no development. The information and communication support during last few decades has been conventional. The agricultural system in Kerala is more complex and the need for information is very high. The extension personnel of the Department of Agriculture disseminated the technological messages to the farmers manually. This approach has not been able to reach majority of the farmers who are spread across the country. This gap remains a challenge for the Extension system even today. To reach over 110 million farmers, spread over 500 districts and over 6000 blocks is an up hill task. The diversity of agro-ecological situations adds to this challenge further. The success of Green Revolution was mainly achieved due to concerted homogeneous extension approach. Now the extension strategy becomes more complex because the farmers' needs are much more diversified and the knowledge required to address them is beyond the capacity of the grass root level extension functionaries. Today it is possible to find a solution to this situation by using the potentials of Information and Communication Technologies (ICTs) to meet the location specific information needs of the farmers. The information and communication networks are expanding very fast. The Internet connectivity has touched almost all the districts in the country and is moving down to the block and Mandal levels.

The ICT Initiatives in Kerala

KISSAN Kerala

KISSAN stands for, "Karshaka Information Systems, Services and Networking". It is co-ordinated jointly by the Indian Institute of Information Technology and Management- Kerala (IIITM-K), State Department of Agriculture and Kerala Agricultural University. It targets the farmers, agriculture and related stakeholders. It is an integrated knowledge empowerment approach. It is based on need and demand driven extension and educational services.

KISSAN Runs on Four Pillars

1. Interactive and Informative Kerala Agriculture Portal
2. KrishideepamTV Serial
3. Secondary Agri. Call Centre
4. Ten Agri-Kiosks in Thiruvananthapuram district and Virtual University for Agriculture Trade jointly managed by IIITM-K, Department of Agriculture and linked with the Kerala Agricultural University.

Kissan's Knowledge Infrastructure for Agriculture & Related Areas

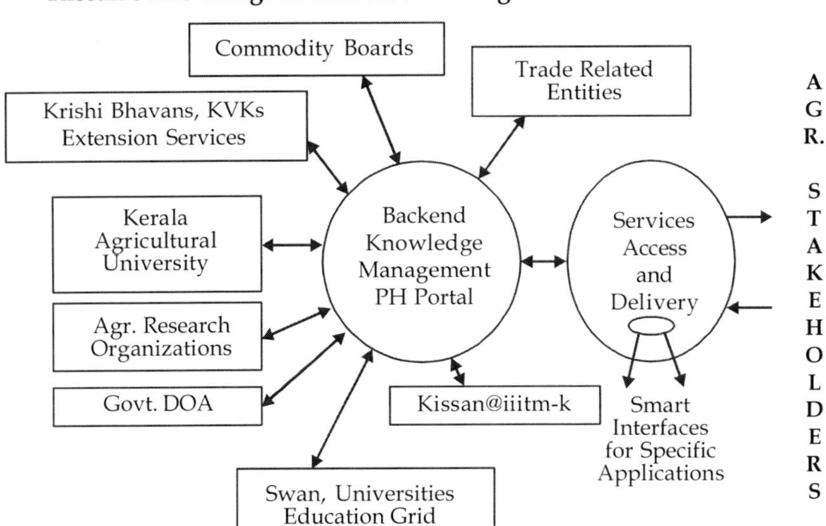

Figure 1 : KISSAN Kerala's Knowledge Network Model

The project's goal is to develop and deploy Agriculture Information Systems and Services to:

i. Address immediate and long term concerns of the farmers
ii. Increase farm productivity
iii. Increase returns for farm produce

The project applies five "fundamental rights" of Knowledge Management in the Agricultural Domain: "Right Information to the Right Persons at the Right Time in the Right Places and the Right Context"

The project was launched on 1st November 2003 in the form of information kiosks at panchayat level to enable the farmers to communicate with agricultural scientists and officials. Such kiosks were set up at ten Krishibhavans of Trivandrum district namely Attipra, Kazhakootam, Managalapuram, Sreekaryam, Pothencode, Nedumangad, Karakulam, Kudappanakunnu, Vattiyoorkavu and Ulloor. These kiosks were equipped with the latest multimedia computers, web cameras, network accessories and dedicated internet connections. This was provided for the farmers and extension workers to interact with experts and researchers across the State and beyond. For this, there is an interactive portal which facilitates knowledge collaboration between the Department of Agriculture, Kerala Agricultural University, Farm Information Bureau and other allied departments. The portal also offers market information, virtual market, weather information, crop information, farming practices availability of planting materials, fertilizers and pesticides, seeds, organic manures and the like. KISSAN's objective is to build effective knowledge management and smart information dissemination systems that link farmers, public research institutions, administrators and private entrepreneurs. The core deliverable of the project is an integrated multi-component, multimodal and geographically distributed agriculture information system that is accessible anywhere and anytime by all concerned.

This project for agriculture is a pilot implementation of application of IT through relevant Information Systems and networking to aggregate, share and disseminate information of importance and interest to the farmers, agriculture workers and officials in ways that enhance the total agriculture development and farmers' welfare in the state. The project aims to establish a farmer centric integrated distributed Information System for agriculture, which would focus on catering to the various requirements and services to the farmers across the State. According to KISSAN news report (Anonymous, 2004), the innovative ICT project in Kerala, received a total number of 1341 hits from kiosks, and the total number of queries answered were' 1531. The total number of emails received back in October 2004 is 26, and the mostly accessed page is on crop information and most accepted page is the market information.

It is a dynamic system focusing on the growing demands of the farmers. The system has introduced an online fertilizer recommendation system (FRS). This FRS provides recommendation on fertilizers, based on the soil test reports provided to the farmers by various soil testing laboratories across the state. The farmer/user has to provide the basic details, details of the soil analysis report and also the preference of crop. Using these data the system will automatically generate a detailed recommendation for the integrated management of farms by providing most accurate information on how to provide balanced nutrition. The report contain such details as the soil amendments to be made, organic manures to be used and also the quantity as well as the frequency of use of chemical fertilizers. The query section provides expert advice to the farmers on the questions posted by them. The system provides information regarding the availability of inputs to the farmers.

It has got vast scope to make it self-financing, generate quality employment in the districts and rural areas. Kissan Krishideepam, an agricultural based Television program under the Kissan - Kerala Project, telecasts the farm programme twice a week. This is a more widely watched programme by all sections of viewers due to its uniqueness in its presentation and technical credibility. The format for presentation includes success stories, technical information, forecast on pests and diseases and various development schemes. Every Friday, there is a phone-in programme in which the farmers can get direct expert advice live from experts in the particular area.

Due to the excellence in services the project has bagged the South Asian Manthan Award -2008 for "Best e-content for Development" and "e-India 2009" Award for the best *"ICT enabled agricultural initiative of the year 2009"* for its successful integration of various technologies like Television, Internet, Telephone and mobile/SMS for delivering useful agricultural services and information to the farming community across Kerala.

Akshaya Project

The Government of Kerala gives a high priority to ICT development; there even exists a separate department for IT, active under the name "Kerala State Information Technology Mission". The IT department's main achievement till then was the 'Friends' project; this project brings government services closer to the people. FRIENDS (Fast, Reliable, Instant, Efficient Network for Disbursement of Services) started in 2000, is an e-governance transaction enabling mechanism which offered integrated computerized payment counters for citizens at each of the district headquarters, for over 1000 combinations of services. The Kerala government has a stated mission of equipping all its local governing bodies (over 1200 in all) with computing facilities, by providing the requisite training to start using these, as well as locally-relevant content and services for their sustenance. The Akshaya project has been a continuation of this.The Akshaya project in the Malappuram district, the government of Kerala has fashioned the nation's first e-literate district, imparting at least one member of every household with basic computing familiarity. This was not done through a district-wide campaign of barefoot computer professionals - instead, the state undertook a massive project to install over 630 telecenters (or subsidized village cybercafés), networked with a broadband wireless IP based backbone throughout the entire area of 38,000 sq. km. The project is a state-private partnership, in which the state agency, Kerala State Information Technology Mission (KSITM), planned the project, set up the connectivity infrastructure, marketed the idea, developed an e-literacy program and created an initial suite of applications.

The Basic Services Provided by the AKSHAYA Centres are Described below :

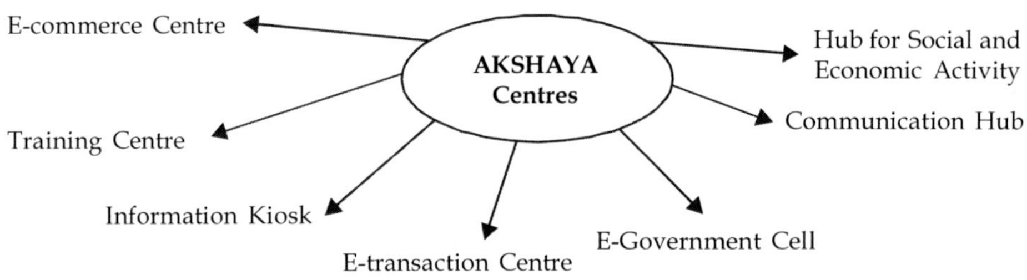

Figure 2 : Akshaya Services

The main goals of the project are:

a) IT-literacy to at least one member of each family of the 65 lakh families in the state.

b) Extend the training initiative into a service delivery mechanism for the local citizen.

Through this project the state envisions

- To bridge the gap between the 'Information Poor and the Information Rich'
- For Creation of a society which is ready for capitalizing on Knowledge for economic and social development
- Akshaya Centres to work as social and economic catalysts for the overall development of the society

- To Empower individuals and communities through enhanced access to information, education and communication facilities
- To integrate communities through creation of e-networks and development of the core sectors like Agriculture, Health, Education, Industry and Resources.

The Five Core Services provided by the project are:

1. *Training centre* : One of the important services at Akshaya Centre is imparting programmes to citizens. Lot of entry and medium level course programmes are offered through this initiative. All the courses are approved by the Government and a certification for merit will also be provided through an online certification mechanism.
2. *Information Kiosk* : Akshaya is envisaged as a one stop information centre. All kinds of information will be made available through the centres. Content has been already generated in five core areas including, Health, Agriculture, Career, Education and Laws and regulations. Government also digitized all applications forms, Government schemes and delivered through Akshaya network. Encouragement for digitizing and updating locally relevant content is also given to Akshaya Centres.
3. *e-Transaction centre* : Extension of FRIENDS Services through Akshaya centres is implemented through this initiative. Presently KSEB and BSNL bills are collected through this package. Entrepreneurs can collect Rs.5 per bill from the customer for offering this service. This facility is now operational in 160 + Akshaya Centres in Malappuram, and will soon be rolled out to other centres in the state.
4. *e-Governance cell* : Government already had taken decisions to convert Akshaya Centres as the last mile units for e-Governance delivery. Akshaya Centre can be used for delivering various e-Governance services to the public, which include, Pubic Grievances Redressal System, Decision Support systems, Online processing of applications, information dissemination services, digital extension of various campaign/ awareness programmes, telemedicine, agriculture intervention *etc.* Many initiatives like e-Parathi (District Collector's Public Grievance Redressal Mechanism, e-Krishi, *etc.,* has been started).
5. *Communication hub* : Akshaya Centres will also be developed as a Communication hub, which will have all ICT based communication facilities to the common man. In Malappuram, all the Akshaya Centres are connected to Wireless internet which enable the centres for various connectivity based services. Intranet based services are also initiated through the network. This is initiative is very important in Kerala as Kerala has lot of population living out side the country.

e-Krishi : e-Krishi is the Market Driven Agricultural Initiative through IT enabled Agri Business Centres in Kerala State implemented by Kerala State IT Mission (KSITM) & Indian Institute of Information Technology & Management - Kerala (IIITM-K) in collaboration with Department of Agriculture. e-Krishi (www.e-krishi.org) was funded by UNDP, monitored by NISG, executed & serviced jointly by the IIITM-K, KISSAN & VUAT projects of the Department of Agriculture, Kerala Agricultural University and Kerala State IT Mission (KSITM).

The e-Krishi is a novel initiative of the State IT Mission to address the gap in agricultural information flow and transaction management. e-Krishi was first launched in Malappuram, as a pilot model, the project could set up a fairly large network of farmers by giving them instant access to a lot of information useful to them, including market demand, price, best practices, and expert advice on quality improvement. According to e-Krishi managers, the success of the project depended a great deal on the solid platform it got from Akshaya in Malappuram. As many as 114 Akshaya centres have been upgraded to e-Krishi centres. They provide all help to the farmer, no matter he is educated or illiterate. Many farmers in Malappuram district has been selling their produce such as paddy, coconut, arecanut, vegetables, banana, rubber, vanilla and medicinal plants through Internet, using the e-krishi portal www.e-krishi.org.

Need for development of Agriculture has been clearly spelt out by the Report of the Commission on WTO concerns in Agriculture (headed by Prof. M.S. Swaminathan). In its report titled "Building a Sustainable Agricultural Trade Security System for Kerala" the commission observes that the substrate conditions essential for building a sustainable Agricultural Trade System relate to the following parameters:

a. Proactive State Policy, which while building a response to the WTO, must defend and extend the economic and developmental gains achieved through state intervention and public action in the past. The State cannot afford to withdraw in the new, WTO-regulated phase of Kerala's development. It must continue to play a role in public investment (for example, in agricultural extension, and infrastructure development) as well as in protecting the lives and livelihoods of those, particularly the poor peasantry and agricultural labour, threatened by the new trade regime.

b. Productivity enhancement by bridging the prevailing gap between potential and actual yields with technologies on the shelf.

c. Quality Improvement through a quality literacy movement for producers and consumers, and by strengthening the infrastructure for sanitary and phytosanitary measures for both domestic and export markets.

d. Profitability Enhancement through concurrent attention to production efficiency and higher productivity, as well as to improve post-harvest technology, value addition and agro-processing. Measures for value-addition will also include organic farming and the production and marketing of organic spices, tea, coffee and fruits.

e. Sustainability Improvement, through attention to the ecological foundations essential for improving productivity in perpetuity, such as soil health care, water harvesting and efficient use and forest and agro-biodiversity conservation.

The key output from the initiative shall be the facilitation and integration of economic activities of all member stakeholders involved in Agriculture thus enabling conversion of under- performing and non- performing agricultural farms into high yielding farms of quality products in demand in the market. Key components of the initiative include Establishment of a robust IT enabled platform where the members can seek information, transact and make or receive electronic payments.

It benefits the member farmers by providing information on technical and marketing aspects. It provides a platform for the input providers and connects to the customers directly.

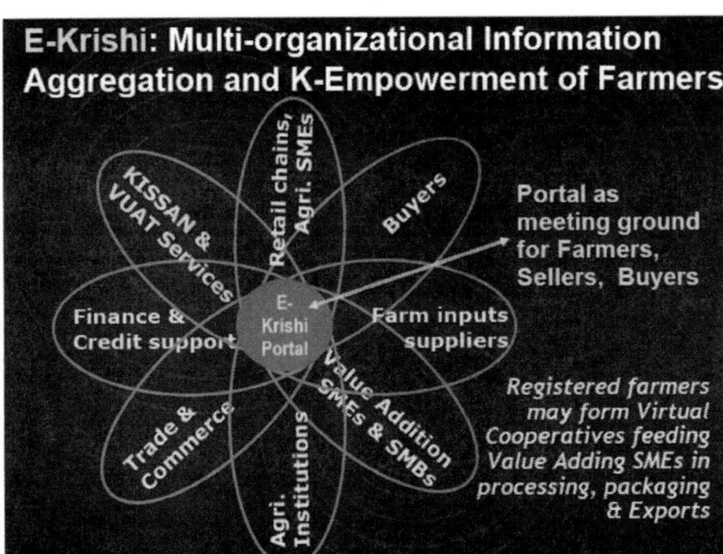

Figure 3 : e-Krishi links

The Kerala State IT Mission is set to introduce its prestigious e-Krishi project in other districts like Kollam, Pathanamthitta, Ernakulam, Thrissur, Kozhikode, Kannur and Kasaragod districts following the instant success of the pilot project in Malappuram. The Vegetable and Fruit Promotion Council Kerala (VFPCK) is joining hands with the Kerala IT Mission to provide everything from seeds to marketing assistance for its member farmers and participants. It has been collecting market information like price changes, stocks and availability, as well as arrivals at 17 major markets in Kerala and four major markets outside the State through its marketing division. The Kerala State Information Technology Mission has launched a toll-free agricultural trade call centre and the revamped e-Krishi Web site.

Impact of the Project

- Over 7000 farmers in Malappuram district have registered.
- Momentum of buyer – seller interactions picking up.
- Word-of-mouth spread of e-Krishi utility.
- Large volume buyers will benefit from placing orders for aggregated supplies from multiple farms.
- Akshaya initiative has made all the farm families Internet savvy.
- '1661' Call Centre services make it easy for farmers to interact with experts.
- Government extension services get more teeth.
- Integration with KISSAN & VUAT (Virtual University for Agricultural Trade) provides farmers support at all stages of the crop cycle and agrarian prosperity chain.
- Being extended to seven more districts.

The Virtual University for Agricultural Trade (VUAT)

Agriculture in Kerala, dominated by plantation crops, heavily depends on foreign trade and hence is highly influenced by WTO regimes. To take advantage of the emerging dynamic market opportunities and to evolve appropriate strategies for minimization of the negative impacts of the trade regimes, the Commission recommended the establishment of a Virtual University on Trade in Agriculture as a consortium of groups having capability in agriculture, trade, information technology and related areas that will make available the right information to the right people at the right time and ultimately serve as an institution for furthering agrarian prosperity. Accordingly the Virtual University for Agricultural Trade (VUAT) was established by the Government of Kerala as a consortium of state and central agencies as per the recommendations of the Commission on WTO Concerns in Agriculture headed by Prof. M.S. Swaminathan. The Kerala Agricultural University and the Indian Institute of Information Technology and Management - Kerala are the twin agencies vested with the implementation of this project with the active partnership of Commodity Boards and export promoting agencies. The Department of Agriculture, Government of Kerala sponsors the project.

The objectives and mandate are:

- To enable farmers for achieving higher income and prosperity through productivity improvement, quality enhancement, value addition and farmer-friendly marketing of agricultural produce
- To enable building up a sustainable agricultural trade security system
- To generate database management systems on trade in agriculture
- To serve as a decision support system to the government, trade and farmers
- To offer short-term courses on specific aspects of trade and IPR in agriculture under the WTO regimes within the framework of a virtual college

Its main functions are to create awareness to all stake holders, to create specific knowledge, market and trade information, advisory and decision support, offering short training programmes *etc.* It has a wide range of consortium partners for data generation, database management, training programmes, short courses, system management, co-ordination *etc.*

The expected benefit of the effort is:

- Knowledge empowerment to the farming community in Kerala to enable improvements in cultivation practices to take advantage of the emerging dynamic market conditions
- Facilitating trade security to farming community
- Promotion of quality revolution among the stakeholders in agriculture
- Enabling traders and exporters to have better access to world market through market intelligence, thereby facilitating higher prices at the farmer level.
- Enabling the Government and the policy makers to provide appropriate policy support to the farming sector
- Assisting farmers to have better access to market and trade information

- Enabling the farmers to achieve sustainable increase in production, productivity and improvement in quality.

Integrating e-Krishi, Kissan & VUAT

An integration of the ICT efforts in the state was found to be effective and fruitful. The e-Krishi run through the Akshya centers, the VUAT and the KISSAN kerala linked with all the serving departments linked was perceived to be effective among the farmers of the State, it can contribute to the total knowledge empowerment of the farmers.

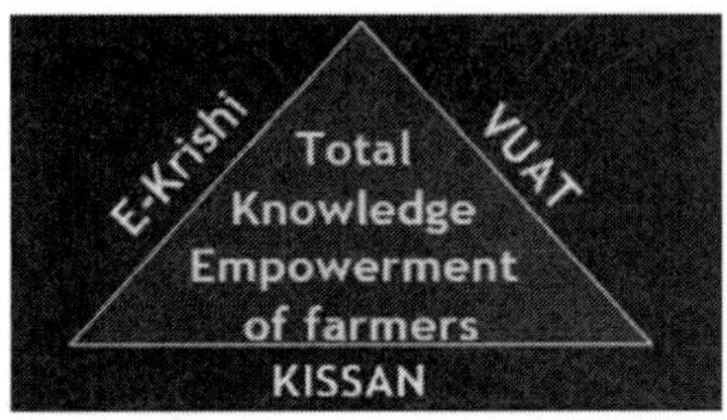

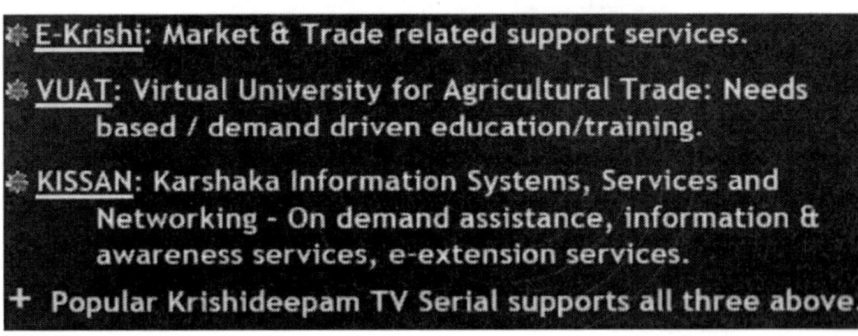

According to a study by Kerala Agricultural University during 2008 the Efficiency of ICT as perceived by the farmers is as follows:

Table 1 : Efficiency perception of the users and non-users

Efficiency perception	User N=270		Non User N=90	
	No	Percentage	No	Percentage
1. Gives personalized and timely information	11	6.11	4	4.44
2. One stop shop information source	41	22.77	8	8.89
3. Saves time and money in interacting	83	46.11	34	37.78
4. Has all information for the farmer wants	39	21.67	34	37.78
5. Supports marketing	6	3.33	10	11.11
Total				**100.00**

The study reveals that the maximum response on the efficiency of the ICT was perceived to be on saving time and money in interacting. We can infer that the farmers need timely information in all aspects with less cost.

Conclusion

Capturing the range of opportunities that ICTs enable is an enormous challenge, but eased by the concept of a smoothly functioning agricultural knowledge system that works to ensure that smallholders can access information and knowledge they need, when they need it, in a language they find accessible and *via* mechanisms that permit feedback, choice, and empowerment. The starting point is to build awareness of agricultural information and other items that can drive demand for access to services, taking advantage of newly available feedback mechanisms give farmers a voice in the creation new services or improvements that can help them make more productive and sustainable decisions. Just as ICTs have transformed developed country agriculture, they are transforming agriculture in developing countries, but the changes underway are not necessarily identical. The rural poor feel the effects of global information networks and global knowledge systems; they may find ways to access it, ultimately empowering the rural poor who are the principal beneficiaries of the efforts. The dawn of the information age has forced Extension to radically change its methods of disseminating research-based information to clientele need based. The abundance of methods for disseminating information creates a need for Extension to know the types of technology its audience owns and/or regularly uses. There appears to be a clear vision about how the agricultural sector should develop in terms of ICT adoption in maximizing benefits for the farming community. Much more coordination is required between agribusiness and the professionals serving the farmer to achieve an integrated and harmonious approach to reap the desired results.

References

Anonymous. 2004. KISSAN NEWS. *Newsletter* of KISSAN Kerala project, November. p 1

12

Taking Agricultural Technologies to the Grassroots on Wheels — A Case Study of the Mobile Agricultural School and Services (MASS) Initiative

Sanjay Kumar Rewani, Kumari Shweta, A.K.Pandey and J.Oraon

ABSTRACT

Jharkhand state is endowed with vast and rich natural resources. It's 80 per cent population resides in 32,620 villages and depends mainly on agriculture and allied activities for their livelihood. Rural population in Jharkhand still suffers from several ailments, weakness and constraints which come in the way of the use of ICT for agricultural development. Low level of literacy, poor socio-economic status, low penetration of extension services, language barrier, lack of electricity, *etc.*, are some of the problems due to which the farmers are unable to access the latest techniques in farming through different ICT services and are using the traditional knowledge and techniques which yields very less. To overcome these problems, Mobile Agricultural School and Services (MASS), has come up with a whole new concept of providing training to the farmers at their doorstep and empowering them with new technologies and skills to foster sustainable agriculture and other allied activities in the villages of Jharkhand. It has designed a well digitally equipped training hall on wheels which move from one village to another and provide a well flavoured training to farmers through self designed, target oriented packages in the form of films and other audio-visual aids including internet facility. It imparts an excellent audio-visual training on each and every aspect of crops, vegetables, medicinal plants and flower production; vermi-compost production; organic farming and lac cultivation. It also provides training on various aspects of different animal husbandry practices like dairy, poultry, piggery, fishery, goat production *etc.* These trainings enhance the knowledge and sharpen the skills of farmers dynamically. The experts of MASS also solve the various problems of the farmers during training and give advice and assistance to farmers to help them improve their methods of production and marketing. Earlier the cost for training was managed by the farmers themselves but now after seeing the success of MASS, several Governmental Organizations, Non-Governmental Organizations and other agencies have come forward to take the financial responsibility. The present study discusses the various aspects of MASS like its objectives, organizational structure, services provided, mode of providing training, coverage area and factors of success. It also deals with SWOT analysis that identifies the strengths, weaknesses, opportunities and threats for MASS. Both primary and secondary data were collected from the office of MASS and Governmental and Non-Governmental Organizations linked to it for the preparation of this article. Field visits were taken up in the villages where services have been provided by MASS to collect the primary data. On the basis of the findings of the study, it is concluded that MASS is very successful in providing sustainable agriculture literacy to the farmers of the villages of Jharkhand and in improving their socio-economic status.

Introduction

India is second most populous country in the world. Majority of its population lives in villages and earns their livelihood through farming. Agriculture is the backbone of Indian economy. It contributes around 26 per cent of the total GDP. Sixty five per cent of Indian population lives mainly in its 600,000 villages. Agriculture is the mainstay of the majority of the villagers as they are employed in agriculture or agriculture related services (www.planningcommission.nic.in).

Farming community in India is facing a lot of problems in maximizing the crop productivity. Inspite of successful research on new agricultural practices, the majority of farmers are not getting upper bound yield due to several reasons. One of the reasons is that expert/scientific advice on crop production and marketing is not reaching the farming community in a timely manner. With the fixed availability of land and water, higher agricultural growth can only be achieved by increasing productivity through effective use of the better technology. The virtual collapse of the agricultural extension system in most states limits farmer's access to better technologies and practices.

It is universally accepted that Information and Communication Technologies (ICTs) offer immense opportunities for the comprehensive social and economic development of the people all over the world. In order to transform agriculture into an information driven, modern and competitive sector, the role of ICT can not be over emphasized. A large number of initiatives have been made and are being made in rural India to deploy ICT as a development tool for creating awareness among farmers and rural artisans and for their betterment. The right information at right time can make a huge difference to farmer's income. Farmers need information from different sources and often need help to integrate them. There is a need to bridge theory and reality at grassroots. Few are prepared to move beyond secondary data. Farmers, in particular small and marginal farmers, must be able to choose agriculture practices and technologies from the full range of approaches available for tackling their problems.

Methodology

Both primary and secondary data were collected from the office of MASS and Governmental and Non-Governmental Organizations linked to it for the present study. Field visits were taken up in the villages where services have been provided by MASS to collect the primary data. With these data SWOT analysis was conducted to ascertain the strengths, weaknesses, opportunities, and threats for MASS.

An Overview of Jharkhand State

The State of Jharkhand was created in November, 2000 after carving 79 lakh hectares geographical area from the State of Bihar for better planning and execution of development activities of a ecologically different zone (Hills & Plateau region compared to Bihar plains). 74.65 per cent of its population resides in rural areas. 54 per cent of its population in rural areas lives below poverty line. The percentage of ST and SC population is 28.81 per cent and 11.35 per cent respectively. According to 2001 census report the literacy of Jharkhand is as such: male 67.94 per cent, female 39.38 per cent and the total 54.13 per cent only (Anonymous, 2006).

Out of 10.13 million working population, 76.86 per cent are engaged in agriculture. The agricultural scenario of the State is not very good since the cropped area is only 22 lakh hectares with average productivity of 1 t/ha (www.jharkhand.gov.in). Rural population still suffers from several ailments, weakness and constraints which come in the way of the use of ICT for agricultural development. Low level of literacy, poor socio-economic status, low penetration of extension services, language barrier, lack of electricity, *etc.*, are some of the problems due to which the farmers are unable to access the latest techniques in farming through different ICT services and are using the traditional knowledge and techniques which yields very less.

Prevailing System of Technology Dissemination in Jharkhand

1. The technology available at the state level is in operation at district level also.
2. Training of farmers is not demand driven.
3. Most services have a curative rather than a problem solving approach.
4. Not much work on group activities.
5. Not much research-extension linkage.
6. Lack of skills and knowledge in village level extension worker.
7. Paucity of funds is a major problem.
8. Practically not much awareness on the "Use of ICT" in technology transfer.
9. Not much concern on food and nutritional security.
10. Training & Visit system is not successful.
11. Approach is crop based & not farming system based.
12. No emphasis on management reforms.
13. No work on institutional build up.
14. Not much emphasis on capacity building & skill upgradation.
15. Not much linkage with local institutions involved in technology transfer.
16. Low level of involvement of small & marginal farmers in technology development and dissemination process.

Mobile Agricultural School & Services (MASS)

MASS is a brainchild project of Mr. Vijay Bharat (M.Sc. Agriculture). It is an institute on wheels which impart training to the farmers at their doorstep. It has changed the whole concept of training. It has designed a well digitally equipped training hall on wheels containing Projector, Screen, Laptop, Electronic Board *etc.* which move from one village to another and provide a well flavoured training to farmers through self designed, target oriented packages in the form of films and other audio-visual aids including internet facility. It can train 50 (Fifty) farmers in one session. The cost for training is around Rs. 13000. These trainings enhance the knowledge and sharpen the skills of farmers dynamically. It re-shapes their aptitude and attitude quite matching the vision of government programme & policies, developmental work and other innovative agricultural activities. It also carries inputs like fertilizers, insecticides, pesticides, seeds and medicines to sell it to the farmers at cheaper cost in comparison to the local market.

Figure 1: MASS Training Bus

Objectives of MASS

1. To provide excellent audio-visual training to farmers, making them watch each and every step of production on screen and giving emphasis over the identification of diseases, insects and pests and their solutions or remedies.
2. Facilitating well equipped training centre with good environment in their respective places.
3. To save precious time of farmers.
4. To save their traveling expenses on training.
5. Providing training through one-to-one method.
6. To teach each and every farmer.
7. To provide on- farm solution.
8. To provide them speedy clinical consultation at the time of natural calamity.
9. Bringing them into modernity from antiquity.

Organizational structure of MASS

MASS has established two different wings to fly successfully as to achieve its set objectives and mission.

A. MASS Training Division : It carries the responsibility to impart training to the farmers at their doorstep as and when they seek for. The officials of this division are well trained to understand the farmer's language and 'psyche' and are able to locate the 'hard spot' first and then chalking out their 'modus operandi' accordingly.

It consists of the following officials:

i. Managing Director (M.D.)
ii. Agricultural scientists/experts
iii. Training In-charge
iv. Software operator
v. Helper-cum-Assistant

B. MASS Film Division : This is another wing of MASS which introspect the problems of the farmers first on the basis of the on-field training feedback. Accordingly, it launches the training packages in the form of films. From shooting to the end product, it takes care of the clarity of the topic.

It consists of the following officials:

i. Director
ii. Script Writers
iii. Cameraman
iv. Team of Actors
v. Editor
vi. Software Developer and Technicians

How MASS Intend to Meet its Objectives?

1. By creating a bridge between the research stations and farmers.
2. By transferring the technical know how to the farmers.
3. By making the farmers business minded.
4. By developing the managerial skill in the farmers.
5. By developing awareness among the farmers regarding their different rights.
6. Strengthening their morals to fight for the rights.

Services Provided by MASS

1. Training on each and every aspect of:
    i. Cultivation of every type of crops, vegetables, medicinal plants and flowers.
    ii. Production methodology on dairy, poultry, piggery, fishery and goat rearing.
    iii. Production of vermi-compost.
    iv. Organic farming.
    v. Lac cultivation.
    vi. Multi-Tier Cropping system
    vii. And several other Agriculture related activities as and when asked from the target group(s).
2. Information about the latest released varieties of crops.
3. Information about new technologies and practices.
4. Facility for on the spot soil testing to check the status of NPK.
5. Sorting out of the current problems with standing crops and animals by visiting the fields of the farmers.
6. News about new policies and existing policies of State Government, Government of India and some leading Banks.
7. Current market rate of different agricultural produces of different cities through internet.
8. Inputs like fertilizers, insecticides, pesticides, seeds and other inputs to the farmers at cheaper cost in comparison to the local market.

Working of the Whole Process

Meeting with farmers of the village

↓

Collection of soil samples and existing facts

↓

Identification of problems

↓

Awareness campaign about innovative training programme *i.e.* MASS

↓

Provision of the need based training programme

↓

Formation of farmers groups

↓

Arrangement of training sessions for farmers

↓

Arrangement of pre-season, on-season and post- season meetings and training programmes

Steps Followed During Training

i. 1st Step : Before training, they start their programme with local folk dances/songs as to bring the farmers in the rhythm of learning.

ii. 2nd Step : Training session begins with Audio-visual training package.

iii. 3rd Step : Experts of MASS reciprocate to the problems and questions raised.

iv. 4th Step : It is followed by a 15 minute Interactive session programme focused on the topic covered during the training.

Figure 2: Farmers under training inside MASS Training Bus

How it is Different from Conventional Method?

- Films give a clear picture to understand the matter.
- Films' final version comes out only after the consultation and suggestions of different scientists and hence the quality of training is of the best quality.
- Every farmer put his full attention during training and hence get maximum benefit.

- This method does not need the proof of methods that has been involved in cultivation.
- The energy of scientists/experts are utilized in the field and hence we get good time to solve the problems on practical level.
- Generally, one week is required to arrange one training programme in conventional method but this method come in action in very short period of time.
- Facilitation of Internet at the doorstep of farmers.

Table 1: The Topics Covered & Number of Farmers Trained through MASS

District	Organized by	Topics covered	No. of trained farmers
Ranchi	HARP, Palandu	Packages & Practices of fruit crops and vegetables	350
Ranchi	DAO, Ranchi (8 Blocks)	Nursery development of vegetables, Cultivation of rice, Multitier system, Cow management & Piggery	1600
Ranchi	Environment Pollution Ban Treaty (EPBT) (8 Blocks)	Packages & Practices of Jatropha, Citrus plant & Integrated pest management	200
Lohardaga	District Horticulture Officer (4 Blocks)	Improved method & hi-tech system of nursery development of vegetables & Cultivation of vegetables & rice	450
Palamu	DAO, Palamu	Packages & Practices of vegetables, Mushroom production & Cow management	200
East Singhbhum	District Horticulture Officer (4 Blocks)	Improved method & hi-tech system of nursery development of vegetables, Cultivation of vegetables & rice	550
Dhanbad	Gangadhar Rastriya Vikas Mission	Packages & Practices of vegetables, Goat rearing, Poultry and Cow management	300
Ranchi	Deputy Director, Horticulture	Production of mushroom	200
Dumka	PD, ATMA, Dumka	Cultivation & Grading of vegetables, Production of maize	240
Ranchi	DAO, Ranchi	Production of mushroom	200

West Singhbhum	PD, ATMA, Chaibasa	Cow management, Organic farming & Composite fish farming	3600
Dumka	DAO, Dumka	Cultivation & Grading of vegetables, Production of maize	1400
Koderma	DAO, Koderma	Watershed management	3500
Garhwa	PD, ATMA	Packages & Practices of gram & wheat	700
Ranchi	NABARD	IPM, Packages & Practices of wheat & maize	300
West Singhbhum	ATMA, Chaibasa (5 Blocks)	Packages & Practices of vegetables, Goat rearing, Poultry and Cow management	1400

SWOT Analysis

A comprehensive SWOT analysis is conducted to identify the strengths, weaknesses, opportunities and threats those related to MASS through primary and secondary data as explained in the methodology. The government officials, trained farmers and local leaders have been interviewed to elicit more information regarding their experiences with transfer of agricultural technologies to the farmers by MASS.

- The study shows that there are many strong aspects of MASS in taking the agricultural technologies to the grassroots.

These SWOT factors are depicted in a SWOT matrix as below:

Strengths	**Weaknesses**
❒ Training of farmers at their doorstep.	❒ Lack of facility for training farmers in regional languages.
❒ Self designed target oriented training materials in the form of films and audio-visual aids.	❒ Insufficient number of training buses.
❒ Financial support by State Government and NGOs for training of farmers.	❒ High cost of training.
❒ On-farm solution of the problems.	❒ Limited technical manpower.
❒ Provision of agricultural inputs and medicines at cheaper cost.	❒ Poor follow-up of the programme.
❒ Facilitation of Internet at the doorstep of farmers.	❒ Limited training materials in print.
Opportunities	**Threats**
❒ Expand its work area in the neighbouring states.	❒ The problem of naxalism in the state.
❒ Virtual collapse of the state agricultural extension system.	❒ Frequent bandh and hartals.
	❒ Deteriorating condition of roads in rural areas.

SWOT Matrix on MASS

Conclusion

One of the main strategies for development of Jharkhand state is to ensure agricultural and rural development on sustainable basis. The disappointing performance of the agricultural extension system in the State, farmers had limited access to better technologies and practices. The study proves that MASS has succeeded in filling this information gap by providing timely advice and guidance on sustainable farming methods to the farmers of rural areas.

References

Anonymous, 2006. Annual Report 2005-06, State Agricultural Management & Extension Training Institute (SAMETI), Jharkhand.

http://www.jharkhand.gov.in/

http://www.planningcommission.nic.in/

13

Village Resource Centres (VRCs) in Uttaranchal

B. Maithili, Neelam Pandey and Paras Kakkar

ABSTARCT

The Himalayan Institute Hospital Trust (HIHT) has been working in the agriculture field since, 1999. At present, it is focusing on a Research-based project on the cultivation of Aromatic/ Medicinal plants aiming to promote Medicinal and herbal cultivation and increase the income and participation of community. Initially the grant -in-aid is being provided to some farmer groups in Shilogi village of Tehri District and Malethi village in Pauri District. These farmers came forward and offered their land for research work. The project supported the community through both, backward and forward linkages. Rural Development Institute, HIHT, is facilitating the whole operation, from choosing the coordinators to the advisory agencies. RDI is trying to make sure a proper people (especially poor) friendly delivery mechanism through village Resource Centre. VRC is a knowledge centre for addressing various needs of the local especially rural communities such as education, health, weather, agriculture, natural resources and livelihood opportunities through IT and space applications. ISRO is implementing this project in the State through its sister concern at Dehradun, Indian Institute of Remote

Sensing (IIRS). HIHT is also performing as a nodal centre for enabling implementation through other NGO's (in Uttarakhand) involved in the VRC project covering 8 districts of Uttarakhand through 16 VRC's and two expert centres. These are VSAT enabled information centres. VRC is also being used for live interaction of farmers with the Scientists and dissemination of other farmer related information like latest technology *etc.*

Introduction

There are more than 6,00,000 villages in India and around 700 million people live in villages. Many villages are relatively deprived in terms of basic amenities and services, especially those related to education, health, sanitation and empowerment. There is a need to improve the quality of life in villages to achieve overall national development.

Indian Space Programme started in early 60's has become largely self-reliant with capability to design and build satellites for providing space services and to launch them using indigenously designed and developed launch vehicles. Department of Space/ISRO has evolved many useful applications emanating from the Space Technology. While Indian National Satellite (INSAT) system continues to provide regular services in the areas of telecommunications, business communication, broadcasting and meteorological services. Several initiatives have been taken-up to expand the application of INSAT to new areas like Tele-medicine and Tele-education. ISRO has also been a champion in demonstrating the use of space technology for societal good and has piloted several socially relevant space application projects.

With seven satellites, Indian Remote Sensing Satellite (IRS)-1C, IRS-1D, IRS-P3, IRS-P4, Resource sat, Technology Experiment Satellite (TES) and CARTOSAT-I, in operation, India has the largest constellation of remote sensing satellites today providing data in a variety of spatial resolutions and spectral bands. The data is used for various applications in the fields of agriculture, forestry, ground and surface water, drought assessment and monitoring, flood mapping, land use and coastal studies *etc.* Space technology, as the powerful enabler, provides alternate route for holistic and rapid rural development. India has been among the world leaders, in the areas of developing end-to-end capability for both remote sensing and communication satellites.

As mentioned above, space technologies - satellite based communication and remote sensing has demonstrated their capabilities to provide services related to health care, education, weather, land and water resources, agriculture *etc.* To provide these space-based services to the rural areas,

Department of Space/ISRO has initiated a programme to set up Village Resource Centres (VRCs) in partnership with concerned state, central agencies and NGOs.

Space based services, emanating from Satellite Communication (Sat Com) and Remote Sensing satellites hold considerable value to transform village society. Remote Sensing provides information on natural resources which could be used to identify prospective zones of ground water occurrence and recharge, wasteland type and extent, watershed attributes, environment, infrastructure related information, alternate cropping pattern, water harvesting *etc.* In order to disseminate the services emanating from the space systems as well as other Information Technology (IT) tools, to the rural communities, a partnership between ISRO and Indian Institute of Remote Sensing (IIRS) and four NGOs of Uttaranchal *viz.*, (i) Himalayan Institute Hospital Trust (HIHT) (ii) Sri Bhuvaneshwari Mahila Ashram (SBMA) (iii) Central Himalayan Rural Action Group (CHIRAG) and (iv) Institute of Himalayan Environmental Research and Education (INHERE) have been formed to set up VRCs in Uttaranchal.

Brief profile of each NGO and expert centres are given below:

1. NGOs and Expert Centres

1.1 Himalayan Institute Hospital Trust (HIHT)

Himalayan Institute Hospital Trust (HIHT) is a non-profit Organization committed to health, education and economic self-sufficiency. The organization is located at Swami Rama Nagar, mid way between Dehradun and Rishikesh. Resources of the trust include a 750 bed large tertiary hospital with super specialty services, a Holistic Health programme, a Medical college offering graduate and post graduate courses along with other Paramedical graduate courses, a School of Nursing, a Rural Development Institute and a Healing centre for Alternate Systems of Medicine. More than a thousand professionals, academicians, researchers and skilled resource including community workers are employed with various facilities of the Trust. The mission of the trust is to develop integrated and cost effective approaches to healthcare and development addressing the local population, which can serve as a model for the country as a whole and for the underserved populations worldwide.

The Rural Development Institute (RDI) is the rural outreach of HIHT, providing direct and indirect services in the districts of Uttaranchal. The primary focus areas of service have been Health, Water, Education and livelihood measures. Since a decade, Rural Development Institute has been engaged in the health field in the areas of primary health, reproductive and child health, school health, adolescent health and alternate systems of medicine. The institute has its outreach in over 1000 villages across the state serving a population of 3,50 ,000.

Besides direct implementation of programs, which continue till date, the institute has been engaged in providing support to other agencies in many aspects of medical care. Designated as mother NGO by Government of India, Ministry of Health & Family Welfare, HIHT been providing support to NGOs in six districts in Uttaranchal in the field of reproductive health.

1.2 Sri Bhuvaneshwari Mahila Ashram (SBMA)

Sri Bhuvaneshwari Mahila Ashram is an independent organization that works with the people of Uttaranchal Himalayas. The mission of the organization is to work with communities, its organizations/institutions, government, technical and academic institutions and other civil society groups.

1.3 Central Himalayan Rural Action Group (CHIRAG)

Central Himalayan Rural Action Group (CHIRAG) is a non-profit grassroots development organisation working with rural communities in the Central Himalayan State of Uttaranchal. The key thrust areas of work by the organization is Natural Resource Management with the objective of improving the quality of these resources and their management to enhance quality of people living in a symbiotic relationship with nature. Major activities are forest management and reforestation, water management, agriculture, horticulture, animal husbandry, alternate energy and integrated watershed management.

1.4 Institute of Himalayan Environmental Research and Education (INHERE)

INHERE was founded at Chokhutia block of Almora district with the objective of holistic development of the community utilizing the vast resources, human and natural resources of the Himalayas. The organization is active both as a support organization and a grass roots implementation agency. INHERE focuses on community work through community organization. It works on the approach of integrated development by using local resources and encouraging entrepreneurship skills in the poor, especially the women. The thrust areas of the organization are Environment, Livelihood & Food Security, Social Service and Human Resource Development.

2.Scope and Objectives

The newly formed state of Uttaranchal in the north-western region of the Himalayas and much distinguished from the parent State of UP., ranks 19th according to a composite index of social and economic development.

Health issues relevant to Uttaranchal can be perceived as being linked to the workings of both patriarchy and poverty, which is a consequence of poor development. Safe motherhood remains the lowest prioritized area of concern and there is high prevalence of nutritional disorders, habit-forming substance abuse, worm infestations, dental problems and anaemia among the Young. Coverage of women with 'Antenatal Package' is less than 50 per cent in the State and only approx 18 per cent of the deliveries are institutional; one third of the currently married women suffer from RTI's and STI's. There is a large unmet need in the State, with limited choices of contraceptives. The Infant mortality rate in Uttaranchal (41), though lower than the national average is significantly higher in rural areas (62). Low birth weight is an important cause of concern in Uttaranchal and 42 per cent of children below three years of age is underweight. With demographic and epidemiological transition, Uttaranchal also has acquired the 'double burden of diseases' - of both communicable and non-communicable pathogenesis. The prevalence of malaria in the state is 450 per lakh and that of Tuberculosis 1225 per lakh; 40% of population in the hilly districts lie in the endemic belt for Goiter. The potential for rapid increase in HIV cases strongly prevails due to migration factor and subcultures of risk practices observed with entertainment and tourism industry.

Migration has created a "money order" economy and poverty rooted in subsistence existence further adds to the creation of conditions leading to high mortality and morbidity. All the hilly districts are characterized by very small landholdings, poor agricultural productivity, lack of infrastructure, low level of industrial development, low level of social services, peculiar geographical conditions and the difficult nature of terrain. Micro village-level studies have shown that about eight percent of the working population is engaged in crop agriculture and animal husbandry. Despite a very high cropping intensity, out of the total expenditure on food, the value of cereals, pulses, oilseeds *etc.* produced indigenously is about 39 per cet only and the balance 61 per cent of the food required is imported from other region.

Telemedicine Started to Accomplish the Following Objectives

2.1 Livelihood Challenges

The primitive nature of agriculture, lack of irrigational facilities and non-availability of employment pose a livelihood problem in many districts. More than one-third of the households across the state belong to extremely food insecure category that fall short of their annual food requirement. Besides seasonal food insecurity, borrowing food directly or money for food, selling livestock and other assets, looking for some alternate employment locally or migration to other places as major coping strategies are hallmarks of livelihood challenges in the state. Small land holdings, lack of local employments, absence of able-bodied working males, a higher percentage of elderly people and the mind set to migrate among the young are important 'push factors' of out-migration and 'post office' economy of the state. Women as the de facto managers of natural resource base acquire equal if not more importance than men in the mountains. They constitute the main agricultural workforce in Uttaranchal - a fact proven by their workforce participation rate of 41 per cent as compared to the national average of 22.7 per cent (1991). Access to loans and credit are linked with ownership of assets and are therefore denied to her. Land rights remain with men, leaving women with limited options. In Uttaranchal, the out migration of men, large-scale unemployment, ecological collapse, and their traditional backbreaking life, have all conspired to worsen the lot of women to degrees unimaginable.

Activities

- Online Community trainings on Income generation and self help group's formation and Agri based information given to the women groups through Village Resource Centre.

2.2 Poor Usage of IT in Rural Transformation

Lack of comprehensive information and technological system on natural resources like soil texture, ground slope, surface flow, drainage system, ground water prospects *etc* never equip farmers to go for cost effective and eco balanced farm interventions. Poor deliveries of agricultural extension services in the mountainous outreach have not given opportunities to masses to adopt value added best practice. The land management, cropping pattern, seed treatment, and pest management are based on traditional practices with no further linkages to the modern science and technology. Lack of capacity building events like farmer's advisories never capacitated the rural masses to bring them at entrepreneurial level through maximizing the economic outputs from the available resources.

A rich heritage of oral tradition in Ayurvedas/Ethno medicine has flourished in the mountains and an evidence-based therapeutics with medicinal flora/ herbs exists. The mountainous area traditionally has been home to local healing practices by people with knowledge of astrology, herbs and ancient Vedic practices. But over time, roots of herbal home remedies in the community have weakened considerably for various historical reasons. Lack of political will and pro-people initiatives for development have perpetuated ignorance and failed to harness best traditional health practices and judicious integration with modern medicine.

The proposed VRCs aim at transferring the benefits of the space and IT directly to the communities at the grass root level to uplift the life and livelihood of rural areas. The proposed project as envisaged is meant to serve as a knowledge centre to address various primary need based requirements of the local communities such as education, health, weather, agriculture, natural resource and livelihood opportunities. One of the main aims of the VRCs is to use development in the field of IT and Space technologies to empower rural communities by enabling greater capacity building and awareness. The VRC programme is meant for reaching the benefits of the Space technology directly to the communities at the grass root level.

The ISRO-HIHT Uttaranchal Village Resource Centre (VRC) Programme is a partnership made to uplift the life and livelihood of rural area. It aims to promote a single window delivery of need-based services in the areas of education, health, nutrition, weather, environment, agriculture and alternate livelihoods to the rural population. The VRCs handle both dynamic and generic information to empower rural communities through spatial information infrastructure.

Table 1 : Major Objectives and Activities of the Project

Major objectives of the project	Activities
Tele-education and tele-healthcare services	Health education/English/Maths tuition classes for adolescents and tele medicine
Access to space based information on land, water, soil *etc.*,	Orientation on soil testing in some of the VRC
Advisories on agriculture, land and water management	Online information and field based activities
Providing enhanced livelihood opportunities	Established Aromatic and Medicinal plants nursery
Providing information on price, market, pests, diseases, livestock, govt. schemes, job opportunities, e-governance related information *etc.*	Through G.B. Pant Institute

3. Locations of VRCs

Sixteen VRCs Established in the following locations in Uttaranchal with Himalayan Institute Hospital Trust as the lead NGO. The details of VRCs are as follows :

Table 2 : VRC Locations

Sl. No,	Location of VRC	Block	District	NGO Responsible
1.	Thano	Raipur	Dehradun	Himalayan Institute Hospital Trust (HIHT)
2.	Nagthat	Kalsi	"	– do –
3.	Lakhamandal	Chakrata	"	– do –
4.	Kwanu	Chakrata	"	- do
5.	Malethi /satpuli	Kaljikhal	Pauri	
6.	Chamba	Chamba	Tehri	– do –
7.	Ghat	Ghat	Chamoli	Sri Bhuvaneshwari Mahila Ashram (SBMA)
8.	Gairsan	Gairsan	"	–do–
9.	Anjansen	Jakhnidhar	Tehri	–do–
10.	Josiada	Town Uttarkashi		Uttarkashi – do –
11.	Gola	Bagheswar	Bagheswar	Central Himalayan Rural Action Group (CHIRAG)
12.	Simayal	Ramgarh	Nainital	–do–
13.	Chinoni	Chakotia	Almora	Institute of Himalayan Environmental Research and Education (INHERE)
14.	Bayeda	Dwarahat	–do–	– do –
15.	Bhikiyasen	Bhikiasen	–do–	– do –
16.	Jamradi	Bhainsiajhana	–do–	–do–

These 16 VRCs will cover around 7,25,000 population. The location details of each VRC are given in Table 2.

Table 3 : Expert Centres for Tele-medicine and Tele- Agriculture

Sl No	Location of Expert Centres / TM Centres	Expertise	Taluk/Block	District	NGO/Agency Responsible
1	Himalayan Institute Hospital Trust (HIHT)	Tele-Medicine	Doiwala	Dehradun	Himalayan Institute Hospital Trust (HIHT)
2	G.B. Pant University of Agril. & Tech.	Tele-Agriculture	Rudrapur	Udham Singh Nagar	G.B. Pant University of Agril. & Tech.

Expert Centre proposed: At Himalayan Institute Hospital Trust for telemedicine & G.B. Pant University of Agriculture ad Technology for tele-agriculture.

4. Mechanisms of VRCs

Village Resource Centres become single window delivery mechanism for tele-medicine, tele-education, natural resources data, agriculture advisories, land and water resources advisories; interactive farmers' advisories; e-governance services and weather advisories.

Tele-Education : Focus is on imparting vocational training at local level - aimed at skill development and capacity building to catalyse alternate livelihood in rural areas; supplementary teaching to rural children; and non-formal and adult education. Adolescent training on general health and life skill also imparting to the adolescents. According to the children's needs mathematics and English Speaking classes were also conducted for the students

Tele-Medicine/ Healthcare : Thrust is on both preventive and curative healthcare at primary level. The Tele-medicine system at VRCs consists of customized medical software integrated with computer hardware, along with medical diagnostic instruments connected to the VSAT at each location. With the help of local doctor/paramedical, expert medical consultation and counselling are provided to the villagers from specialty hospitals. Generally, the medical record/history of the patients is sent to the specialist doctors, who will in turn study and provide diagnosis and treatment during video-conference with the patients end. Tele-medicine helps patients in rural areas to avail timely consultations of specialist doctors without going through the ordeal of travelling long distances.

Land & Water Resources Advisories : VRCs providing spatial information on various themes such as land use/ land cover, soil, groundwater prospects, and also enable the farmers to get query based decision support.

Interactive Farmers' Advisory Services/ Tele-agriculture & Demonstrations at field level : Online interactions between the local farmers and agriculture scientists working at Scientific Institutions. The advisories covering a wide range of subjects starting from alternate cropping systems, optimisation of agricultural inputs - seeds, water, fertilizer, insecticides, pesticides and producer oriented marketing opportunities. Community centred advisories on soil and water conservation, on adopting water efficient cropping patterns, on practices related to rainwater harvesting/ground water recharge, information on market/ price, pests & diseases *etc.*, also made available. Farmers/ villagers of one VRC can interact with farmers/ villagers of other VRCs on crops, pesticides, *etc.*

e-Governance services : These services includethe formation of the Self Help Group (SHGs) and micro credit planning.

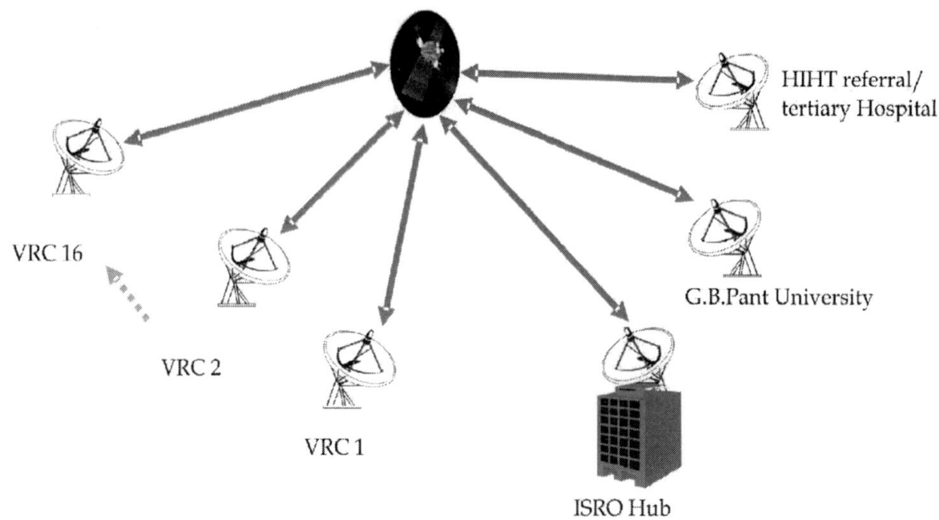

5. Responsibilities of ISRO

Major responsibilities of ISRO/DOS are to provide the satellite (VSAT) based communication back bone/connectivity, customisation of available natural resources database/advisories, providing minimal tele-medicine and tele-education facilities, software for accessing, displaying, querying natural resource database/advisories.

The Village Resource Centre (VRC) Programme will provide VSAT connectivity at 16 locations, spread in 8 districts of Uttaranchal State.

The proposed network is capable of catering to three types of requirements. The first one is the connectivity for telemedicine services wherein any VRC node can establish the connection with selected specialty/large/district hospitals. The second type of the connectivity is for training and data services where all the services will be provided from a central location. The third type of connectivity is for the services, which are exchanged among the VRCs, and interaction between VRCs and few expert centres. For delivering the planned services, the technology options available are audio tele conference and video-conference. The beneficiaries of VRC could access the services through broadcasting, point-to-point connectivity and intranet on a time-shared basis.

The communication system in the network of VRC programme has three components namely, the satellite, a hub or the uplink facility and the remote terminals. The remote terminals should be capable of having two-way communication to facilitate interaction.

6. Responsibilities of NGOs

NGOs provide building (a hall with two/three rooms) for setting up of the VRCs with necessary furniture, lighting and space for the installation of VSAT antenna and for tele-medicine, tele-education/ tele-agriculture facilities in each location. Computer Systems should be preferably, Pentium IV, for accessing the natural resource database/ querying and other services *etc.*, provided by them. Contents for tele-education/tele-agriculture programme will be generated/ arranged by them. The NGOs arranged for doctors (preferably from PHCs) at VRCs for interaction with expert doctors through telemedicine facilities. The priorities, perceptions, aspirations and requirements of the people are understood by conducting Participatory Rural Appraisal in all the locations by respective NGOs. This is helpful for deciding on the deliverables and various advisories. Census data on population and other details are collected and suitably integrated by respective NGOs. The information collected through PRA/ Census used to arrive at suitable agriculture/ land/ water resources advisories and to identify the health/ education needs. All associating agencies collect all the details on governmental schemes, pests & diseases, market/ price, livestock *etc.*, from various sources and these information are provided through respective VRCs.

The trained manpower for manning the VRCs is provided by the respective NGOs. The person responsible for handling the VRCs should have a basic knowledge of computer operations. The NGOs bear the cost of running (including staff salary, training *etc.*,) and maintenance of the VRCs.

14

Community Information Centre EmpoweringRural Poor through ICT

Dillip Pattanaik

ABSTRACT

Information and Communication Technologies (ICTs) are for everyone and that does not mean for urban or rural communities. Recent developments in the fields of information and communication technology are indeed revolutionary in nature. By definition, Information and Communication Technologies (ICTs) are a diverse set of technological tools and resources to create, disseminate, store, bring value-addition and manage information. Knowledge thereby becomes the fundamental resource for all economic and developmental activities in the knowledge society, of which rural poor form an equal part. The process of synthesis of knowledge possessed across communities with the global pool of knowledge and with the scope for further enrichment provides the genesis for information sharing. Isolation of rural communities from the mainstream economy and their lack of access to information because of societal, cultural and market constraints have let them to become distant from the global pool of information and knowledge. The one resource that liberates them at large from poverty and encompasses them is knowledge and its widespread dissemination. The relevance of ICT to rural communities depends upon its ability to provide need based and location-specific information on health, nutrition, education, natural resource management, livelihood, issues relating to weather, water, energy and agriculture, markets, information on different development programs *etc.* These isolated rural poor have to be an equal beneficiary to advantages offered by the ICT and the products and processes, which emerge from their use. This paper is based on action based research. IRMA-India has established four community information centres in a cluster of rural villages where these centres are working as knowledge/information provider to the communities. These centres are the sources of knowledge, skills, information for social and economic empowerment of rural families towards reaching the unreached and voicing the voiceless.

Introduction

The Indian sub-continent covers an area of 3.29 million sq. kms. It is the 7th largest country in the world. India's population as on 2001 was 1,027 million. India accounts for a meager 2.4 per cent of the World surface area and yet it supports and sustains a whopping 16.7 per cent of the world population.

About 75 per cent of the country's population lives in rural areas in over 6,60,000 villages. Known for its diversity, the country accommodates about 25 per cent of its population from indigenous and disadvantaged communities (Tribals and Dalits) known as scheduled tribes and scheduled castes.

Agriculture is the mainstay of Indian economy and about 60 per cent of the labour force and contributes nearly 26 per cent of the GDP and accounts for about 18 per cent share of the total value of the country's exports.

The literacy rate in the country is 65 per cent for males and 51 per cent for females. Twenty seven per cent of the rural population lives below poverty line and in some states like Orissa; the poverty rate is 48 per cent in rural areas.

ICT for Development

Long before 'ICT for Development/Poverty Reduction' became fashionable, 'old' technologies like radio and television had been used to disseminate information among the rural and isolated poor. As a medium for delivering information directly to the poor, these technologies can achieve far wider outreach at much lower cost than internet-based approaches. However, comparison of different ICTs should not be restricted to outreach and cost considerations. An important area of innovation in ICT for Poverty Reduction is to exploit the particular strengths of different ICTs by combining them to deliver a more complete communication package.

The world is in the midst of a knowledge revolution, complemented by the opening up of entirely new vistas in information and communication technology. Recent developments in the field of information and communication technology (ICT) are indeed revolutionary in nature. The beneficial impact of ICT on the rural economy and quality of life is now widely recognized. ICT in convergence with other forms of communications have the potential to reach the rural and tribal communities as well, thereby, empowering them to participate in the economic and social progress of the country. The relevance of ICT to rural communities depends upon its ability to provide need based and location-specific information on health, nutrition, education, natural resource management, livelihood, issues relating to weather, water, energy and agriculture, markets, information on different development programs *etc.* Women have to be an equal beneficiary to advantages offered by the ICT, and the products and processes, which emerge from their use.

The most critical development issues relating to ICT and evolution of knowledge societies must be approached from global and local perspectives through the joint participation of the public, private, the Panchayati Raj Institutions (the three-tier local government system), non-governmental organizations as well as academia and the members of the civil society. Gender mainstreaming becomes the crosscutting theme in all these issues.

ICTs are defined as technologies that facilitate communication and the processing and transmission of information by electronic means. ICT is a tool for poverty reduction when it is applied to meet the information and communication needs of the poor and isolated communities. However, the specific technology that is applied is less relevant than questions about freedom of expression, social and economic impact, inclusiveness, outreach and sustainability.

ICT models thrive on innovations, customization and people's participation. Innovative approaches to the use of ICT in the areas of local governance, e-commerce, e-advocacy, e-income generation activities could accrue unlimited benefits to the community. Equitable access to ICT and autonomy to receive and produce the information relevant to their concerns and perspectives are therefore critical issues. ICT strategies can succeed in bridging the poverty gap only if there is a concrete effort towards formulation of enabling policy frame works and avenues which create opportunities and incentives to participate and benefit.

The Indian Scenario Relating to Promotion of ICT

India is distinguished for its geographical diversities, mostly dotted by mountains, hills, rivers, forests, wetlands, deserts and scattered habitats extensively existent in remote areas. With a sizable population of the country's population living in isolated and scattered habitations, besides small and large villages, ensuring universal access through roads and other means of connectivity has been indeed difficult.

About 50 per cent of villages still lack all-weather connectivity. However, the telecommunications and power services in India have improved significantly. Presently, India is the 5th largest telecom network in the world. The Government have plans and programs to provide telecommunication facilities and power to all villages.

The Ministry of Communications and Information Technology of Government of India and the Telecom Regulatory Authority of India as well as several State Governments have already developed the strategies for accelerating the growth of Internet and broadband connectivity in rural India. The National Informatics Center (NIC), and other institutions have considerable reach in the country. The Mission 2007, a national alliance of public and private sectors with the Village Knowledge Center movement to bring knowledge, livelihood and prosperity is also an important initiative for promoting ICT to the benefit of rural communities. The Government of India's new initiative on right to information is another demanding factor for the need and provision of ICT services countrywide.

A. Community Information Centre in a Cluster of Rural Villages in Orissa

1. The Initiative

Orissa is the poorest State of India having 48 per cent of its 40 million people living below poverty line. 85 per cent of its population lives in rural areas spread over 60,000 villages. Population of marginalized and unprivileged communities like tribals (ST) and dalits (SC) forms 38 per cent of total population of the State. About 60 per cent of the villages are not yet connected with all-weather roads. Agriculture is the mainstay of the economy. Lack of access to information, basic, social and economic services, facilities and goods are major contributing factor to the backwardness and poverty of the rural communities.

Based upon the above conditions, IRMA-India with support and collaboration from its parent organization and donors, established four Community Information Centres in two clusters of Orissa State. One cluster is in a remote tribal areas and the other one is in cluster of rural villages.

2. Location

(1) Daspalla Block

The project area covers 17 villages, 517 households and 2544 population. Out of it there are 269 tribal households (52 per cent), 117 Dalit households (23 per cent) and 128 backward community households (25 per cent). Percentage of illiterates is more than 71 per cent and 95 per cent of families live below poverty line. Two community information centres is located at Baliberena and Bam and caters to all these 17 villages.

(2) Khallikote Block

The project area is under Khallikote Block of Ganjam District in Orissa State, covering 5 Gram panchayats (local government unit) with 32 villages,

Project Location and Coverage

(1) Badhinuapalli CIC

- Covers 3 Gram Panchayats
- 22 villages
- 2952 households
- 15224 populations

(2) Kanheipur CIC

- Covers 2 Gram Panchayats
- 10 villages
- 2266 households
- 12170 populations

27394 population and 5218 households. Two community information centres is located at Badhinuapalli and Kanheipur and caters to all these 5 Gram Panchayats.

Each CIC has been equipped with a television, audio-video facilities, power backup, telephone/internet connectivity and computers with hardware like printer, scanner, and web cameras and a small library with books and reading room facility.

(a) Deliverables

The Community Information Centre (CIC) deliverables encompass in terms of social, economic and political which has to generate benefit for the groups like women, disables, scheduled caste, scheduled tribe and people living below poverty line. CIC is a base to generate awareness among the local communities regarding the benefit of modest ICTs in improving their socio-economic well being and in terms of alleviating poverty.

The following services are being delivered at CICs:

- Ready availability of information, procedure, system and guidelines about pro-poor government schemes, plans, policies, programmes and services.
- Easy access to information on health care, education, water and sanitation, animal husbandry, agricultural practices and inputs *etc.*
- Easy access to timely information on markets, examination results, employment opportunities, weather and disaster related early warnings.
- The communities would be well aware of their rights, entitlements to public services, facilities, Acts, rules and regulations.
- Easy, cheaper, faster and timely electronic communication can be made available to the community.
- Availability and application of new technologies which help improving their quality of life and standard of living.
- Computer literacy training to the community (7-50 years) involving children, youth, women and men.

(b) Key Principles of CICs

- It is a people-centred programme based on community ownership.
- It is taking into account the local context and the information needs of the local people and provides useful demand-driven services. The program is not meant to demonstrate the power of technology. Usefulness is more important than the use of latest technology.
- The program is inclusive and to ensuring social inclusion in access.

(c) Partnership

The program is establishing active partnerships with all key stakeholders: the local community, Panchayati Raj Institutions (PRIs), the local Gram Panchayat (local government unit at lowest level) besides Government Ministries/Department/ Agencies, technical and academic institutions, professional bodies, corporate and private sectors *etc.*

PRIs : With the 73rd Constitutional Amendment Act, 1992 the three-tier Panchayati Raj Institution system has emerged as an effective network of local institutions for local governance. The lowest democratic unit under the PRI system is the Gram Panchayat, which is mandated to plan and implement projects for economic development and social justice. A Gram Panchayat constitutes with a population between 2000 to 10,000 taking a cluster of villages. In addition to the local community, the Gram Panchayat is a key partner of CIC project.

(d) Community Participation

Community participation is vital in all rural and community-based projects. It is broad-based and representative and cut across social and economic status. The local Gram Panchayat is also an active part of the local community. Different kind of meetings at village level, cluster level and area level has conducted to bring their attention on the use of information center for their access to information and to reach the communities globally.

The local communities are involved with the CIC process since beginning as it's a participatory community development programme. A CIC Development Committee have been formed for each CICs to look after the CIC's sustainability as well as convergence of other Government programmes by taking of 7 members including Sarapanch (local government elected representative), school teacher, local leader and learned people of the programme area. Project coordinator of IRMA-India has followed up them with basic inputs. The committee is meeting frequently and also helps during the CIC awareness campaigns which are being organized in villages.

Community Information Awareness Campaign

As a continuous process, Community Information Centre Awareness Campaigns are being done on a regular basis to bring awareness among the local community on the objectives of the project and CIC. We have adopted different methodologies to conduct these awareness programmes so that we reach at the remote places in the project area and people to know about the usefulness of the information that are available at the CIC. Following are the methodologies adopted for generating awareness:

- One to one communication between the people already used the CIC and or the people, who ever heard about the center and its function, Staff members in the community.
- General and cultural group discussions, whenever they are gathered.
- Village level meetings.
- Panchayat (local government) level public meetings.
- Through school students to their parents and relatives.
- Being a part of the local functions and playing role as a facilitator to the agriculture and health programmes conducted by the local Government institutions.
- Involving CBOs and youth clubs in villages.

CIC Awareness Campaign at Community Temple, Barapadar (Badhinuapalli CIC)

CIC Awareness Campaign at Community Temple, Barapadar (Badhinuapalli CIC)

CIC Coordinator speaking at CIC Awareness Campaign, Kanheipur (Kanheipur CIC)

Speaking to the local people at Kanheipur about CIC Vision, Mission and Goal

Women from the project area participated Community Awareness Campaign on CIC

Women from the project area participated Community Awareness Campaign on CIC

(d) Connectivity

The most feasible and cost-effective system has been used for connectivity. Connectivity and content is receiving concurrent attention of CICs. They are used to satisfy the local community.

Telephony is still very relevant for rural access. Internet technologies offer new options to provide cheaper and more flexible services.

(e) Content

Creation and updating of relevant content to suit local needs is a key factor in the CIC program. The information provided is demand driven and relevant to the day-to-day life and work of rural men and women. Also semi literate women are being accorded priority in training to operate the center.

Contents are being delivered both by conventional and electronic means. Merely providing information is not enough. Knowledge dissemination is linked to access to the inputs needed to apply the knowledge for economic activities. The convergence between printed materials, radio, TV and Internet provides useful examples of how to create local content, relevant to local needs but also to local culture and provide this content in local languages.

CIC is providing mainly three types of contents namely, generic (health, government schemes, local news, employment news *etc.*), dynamic (market, school, examination results, weather *etc.*) and timely information (wave height, potential fishing zone details, *etc.*).

This model is designed to empower rural families with new knowledge and skills where the empowerment starts with the poorest and the most unprivileged men and women. The local population has a sense of ownership of the CIC. It is client managed and controlled. The local population including the local Gram Panchayat to make contributions towards the running expenses, so that long-term economic sustainability is ensured. But, it may take some time.

(f) Management, Monitoring and Evaluation

The CIC management committees have been established consisting of experts, local NGOs and Self-Help Groups, local Gram Panchayat, and members from the community including potential women volunteers. They are being oriented always with capacity building training and time to time cooperation to run the CIC smoothly.

(g) Sustainability

- The CIC is creating long-term, self-sustaining solutions, which reflects local needs and require local entrepreneurship. So, it helps to fuel the creation of additional local business and community enterprises.
- Local entrepreneurs and communities finding new and innovative ways for increasing production and economic avenues. Business at local public access points is being demonstrably increased.
- The CIC is vital in developing a rural urban e-Commerce service network.

(h) Key features of the CIC

- People focus and community ownership and involvement of women communities.
- Building multi-stakeholder partnership and networks.
- Managing content creation, gathering, validation and dissemination
- Managing technology and connectivity.
- Managing delivery of content on time.
- Managing for improving the well-being of the community.

(i) Convergences of other Programmes and Supports

- *Krishi Mela (Agricultural Fair)* : An agricultural fair organized at Badhinuapalli CIC on 14 July 2008, with the participation of 286 participants from project villages. The fair was organized by IRMA-India in cooperation with the Department of Agriculture, Govt. of Orissa as well as Department of Horticulture. The block level and district level government officials attended the fair. A video on improved paddy cultivation shown to the participants produced by the Agriculture Department. A small booklet on technical questions and answers on agriculture also distributed.

- *Education Fair* : An 'Education Fair' organized on 5th August, 2008 at our Badhinuapalli CIC. The students from the nearest schools were paricipated in the fair. Session wise computer literacy classes were conducted for them. During the fair, five students were identified for a continuous free computer literacy course for about three months. Two school teachers were also identified to take this training alongwith the students.

End Note

IRMA-India is on its mission on development, promotion and dissemination of Information and Communication Technology for empowering the rural communities both men and women and strengthening the rural livelihood system.

15

Expert System for Pest and Diseases Diagnosis in Rubber

N. Balasubramani

ABSTRACT

Expert Systems could be incorporated into extension education programmes to assist in solving problems using knowledge gathered from experts. During the process of problem solving, the Expert System becomes a source of educational material for the user. Hypertext may be incorporated into an Expert System as a means of providing the user with additional information. A hypertext approach to system development involves attaching information to specific phrases used in the system. The form of attached information can be textual, graphical or procedural. The methodology of developing expert system consists of two parts *viz.*, developing computer-based Expert System and testing its effectiveness. The first part includes selection of most relevant subject matter area and procedure used in the development of an Expert System. The second part deals with the research design and technique adopted to test the effectiveness of computer-based Expert System versus human expert.

Introduction

The information need of farmers on new crops, pesticides, farm machines and farming techniques is keep increasing over the years. In order to make accurate decisions, farm managers/ extension workers/ farmers need speedy access to advice on agricultural problems which could be timely, reliable and consistent. Hence, they have to resort on the subject matter specialists like agronomist, entomologist, plant pathologist, soil scientist, horticulturist, farm scientist *etc.* Though the subject matter specialists extend a willful service, the availability of resource persons, affordability of information in time and accessibility, all that matters a lot. It is in this juncture the hallmark in use of telecommunications and computer-based information technology in the era of globalization probably would be the best alternative and rather means for sea change in extension.

Development and promotion of Information and Communication Technology (ICT) in agriculture may supplement and compliment the extension system. Our extension strategy should not only be depending on conventional extension methods like, trainings, demonstrations, Radio and TV broadcasts. It should make a radical drift into computer-based information technology in agricultural extension (FAO, 1993; Zijp, 1994). This will bring new information services to the farming community and would bring about a significant change in agriculture in terms of reduced costs, increased storage of information, easy usage, rapidity *etc.*

Computer technology in its recent advancement focuses on software programmes that are available to assist in filling the knowledge of experts for analysis and design of complex problems. Knowledge based computer programmes or Expert System containing "expert knowledge" would help access information to solve these problems. It would further improve the access to scarce human expertise. It would also be a promising means of providing information.

The branch of science known as artificial intelligence covers a number of field application. Expert System is one such field, which has attracted significant attention in recent years. However, agricultural research and extension have been devoting too little attention to this field. There is great potential for research in the field of artificial intelligence and this should not be completely ignored by those engaged in agricultural research and extension. In the changing agricultural scenario, agricultural field has not been computerized so far in a way it is demanded. However, in the last decade, artificial intelligence based computer programmes called Expert System received a great deal of attention throughout the world, due to its impressive problem solving capability in a variety of fields.

The Expert System (ES) has mostly been confined to USA, Canada and Europe, whereas in developing countries, the application of ES technique in agriculture is insignificant. The application areas of Expert System will be plenty. It has enormous scope to help the farming community by enabling them to take sharp decision.

Meaning of Expert System

An Expert System (ES) is a computer programme that is designed to emulate the logic and reasoning processes that an expert would use to solve a problem in his/her field of expertise, using artificial intelligence technology (Waterman, 1986). It performs many functions as an expert does, such as posing relevant questions and explaining its reasoning process. It further interacts with human beings that the content matches with the comprehension and good understanding.

An ES tends to behave as human experts in decision-making and is highly interactive. It has the ability to capture human decision-making expertise, interactive and represent this expertise as a series of rules and facts. It will minimize or avoid errors in complex tasks, protecting the perishable knowledge of experts, and make it available when and where required. It will consider all possible alternatives, display unbiased judgments, available readily for use unlike human experts and less expensive to consult than human experts. One of the most exciting features of ES is the availability of this very sophisticated computer technology for immediate practical use by the entire agricultural community.

Need and Application of Expert System

The potential for application of ES in agriculture is infinite as the vast network of scientists working in India at Central Institutes, Project Directorates, National Research Centers, All India Coordinated Research Projects, Krishi Vigyan Kendras, Agricultural Universities and so on so forth. The experience and expertise of these vast networks of scientists may be captured by Expert system and applied in a broader range of problems and situations in agriculture.

The conventional systems may be substituted and supplemented with Expert System, as because it is a blend of information catering to the diverse needs of farmers facilitating for all possible questions and relevant answers in an inbuilt mechanism. Hence, Expert System would help in a big way to solve the problem reposed by the farming community as a holistic service device.

Information gap could be possibly reduced by Expert Systems which can emulate the problem-solving logic of human experts are potential tools to aid in training, particularly in diagnosis and pest management decision-making.

Expert Systems could be incorporated into extension education programmes to assist in solving problems using knowledge gathered from experts. During the process of problem solving, the Expert System becomes a source of educational material for the user. Hypertext may be incorporated into an Expert System as a means of providing the user with additional information. A hypertext approach to system development involves attaching information to specific phrases used in the system. The form of attached information can be textual, graphical or procedural.

Developing an Expert System

Expert's knowledge is the central and key component of developing an Expert System. The knowledge acquisition must be regarded as much as an art as a methodical and scientific procedure. One approach, however, that often seems to be ignored is the collection, integration and use of research results, while an expert should clearly build research results into their expertise.

Experience suggests that, in addition to potential users contributing to the knowledge base, their involvement produces a valuable, lasting outcome. The users need to be involved at several stages of development of Expert system. The group discussion, consultation, and gathering of attitudes of potential users on an individual basis would be more essential in developing an expert system. It is essential to ensure that each user is personally 'heard' while developing of ES.

There are two ways of building Expert System: one is to develop "from scratch: *i.e.* to code the Expert System as a normal computer programme for each domain using Artificial Intelligence (AI) programming languages like CLIPS, PROLOG, LISP *etc.*, and the other is to use an Expert System shell *i.e.* to build an Expert System with the help of specially designed programmes that are commercially available which may be used for a particular domain. The shell enables the user to build their own Expert System with little or no help from knowledge engineers. Thus shells can make considerable saving on programming time. Because of this, building Expert System can be faster and more economical. Hypertext proved to be a useful method of providing additional educational information to the Expert System user and should be considered in the development of Expert Systems for extension education.

Survey of Expert System in Agriculture

The works carried out on Expert System in agriculture and allied sectors worldwide were collected and presented as follows:

Sl.No.	Authors	Name of Expert System	Utility	Software/ Shell used
1.	Fermanian *et al.* (1985)	PLANT/tm	Diagnosis of weed in turf	-
2.	Jones and Haldeman (1986)	CHAMBER	Management of environmentally controlled crop research facility	-
3.	Lemmon (1986)	COMAX	ES for cotton crop management	-
4.	Palmer (1986)	COMAX	Soybean crop variety selection	PROLOG
5.	Shroyer *et al.* (1987)	WHEAT WIZ	Cultivator selection tool	-
6.	Bennett and Sneed (1988)	COMAX	Planning, design and evaluation of irrigation systems	PASCAL
7.	Floris *et al.* (1988)	COMAX	Real-time operation; real-time meteorological data handling	PASCAL
8.	Getforth and Macvicer (1988)	OASIS	Operation of control structures; real-time meteorological data handling	PASCAL
9.	Haie and Irwin (1988)	EXSYS	Drainage diagnosis	PASCAL
10.	Halterman *et al.* (1988)	ES	Double cropping management	-
11.	Boggess *et al.* (1989)	FinsARS	Financial analysis for farm business management	-
12.	Stone and Toman (1989)	COT FLEX	Cotton crop management; coupled with SOYGRO model	PASCAL
13.	Batchelor *et al.* (1989)	SMART SOY	Soybean crop management	Insight 2+
14.	McClendon *et al.* (1989)	SMART SOY-IRRIG	Soybean irrigation	Insight 2+
15.	Morgan *et al.* (1989)	CUE	Crop variety selection	SELECT
16.	Hart *et al.* (1989)	CUE	Irrigation system selection	LISP
17.	Hershaeur *et al.* (1989)	CUE	Canal water distribution; canal network incorporated	LISP
18.	Bhatty (1990)	RESEXP	Reservoir operation; DP model integrated	PROLOG
19.	Helms *et al.* (1990)	CIRMAN	Crop insurance strategies	-
20.	McGregor and Thornton (1990)	CVSES	Wheat crop variety selection	CRYSTAL

contd......

21.	Oswald (1990)	TANK	Tank systems diagnostic analysis	PROLOG
22.	Han *et al.* (1991)	ES	Sprayer diagnostics	-
23.	Hasbini *et al.* (1991)	PUMP	Operational guidelines for center pivot systems	PASCAL
24.	King *et al.* (1991)	MKBS	Fertilizer and irrigation applications	Turbo C
25.	Nevo and Amir (1991)	CROPLOT	Multiple crop selection	Rabbi
26.	Srinivasan *et al.* (1991)	ESIM	Delivery system operation; canal network incorporated	EXSYS
27.	Clarke *et al.* (1992)	IRRIGATOR	Irrigation scheduling; ET method selection	PC PLUS
28.	Elango *et al.* (1992)	BDM-EXPERT	Drought management integrated with CASIMBOL model	IITM RULE
29.	Kumar *et al.* (1992)	KBS	Economic feasibility of irrigation system selection	Level 5
30.	Nakamura and Tsukiyama (1992)	ES	Irrigation canal renovation project planning	-
31.	Plant *et al.* (1992)	CALEX/cotton	Cotton irrigation scheduling	CALEX
32.	Raman *et al.* (1992)	BDM-EXPERT	Crop planning under droughts; LP model inferencing	Insight 2+
33.	Bralts *et al.* (1993)	ES	Hydrologic analysis of micro irrigation system	-
34.	Mohan and Arumugam (1994)	CROPES	Multiple crop selection	IITM RULE
35.	Nevo *et al.* (1994)	CROPLAN	Optimal crop planning; LP model integrated	PROLOG
36.	Pasqual (1994)	ES	Identification and control of weeds in wheat, barley and oats	-
37.	Arumugam (1995)	TANKES	Tans system operational guidelines; real-time operation	VP-EXPERT
38.	Mohan and Arumugam (1995)	ETES	ET estimation method selection	VP-EXPERT
39.	Nuthall and Bishop-Hurley (1996)	-	ES for animal feeding management	VP-EXPERT

contd......

40.	Yialouris *et al.* (1997)	VEGES	A multilingual Expert System for the diagnosis of pests and diseases and nutritional disorders of six greenhouse vegetables	AUA-ES
41.	Ganesan (2002)	AGRES	Diagnosis of pests and diseases of major crops of Kerala	-
42.	Balasubramani (2004)	RUBEXS - 04	Disease Diagnosis in rubber plants	VB

Many computer-based decision support systems have been developed across the globe to help farmers to plan and manage various aspects of their farming operations. From the above table, we could understand that, the expert systems that address wide ranges of tasks such as nutrient management, pest management, soil conservation, energy conservation, financial management, water quality protection *etc.* Some of these systems address more than one dimension of the management task, but few allow simultaneous consideration of resource conservation, environmental protection, productivity, and profitability. The Expert systems developed by various experts clearly indicates that, different softwares were used to develop computer-based Expert System. Further, these were used as an effective tool in various fields of agriculture. But, It is essential to develop a user friendly computer based Expert System considering the flexibility, simplicity, nature of problem and familiarity of the software.

Methodology

The methodology consists of two parts *viz.*, developing computer-based Expert System on rubber plant and testing its effectiveness. The first part includes selection of most relevant subject matter area in rubber cultivation and procedure used in the development of an Expert System. The second part deals with the research design and technique adopted to test the effectiveness of computer-based Expert System versus human expert in terms of (i) Knowledge gain, (ii) Knowledge retention, (iii) Knowledge gain related to skill, (iv) Symbolic adoption and farmers perception on computer-based Expert Systems as a tool for technology transfer.

Selection of Farm Technology for the Development of Expert System

Based on the familiarity, rubber crop had been selected for the development of Expert System. Among rubber cultivation, a list of 10 important subject matter areas were identified from literature, discussion with scientists and extension personnel working in rubber cultivation and promotion. These 10 subject matter areas were grouped into three broad categories *viz.*, (i) rubber production technologies, (ii) protection technologies, and (iii) harvesting and post harvesting technologies. Thirty judges, 10 each from scientists, extension personnel working on rubber and rubber growers of non-sampling areas were identified. The judges were asked to rate these 10 subject matter areas on a three point continuum *viz.*, most needed, needed and not needed. The judges rated these subject matter areas based on their experience and expertise in the field of rubber as they felt necessary to develop Expert System for the rubber growers. The scores of the judges were tabulated and analysed. The rubber protection technologies with highest mean scores were selected for developing computer-based Expert System.

Designing of Expert System

The methodology of developing computer-based Expert System on rubber crop is divided into two main parts: knowledge engineering and software engineering. The knowledge engineering methodology included: acquiring the knowledge, analysing, modeling the acquired knowledge and verifying the modeled knowledge. An exhaustive knowledge base on plant protection technology of rubber including leaf diseases, stem diseases, root disease, non-microbial maladies and pests were acquired from various sources. The documented knowledge was analysed, grouped in a common knowledge base and it was represented in the form of flow chart, with the purpose of finding the domain knowledge which the experts were using to reach a conclusion from specific components in the domain layer. So, this inference analysis was aiming at modeling the acquired knowledge. The task analysis was done to find the sequence or the procedure, which the experts used to reach a final conclusion. The knowledge base was verified at the knowledge acquisition stage, analysis stage and implementation stage and got approved by the expert as a valid way for solving the problem.

Software engineering methodology : Different types of tools have demonstrated the enormous range of possible Expert System architectures that could be build around different representations and control regimes. However, flexibility, simplicity, nature of problem and familiarity of the language to the researcher were considered for selecting the Expert System architecture. With that, the programming language Visual Basic 6.0 was chosen for designing and developing the Expert System on rubber. The main advantage of Visual Basic is that it is a rapid application development tool. This is the main reason for Visual Basic's extensive use as a prototyping language. Visual Basic applications insulate from having to deal with low-level message handling.

Procedure Followed in Designing RUBEXS-04

The computer-based RUBber EXpert System was abbreviated as RUBEXS-04. The numerical value 04 indicates, the year of designing *i.e.* 2004. The diagram given below represents the process of development of Computer-based Expert System RUBEXS-04 (Figure 1). The RUBEXS-04 was programmed using Visual Basic 6.0. It is a graphical user interface, which uses graphics to organise workspace. It also uses event-driven architecture. The interface comprises a form with controls. The step by step procedure followed in designing RUBEXS-04 is given below.

Steps Followed in Designing RUBEXS-04

- The user interface was created by drawing controls such as text box, labels, option button, command button *etc.*, on a form. Each form was carrying single question of a cause or about a situation coupled with dichotomous options of Yes or No. A form is one of the main building blocks in this application. Users interact with the controls on a form to obtain the desired result. The form becomes the first on-screen window a user sees while running a programme. In order to make the programme easier to read and debug, appropriate names were assigned to the objectives.
- The properties were set for the form and controls to specify values such as the caption, colour, size *etc.*
- Codes were written to bring the application to life or after setting the initial properties for the form and each object, codes were added that executes in response to events.
- HTML template - An Hypertext Markup Language (HTML) page is designed to act as a template; some parts of the page are replaced with dynamic content before the page is displayed especially, advice and more details of the cause are designed to act as a template.
- Describing symptoms in words is very difficult and sometimes is very confusing. Therefore, necessary images were inserted into relevant forms, with the main purpose of describing a disorder symptom, and confirming the diagnosis of the cause of a certain disorder by the user.
- The voice is recorded and edited using a software sound forge. It is one of the best sound editing and recording software, and the background sound was given to add flavour to the multimedia presentation. It was developed both in local language and also in English.
- After the project has been created, it has been saved using the 'Saved Project As' menu item on the 'File' menu option. Saving a project also saves its form and code modules.

- Debugging tools were used to check for errors and modified the code whenever necessary.

Development of Expert System

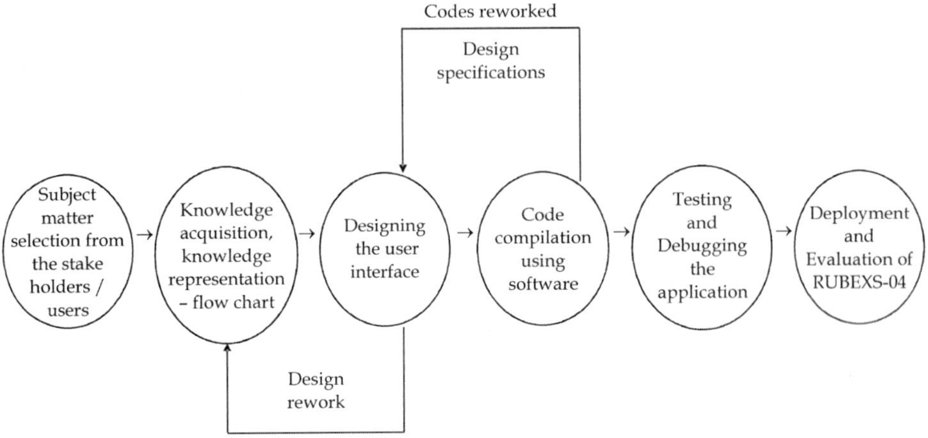

Figure 1 : Development Process of RUBEXS-04

Functioning of RUBEXS-04

The RUBEXS-04 will run form by form. Each form contains one question related to diagnosis of pest and diseases of rubber crop. Under each question, there will be two control buttons with the option of Yes or No. User has an option to click either or command buttons based on their experience and prevailing situation about pest and disease in their rubber plantations. Based on their answers, RUBEXS-04 takes the users to next relevant questions. Likewise, it keeps on interact with the user and finally gives the diagnostic result. In the diagnostic result form two more additional control buttons are available with the caption and Click on "Advice" button tells about the measure to control the problem diagnosed and click on "More" detail button gives more information about the cause diagnosed by the RUBEXS-04.

Locale of the Research

The traditional rubber growing area and familiarity of the locality to the researcher is considered for selecting the locale of the research in the deployment of RUBEXS-04 and testing its effectiveness.

Sampling Procedure

For the effective transfer of technology and for empowerment of the small rubber growers the Rubber Board has promoted formation of grass root level organisation/ institution building at the village level to be known as Rubber Producer's Society (RPS). All small rubber growers having their holding located within the operational area of the concerned RPS can be members. The members in each RPS varied from 50 to 200. At the end of 2000-01 there were 2,100 RPSs registered in Kerala (Rubber Grower's Companion, 2004). There were 60 RPSs functioning under the Rubber Board, Regional Office, Mannarkkad.

Since, RPSs were envisaged to function as extension arms of Rubber Board and to facilitate a meaningful devolution of extension functions to render effective service to the small growers, the RPSs were considered as appropriate and potential center for exposing the treatments. Out of 60 existing RPSs under Mannarkkad Regional Office, the three RPSs were randomly selected. These were also active, well functioning and representing the entire Regional Office.

Selection of Respondents

RPSs were registered under charitable societies and formed based on a model by-law drawn by the Rubber Board, Kerala. Only rubber growers were permitted to become member in RPS. Hence, membership registers were maintained at each RPS. Based on the membership registers, wide publicity were given to all the members of the selected RPSs to participate in the "Special diagnosis campaign" organised in the stipulated date, time and venue of the respective RPSs.

Accordingly from each RPS, rubber growers ranging between 60 and 75 had participated in the campaign. From these participants a sample of 40 rubber growers from each RPS concerned were selected randomly. Hence, a sample size of 120 rubber growers was the total sample for the study.

Selection of Experimental Design and Treatments

Multiple Group Randomised (MGR) design was adopted in this study. This design could enable the researcher to establish adequate relationship between the dependent and the independent variables. Moreover, this is the classical experimental design containing procedures that are more accurate.

Keeping the objectives and hypothesis in mind, four different treatments such as human experts without discussion, human experts with discussion, RUBEXS-04 without discussion, and RUBEXS-04 with discussion was selected. These treatments were tested for their relative effectiveness using the MGR design.

Each treatment was replicated thrice. Considering 10 respondents per replication, there were 30 respondents per treatment. Thus a total of 120 respondents were the subjects for the four treatments.

Accordingly, 12 randomised groups were selected from the three RPSs to find out the relative effectiveness of the four treatments in terms of (i) knowledge gain, (ii) knowledge gain related to skill acquisition, (iii) symbolic adoption and, (iv) knowledge retention of the subjects.

'Before-After' technique of measurement was used to find out the effect of a particular treatment and hence a separate control group was not taken into consideration.

In this design there may be a possibility for the occurrence of heterogeneous nature of the subjects among the experimental groups. Hence, it was decided to apply Kruskal-Wallis 'H' test, to find out whether there existed significant difference among the subjects selected for the four treatments in respect of their profile characteristics. It was observed that, there was a non-significant 'H' value which indicated the existence of homogeneity among the subjects selected for the experiment from the three different RPSs. In other words, the sub-groups involved in this study were drawn from the same population and so the experimental groups selected by multiple random group design were justified. Thus, it was evidenced that the different groups involved were matching in all respects.

Selection of the Treatments and their Assessment

It was decided to have four treatments comprising of 12 experimental groups. Each group consists of 10 randomly selected subjects. All the four treatments were replicated thrice in the three selected RPSs. Thus, four treatments were allotted randomly to 12 experimental groups to find out the relative effectiveness of four selected treatments in terms of (i) knowledge gain, (ii) knowledge related to skill acquisition, (iii) symbolic adoption and (iv) knowledge retention.

Accordingly, the selected treatments for the study were as follows:

Treatment-1 (T_1) : Human Expert without discussion

Treatment-2 (T_2) : Human Expert with Discussion

Treatment-3 (T_3) : RUBEXS-04 without discussion

Treatment-4 (T_4) : RUBEXS-04 with discussion

In each RPSs, the four treatments were assigned randomly to the four groups comprising 10 subjects each, thus adopting the principle of randomization to increase the precision/ validity of the experimental results.

Details of Number of Contacts Made

i)	At pre-exposure stage (knowledge gain, knowledge related to skill acquisition)	:	120
ii)	At immediate post exposure stage (knowledge gain, knowledge related skill acquisition and symbolic adoption)	:	120
iii)	At knowledge retention after 15 days	:	120
	Total number of contacts made		360

Control group : Since the base level knowledge was assessed, there was no need to have control groups

Experimental Procedure

Exposure of Treatment (T_1): Human Expert without Discussion

After deciding the treatments, the research programme was fixed with senior level scientists of Rubber Research Institute of India (RRII), Kottayam. An Entomologist and Pathologist, had more than 20 years of experience as scientists in their respective field, to serve as resource persons for the selected treatments (T_1 & T_2) was arranged. The scientists here after would be referred as Human experts. The human experts have been apprised about the treatments and number of pest and disease to be covered for diagnosing the pest and disease of rubber crop. After briefing about the treatments to the human experts, the initial knowledge level, knowledge related to skill and other data regarding independent variables of the subjects were secured. Then the subjects were exposed to T_1. The experts had explained the subjects step by step in their usual way of diagnosing the major pest and disease of rubber crop based on its symptoms and other factors. They also explained the suitable measures to be taken up to control the pest and disease. In treatment (T_1), the subjects were not entertained to discuss and clarify their doubt. This section was followed by the assessment of gain in knowledge and knowledge gain related to skill by the subjects. The difference in the knowledge and skill levels before and after exposure was taken as knowledge and skill gained by the subject. The mean knowledge and skill gained by subjects for each treatment were compared by employing analysis of variance technique to draw inference.

After the exposure of the subjects to the treatments pertaining to knowledge and knowledge related to skill level, the subjects were assessed for their symbolic adoption behaviour. The subjects were also assessed for quantum of technical information retained after 15 days of exposure to T_1.

Exposure of Treatment (T_2): Human Expert with Discussion

The same methodology was followed as that of T_1. But, immediately after exposure, the subjects were allowed to participate in the discussion forum with the human experts for a period of 15 to 20 minutes.

Exposure of Treatment (T_3): RUBEXS-04 without Discussion

Pre-exposure knowledge level, knowledge related to skill and other data regarding independent variables of the subjects were assessed. The purpose of RUBEXS-04 and procedure to be followed in operating RUBEXS-04 were explained to the subject. After orientation, the treatment was exposed to the group of subjects through Liquid Crystal Display (LCD) projector, form-by-form covering the major pest and disease which affecting the rubber crop. It was strictly monitored that no interaction took place between subjects as well as with researcher. After exposure of the subjects to the treatment, the respondents were assessed for knowledge gain, knowledge gain related to skill and symbolic adoption. The subjects were also assessed for their quantum of knowledge retained after 15 days of exposure to treatment.

Exposure to Treatment (T_4) : Computer-based Expert System with Discussion

The methodology followed to expose the treatment (T_4) was similar to that of treatment (T_3). But, here the researcher allowed the subject to discuss among themselves and also with the researcher to clarify their doubts for a period of 15 to 20 minutes after exposure. Pre and post exposure of the subjects to the treatment pertaining to knowledge gain, knowledge gain related to skill and positive decision taken by them to adopt for effective control of pest and disease in their plantation were assessed. Similarly, the extent of knowledge retained after 15 days of exposure to treatment were also collected from the subjects.

Tools and Techniques of Data Collection

A pilot study was conducted to pre-test the interview schedule to test verify the applicability and understandability of the content. Based on the response, the interview schedule was perfected by incorporating the relevant and deleting the irrelevant items to meet the objective of the study. The data regarding the profile of the participant farmers, perception and familiarity about modern information communication technologies were collected from the respondents with the help of a well-structured and pre-tested interview schedule. Two interview schedules containing the standardised knowledge test were used among the participant-respondents twice *i.e.*, before and after exposure to the knowledge and skill level of treatments for collecting the relevant data. The interview schedule containing the same knowledge test was used after 15 days of exposure to treatments to collect the data regarding the knowledge retention among the participant farmers.

The study was conducted as per the methodology described earlier. The effectiveness of four selected treatments in terms of knowledge gain, knowledge gain related to skill, symbolic adoption and knowledge retention were studied on two dimensions, namely, the significant change, and the relative effectiveness of the chosen treatments on the subjects as a result of exposure to the treatments.

Effectiveness of Treatments in Terms of Knowledge Gain

The information on major pest and disease of rubber affecting the rubber crop and its control measures was chosen as a subject matter area and were exposed to the experimental groups with four different treatments. The knowledge level of the subjects before and soon after the exposure to the treatments were analysed to find out the knowledge gain. The maximum attainable knowledge score for each treatment was 21. Paired 't' test was applied to find out, whether there was any significant difference between the pre-exposure and immediate post exposure gain due to the four treatments.

It was evident from the result that all the four selected treatments namely, human expert without discussion (TK1), human expert with discussion (TK2), RUBEXS-04 without discussion (TK3) and RUBEXS-04 with discussion (TK4) had highly significant 't' value indicating that all the four treatments were effective in terms of knowledge gain.

The mean knowledge gain was maximum with a score of 6.80 in RUBEXS-04 with discussion which indicated 32.38 per cent of knowledge gain. This was followed by human expert with discussion with a score of 3.97, human expert without discussion (2.74) and RUBEXS-04 without discussion (2.67) which resulted in 18.90 per cent, 13.04 per cent and 12.71 per cent knowledge gain respectively.

It was observed that RUBEXS-04 with discussion (TK4) was found to be most effective and superior in imparting knowledge. This was followed by human expert with discussion (TK2), human expert without discussion (TK1) and RUBEXS-04 without discussion (TK3) which were found to be less effective with regard to imparting knowledge to the subjects in that order. In RUBEXS-04, the delivery of information through text, pictures and audio were tailored to its users to learn at their pace. The picture was an 'eye catcher' which would have made instantaneous and strong emotional appeal, would have attracted the attention of the subjects and made them more receptive to the idea, which was exposed. The principle of "seeing is believing, holds good, because one picture was worth more than thousand words". The RUBEXS-04 enabled the subjects to visually observe as well as to hear messages not only aroused interest among the learners but also enriched learning situation by sustaining interest, thus promoting better learning in terms of mean knowledge gain. The RUBEXS-04 followed with discussion forum helped the subjects to confirm the field problem such as tapping panel dryness, pink disease *etc.*, which were observed in their own situation when the same were exposed through RUBEXS-04 and by way of holding discussion with the extension officer present in the situation. Hence, discussion forum had paved way for clarification of doubts, better comprehension of the subject matter and conviction over the ideas too. This might be because of the higher number of senses of the learners involved in combination and hence the result was better compared to other treatments.

Effectiveness of the Treatments in Terms of Knowledge Gain Related to Skill Practice

It is observed that the mean knowledge gain related to skill was maximum with a score of 4.66 which accounts for 46.60 per cent of the knowledge gain related to skill when the subjects were exposed to RUBEXS-04 with discussion (TS4). This was followed by the treatments namely human expert with discussion (TS2) with the mean score of 4.43, human expert without discussion (TS1) with the score of 3.83 and RUBEXS-04 without discussion with the score of 3.70 which accounted for 44.30 per cent, 38.30 per cent and 37.00 per cent of the knowledge gain related to skill respectively.

It was also noticed that more number of subjects (96.67 per cent) had gained knowledge related to skill when they were exposed to RUBEXS-04 with discussion (T_S4) compared to other three treatments. In the treatment T_S2, the percentage of subjects who changed their knowledge level was 90.00, followed by T_S1 with 83.33 per cent. Whereas the least knowledge related to skill (73.33 per cent) was gained when the subjects were exposed to T_S3. This may be the reason that the RUBEXS-04 was exposed through computer in combination with LCD projector these two were the novel media to most of the subjects coupled with the appropriate pictures of disease and pests in the RUBEXS-04 followed by discussion which might have attracted the attention of the subjects and motivated them to learn more.

Relative Effectiveness of Different Treatments in Terms of Knowledge Gain Related to Skill Practice

It was observed that among the four treatments, RUBEXS-04 with discussion was found to be the most effective and superior to impart knowledge related skill about plant protection aspects of rubber crop, followed by human expert with discussion (T_S2), human expert without discussion (T_S1) and RUBEXS-04 without discussion (T_S3) in that order of succession.

The reason may be due to fact that the user friendly and less time required to acquire the operational skill of RUBEXS-04 along with discussion to clarify the doubts about their real problems might have attracted the subjects to involve themselves in learning activities and thereby they would have gained more knowledge related to skill compared to other treatments.

Effectiveness of Treatments in Terms of Retention of Knowledge

The retention of knowledge after 15 days of exposure was assessed and those scores were compared with the quantum forgotten. It was observed that the treatment T_r4 had the highest retention of knowledge gained (72.35 per cent) followed by treatment T_r3 (65.64 per cent), treatment T_r2 (46.85 per cent) and treatment T_r1 (28.47 per cent). This is in other words the quantum of forgetfulness was higher in treatment T_r1 (Human expert without discussion) and lowest in treatment T_r4 (RUBEXS-04 with discussion).

These results clearly indicate that the treatment T_r4 was found to be superior and most effective in terms of knowledge retention compared to all other three treatments.

Relative Effectiveness of the Treatments in Terms of Symbolic Adoption

Result indicated that all the four treatments had influenced the subjects evenly for symbolic adoption of plant protection of rubber crop.

"SWOT" Analysis of Expert System

SWOT analysis was used to assess the strengths, weaknesses, opportunities and threats of using computer-based Expert System in agriculture. The subject's responses were quantified using percentage analysis and the results of the SWOT analysis is furnished under four sub headings.

Strengths of Expert System

The subjects responses with regard to Expert System were collected, analysed based on their perceived importance and presented in the Table 1.

It could be inferred from the results presented in Table 1 that an overwhelming 86.67 per cent of the subjects reported that practical experience and knowledge of human experts could be stored and used for longer period of time was the primary strength of Expert System. It is possible to accumulate and integrate the knowledge and expertise from diverse sources like agricultural specialists and advisors especially the yeomen experiences of the retired experts and their problem solving behaviour could be well simulated in agriculture.

Table 1 : Strengths of Expert System

(n=60)

S.No.	Items	Number	Per cent
1.	Practical experience and knowledge of human expert can be stored and used for longer period of time	52	86.67
2.	Helps for quick decision making	49	81.67
3.	Self learning is possible	46	76.67
4.	Highly convinced about diagnostic result	44	73.33
5.	Highly interactive	39	65.00
6.	Saves money	36	60.00
7.	Saves time	36	60.00
8.	Reduces the dependency of human expert	23	38.33

Unfortunately agricultural specialist's assistance would not be always available to the farmers whenever required. In order to alleviate this problem and remain competitive, 81.67 per cent of the respondents identified Expert Systems as powerful tool with extensive potential to help in quick decision making in agriculture.

Slightly more than three-fourth (76.67 per cent) of subjects were generally regarded that self learning had been possible through Expert System. Moreover, they believed that the navigation tools available in the Expert System could be successfully used and could operate the system with little support due to its user friendly nature.

It is clear from the Table that nearly three-fourth (73.33 per cent) of the subjects in general convinced and agreed with the advice and conclusions of the Expert System, and similarly the advice and more details offered on plant protection aspects of rubber crop. Slightly less than two-third (65 per cent) of the subjects expressed the strength of Expert System as "highly interactive". Because it made the subjects as active participants by keep on asking the questions related to the problem with regard to plant protection aspects of rubber till the diagnosis.

The result indicates that more than half of the subjects (60 per cent) expressed that with the correct use of RUBEXS-04, the farmers will be able to solve simple disorder, pest and disease of rubber crops themselves and so call upon the services of the specialists and extension officers only in the case of doubt or more serious difficulty. They also indicated that it may help to reduce the time and expense involved in extension visits to deal with relatively easily identifiable problems.

It is also evident from the Table that slightly more than one-third of the subjects expressed that Expert System reduces the dependency of human experts. The remaining 61.67 per cent of the subjects felt that Expert System cannot totally replace the human experts. They further explained that the Expert System would support them with additional information and knowledge.

Weakness of Expert System

The weaknesses expressed about the Expert System were analysed and is furnished in Table 2.

Table 2 : Weakness of Expert System

(n=60)

S.No.	Items	Number	Per cent
1.	Required computer and other accessories	60	100.00
2.	Needs regular update, whenever situation changes	59	98.33
3.	Skilled personnel required to update the knowledge base of the Expert System	58	96.67
4.	Computer literacy is required to operate	18	30.00
5.	Useful for only literate farmers	17	28.33
6.	Need power supply	16	26.67
7.	Language barrier	13	21.67
8.	Low credibility in the diagnostic result	12	8.37

It is noticed from the above Table that cent per cent of the subjects expressed that, computer and other accessories are required to utilize Expert System. It was followed by 98.33 per cent of the subjects felt the next most weakness as it required regular update, whenever situation changes. Because the Expert System was designed considering the present situation further it calls for updating to meet the changing situations.

The other weakness of Expert System was identified by 96.67 per cent of the subjects, who opined that requirement of skilled personnel to update the knowledge base of the Expert System. Less than one-third of subjects identified other weaknesses to Expert System in the order of succession: computer literacy to operate, useful only to literate farmers, need power supply, language barriers and low credibility in the diagnostic result.

Opportunities in using Expert System

The subjects had expressed some of the opportunities in using Expert System. The results are analysed and presented in Table 3.

Table 3 : Opportunities in using Expert System

(n=60)

Sl.No.	Items	Number	Per cent
1.	Steady growth of information communication would help to use the Expert System	49	81.67
2.	Declining trend in the cost of computer and other accessories	47	78.33
3.	Possible to bridge the knowledge gap between scientist and farmers	46	76.67
4.	Widespread computer literacy noticed among children's of farmers in rural area	40	66.67
5.	Literacy level of Indian farmers is at an increasing pace	38	63.33
6.	Potential to develop Expert System in local languages	32	53.33
7.	Solar power/battery operated back ups can be used to run the computer system	29	48.33
8.	Possible to update the knowledge base of the Expert System	22	36.67

It was observed from the Table 3 that 81.67 per cent of the subjects were expressed the foremost opportunities, as the steady growth of information communication would help to use the Expert System in future. It was followed by declining trend in the cost of computer and other accessories were the next important opportunity expressed by the 78.33 per cent of the subjects. This may be due to the reason that the information technology (IT) continues to develop at an ever increasing rate, access to IT technology by farmers becomes easier and some products and services like personal computers, cell phones and internet, becomes little cheaper. Hence the subjects might have realised that the benefit of IT sooner than later would be reaped even by the illiterates and disadvantaged.

Slightly more than two-third of the subjects (76.67 per cent) expressed that there is an opportunity to bridge the gap between scientist's knowledge and farmers through Expert System. This might be due to the fact that the opportunities in applying Expert System in agriculture is almost unlimited as the vast network of agricultural scientists working at SAU, Zonal Research Stations, Regional Stations, Central Institutes, Project Directorates, National Research Centres, All India Coordinated Research projects, KVKs and other Research Centres have been generating knowledge in their respective field, their experiences and expertise could be captured and used through Expert System and thereby the gap between scientists knowledge and farmers could be possibly narrowed down.

The other opportunities in using Expert System were widespread computer literacy observed among childrens' of farmers in rural area, literacy level of Indian farmers is at an increasing pace, potential to develop Expert System in local languages, solar power/battery operated back ups can also be used to run the experts system and it is possible to update the knowledge base of the Expert System, in that order.

Threats in using Expert System

The subjects had expressed some of threats in using Expert System. The results are presented in Table 4.

Table 4 : Threats in using Expert System

(n=60)

S.No.	Items	Number	Per cent
1.	Wrong knowledge gathering and representation lead to wrong conclusion	52	86.67
2.	Can't rely upon the ICT, sometimes makes operational problems	26	43.33
3.	Viruses causes file corruptions and hang over of system	11	18.33

It is interesting to note from the Table 4 that a majority (86.67 per cent) of the subjects had expressed the greatest threat in Expert System was wrong gathering of knowledge and representation lead to improper diagnosis of the problem. This may be due to the fact that the collected knowledge might include many facts, principles, prejudices, beliefs and concepts that are not easily and readily available for decision-making. This specialized knowledge is stored in the knowledge base which is the heart of an Expert System. Hence, the correct decision-making totally depends on the appropriate representation techniques used for representing this principal knowledge.

The other threats identified by the subjects were could not rely upon the information communication technology. Operational problems like viruses causing file corruption and hang over of system. This may be due to the fact that at any time system get stuck and hang. The subjects had experienced this problem when they were exposed to treatments.

Farmers' opinion about feasibility of Expert System

Through their general acceptance of the Expert System the subjects inherently gave their preference of using Expert System in various fields as a tool. The data with regard to their opinion is analysed using percentage and the results are furnished in Table 5.

Table 5 : Farmers Opinion Regarding the Feasibility of Using Expert System in Agriculture (n=60)

S.No.	Items	Number	Per cent
1.	As diagnostic tool	58	96.67
2.	As training / education tool	55	91.67
3.	As extension tool	53	88.33
4.	As decision support system	51	85.00
5.	As management tool	49	81.67

* Multiple responses

It is explicit from the Table that, 96.67 per cent of subjects opined that, the Expert System could be used as a diagnostic tool. This may be due to the fact that, the RUBEXS-04 was reliable in diagnosis, the data seemed to be adequate in reaching a well-accepted diagnosis.

A majority of the subjects felt that the Expert System could also be used as training/education tool. This may be due to the fact that the subjects might have had a favourable experience with RUBEXS-04 which was augmented with multimedia capabilities which would have enhanced the understanding of the subjects.

More than three-fourth of the subjects expressed that there is a feasibility to use the Expert System in extension, decision-making and as management tool in agriculture. This is because, the hyper text may be incorporated into Expert System which has been proved to be a useful method of providing additional educational information to the user of Expert System and also helps the field extension officers to disseminate technology with confidence.

Suggested Model for Design and Effective Utilization of Expert System

Based on the experiences and empirical evidence evinced in this experimental study the following stages are proposed for design and effective use of Expert System in Agriculture.

Stage I : Identify the major and location specific problems in agriculture with its stakeholders through participative approaches.

Stage II : Collect the practical experience of the experts in the concerned field and available literature on the particular subject under study.

Stage III : Represent the collected knowledge base appropriately according to the various branches of subject matter.

Stage IV : Choose the simple and user friendly software from the bundle of available softwares.

Stage V : Consider the technical components of expert system such as sequential and logical diagnostic path, field reality of the problems, use of local language and simple words, use of adequate and appropriate photographs and animation, insertion of audio *etc.*

Stage VI : Validate the developed Expert System with its clients for its refinement.

Stage VII : Use the developed Expert System in an interactive mode with its clients with preliminary orientation about the Expert System.

Stage VIII : Place the Expert System in a common venue like office of the block level agricultural offices/ Krishi clubs. This Expert System may be put to operation and execution by the self help groups/ commodity groups/ farmers organisations.

Stage IX : Identify and train local diploma / graduates in agriculture/ Extension officers about the usage of expert system

Stage X : Dissemination of farm technologies using Expert System by the Officers of agriculture and allied departments or trained individuals to achieve effective transfer of technologies in terms of better knowledge gain, retention, knowledge related skill, symbolic adoption and thereby adoption.

Stage XI : Based on the feedback and emerging problems in the agro-ecological system the stages may be reconsidered to design new Expert Systems to suit the changing needs.

Conclusion

There is an ample scope for computer based Expert System to be widely developed and used in the field of agriculture. The application of Expert System had already demonstrated the value across the globe as diagnostic tool, training/education tool, extension tool, decision support system, and as management tool. The expert system do not replace people, but serve as intelligent assistant, improve the quality and productivity of decision-making in farmer's field.

References

Arumugam, N., 1995. Expert Systems for Irrigation Management. *Ph.D. Thesis (Unpublished)*, Department of Civil Engineering, Indian Institute of Technology, Madras, India.

Balasubramani, N., 2004. Designing and Testing the Effectiveness of Computer- Based Expert System on Cognitive and Conative Domains of Rubber Growers, *Ph.D. Thesis (Unpublished)*, TNAU, Coimbatore.

Batchelor,W.D., M.E.Wetzstein and R.W.McClendon, 1989. Economic Theory and Expert System Information Technologies in Agriculture. *European Review of Agricultural Economics*, 18(2): 245-261.

Bennett, T.B. and R.E.Sneed., 1988. An Expert System for Irrigation Planning and Design. ASAE Paper No.88-5021. American Society of Agricultural Engineers, St. Joseph, MI.

Bhatty, M., 1990. Hybrid Expert System and Optimization Model for Multi-purpose Reservoir Operation. *Ph.D. Thesis*, Dept. of Civil Engineering, Colorado State University, Ft. Collins Co.

Boggess, W.G., P.J.Van, Blokland and S.D.Moss., 1989. FinARS : A Financial Analysis Review Expert System. *Agricultural Systems*, 31: 19-34.

Bralts, V.F., M.A.Driscoll, W.H.Shayya and L.Cao., 1993. An Expert System for the Hydraulic Analysis of Micro Irrigation Systems, 9(4): 275-285.

Clarke, N.D., C.S.Tan and J.A.Stone, 1992. Expert System for Scheduling Supplemental Irrigation for Fruits and Vegetable Crops in Ontario. *Can. Agric. Eng.*, 34: 27-31.

Elango, K., R.Honert, C.N.Kumar and V.Suresh, 1992. PC-based Management Game for Irrigated Farming. *Micro Computer Civil Engineering*, 7: 243-256.

FAO Production Year Book, 1993. 43: 8.

Fermanion, T.W., R.S.Michalski and B.Katz, 1985. An Expert System to Assist Turf Grass Managers in Weed Identification. In: Summer Computer Conference, 22-24 July, Chicago, IL American Society of Agricultural Engineers, ST Joseph, MI: 499-502.

Floris, V., D.Simons and R.Simons, 1988. Development of an Expert System for Mark Twain Reservoir Operation. In: Computerized Decision Support System for Water Managers. American Society of Civil Engineers, NY, USA.

Ganesan, V., 2002. Development and Field Testing of Expert System as an Aid to Agricultural Extension Work. Leaflet from Kerala Agricultural University, Vellanikkara, Thrissur.

Getforth, G. and T.Macvicer., 1988. An Operation's Advisor for Regional Water Management. In: Critical Water Issues and Computer Applications. American Society of Civil Engineers, NY, USA.

Haie, N. and R.W. Irwin, 1988. Diagnostic Expert Systems for land drainage decisions. *Irrigation and Drainage Systems*, 2(2): 139-146.

Halterman, S.T. J.R.Barrett and M.L. Swearingin, 1988. Double Cropping Expert System. *Transaction of ASAE*, 31(3): 234-239.

Han, Y.J., J.Boomaerts, S.Nugroho, M.G.D.Christenbury and F.J.Wolak, 1991. Development of an Expert System for Sprayer Diagnostics. *Applied Engineering in Agriculture*, 7(5): 22-27.

Hart, W.E., B.A.Ekholt and T.G.Kim, 1989. Irrigation System Selection. ASAE paper No.89-7042. American Society of Agricultural Engineers, St. Joseph, MI.

Hasbini, B.A., G.W.Buchleiter and H.R.Duke, 1991. Expert System for Improved Irrigation Management. Proc. Int. Summer Meet. American Society of Agricultural Engineers, Albuquerque, New Mexico, June 23-26, 1-17.

Helms, G.L., J.W.Richardson, M.J.Cochran and M.E.Rister, 1990. A Farm Level Expert Simulation System to Aid Farmers in Selecting among Crop Insurance Strategies. *Computers and Electronics in Agriculture*, 4(3): 169-190.

Hershaeur, J., A.Karim, H.Owens and A.Philipakis, 1989. A Field Observation Study of an Expert System Prototype Development. *Inform. Manage.*, 17: 107-116.

Jones, P. and J. Halderman, 1986. Management of Crop Research Facility with a Micro Computer-based Expert System. *Transaction of ASAE*, 29(1): 235-242.

King, J.P., I.Broner, R.L.Croissant and C.W.Basham, 1991. Malting Barley Water and Nutrient Management Knowledge-based System. *Transaction of ASAE*, 34(6): 2622-2630.

Kumar, D., C.D.Heatwole, B.B.Ross and T.A.Dillaha, 1992. A Knowledge-based System for Preliminary Selection and Economic Evaluation of Sprinkler Irrigation Systems. *Transaction of ASAE*, 35(4): 441-447.

Lemmon, H.E., 1986. COMAX : An Expert System for Cotton Crop Management. *Science*, 233: 29-33.

McClendon, R.W., W.D.Batchelor and J.E.Hook, 1989. An Expert Simulation System for Irrigation Management. Proc. Int. Winter Meet. American Society of Agricultural Engineers, New Orleans, LA, 12-15 December 1989.

McGregor, M.J. and P.K.Thornton, 1990. Information Systems for Crop Management: prospects and problems. *Journal of Agricultural Economics*, 41(2): 172-183.

Mohan, S. and N. Arumugam, 1994. CROPES: A Rule-based Expert System for Crop Selection in India. *Transaction of ASAE*, 37(4): 1355-1363.

Mohan, S. and N. Arumugam. 1995. An Intelligent Front-End for Selecting Evapotranspiration Estimation Methods. *Computers and Electronics in Agriculture*, 12(4): 295-310.

Morgan, O.W., M.J.McGregor, M.Richards and K.E.Oskouri, 1989. SELECT : An Expert System Shell for Selecting amongst Decision or Management Alternatives. *Agricultural Systems*, 31: 97-110.

Nakamura, R. and H. Tsukiyama, 1992. Analysis by an Expert System of Initial Irrigation Canal Renovation Project Planning. *Water Resource Research*, 6(3): 223-233.

Nevo, A. and I.Amir, 1991. CROPLOT : An Expert System for Determining the Suitability of Crops to Plots. *Agricultural Systems*, 37: 225-241.

Nevo, A., R.Oad and T.Padmore. 1994. An Integrated Expert System for Optimal Crop Planning. *Agricultural Systems.*, 45: 73-92.

Nuthall, P.O. and G.J. Bishop-Hurley, 1996. Expert Systems for Animal Feeding Management - Part II - Farmer's Attitudes. *Computers and Electronics in Agriculture,* 14: 23-41.

Oswald, O.,1990. An Expert System for the Diagnosis of Tank Irrigated Systems: A Feasibility Study. *Ph.D. Thesis,* Center for Water Resources, Anna University, Madras, India.

Palmer, R.G., 1986. How Expert System can Improve Crop Production. *Agric. Eng.,* 67(6): 28-29.

Pasqual, G.M., 1994. Development of an Expert System for the Identification and Control of Weeds in Wheat, Barley and Oat Crops. *Computers and Electronics in Agriculture,* 10(2): 117-134.

Plant, R.E., R.D.Horrocks, D.W.Grimes and L.J.Zelinski. 1992. CALEX/Cotton: An Integrated Expert System Application for Irrigation Scheduling. *American Society of Agricultural Engineers,* 35(6): 1833-1838.

Raman, H., S.Mohan and N.C.V.Rangacharya, 1992. Decision Support for Crop Planning During Droughts. *Journal of Irrigation and Drainage Engineering,* 118(2): 229-241.

Shroyer, J.P., S.C.Young and T.S.Cox, 1987. WHEAT WIZ: A Computer-based Cultivar Selection Tool. *Applied Agricultural Research,* 2(4): 242-247.

Srinivasan, R., B.A Engel and Pandyal. G.N., 1991. Expert System for Irrigation Management (ESIM). *Agricultural Systems,* 36: 297-314.

Stone, N.D. and T.W.Toman, 1989. A Dynamically Linked Expert-Data base System for Decision Support in Texas Cotton Crop Production. *Computers and Electronics in Agriculture,* 4: 139-148.

Waterman, D.A., 1986. A guide to Expert Systems. Addison-Wesley, Reading, MA.

Yialouris, C.P., H.C.Passion, A.B.Sideridis and C.Metin, 1997. VEGES. A Multi-lingual Expert System for the Diagnosis of Pests, Diseases and Nutritional Disorders of Six Green House Vegetables. *Computers and Electronics in Agriculture,* 19: 55-57.

Zijp, W., 1994. Improving the Transfer and Use of Agricultural Information: A Guide to Information Technology. *World Bank Discussion Paper,* No. 247, 105 p.

16

Development of "Citrus e-Clinic" Software

Sajeed Ali and Subrata Manna

ABSTRACT

Mandarin Orange (*Citrus reticulata* Blanco.), known as 'Suntala' to the local populace, is a major cash crop and was once a pride of Darjeeling hills. It is grown in approximately 930 ha with an annual production of 148.224 metric tones and annual turnover of Rs. 25 crores approximately. Since early 1970s, orange orchards in the Darjeeling hills started to be unproductive inflicting huge economic loss showing severe citrus decline. Gradually the malady assumed alarming proportions and threatening the cultivation of the crop. The major causes for the bottleneck of the citrus industry in Darjeeling hills is the heavy incidence of insect pests, diseases and nutritional deficiencies. The rural India has started to realize that access to electronic communication is now not luxury but it is necessary for better communication. Keeping this in mind, computer literacy among the citrus growers (any member of the family) of Darjeeling hills and Sikkim were studied and 53 per cent of the growers found to be computer literate. The present software has been developed to help the Citrus growers to diagnose and solve their problem to increase the production of mandarin orange using the software. The software has been developed in visual basic language and is very easy for farmers to use. Those who have only basic knowledge of computer can easily use this software. The user just has to click the given caption. The software is available both in English as well as in Nepali (Local language). This software provides variety of option to the farmers. It not only delivers electronic information about their problems but also recommends solutions. This software has been a good example of electronic knowledge delivery to the Citrus growers of Darjeeling and Sikkim Himalayas.

Introduction

Darjeeling and Sikkim Himalayas are primarily rural. About 75 per cent of the population of the hill area live in villages and depend on agriculture, horticulture and animal husbandry for their livelihood. A considerable number of population in Darjeeling hills are engaged as labourers in the Tea gardens. Amongst the horticultural crop, Mandarin Orange (*Citrus reticulata* Blanco.), known as '*Suntala*' to the local populace, is a major cash crop and was once a pride of Darjeeling hills. It is grown in approximately 930 ha with an annual production of 148.224 metric tones and annual turnover of Rs. 25 crores approximately. Since early 1970s, orange orchards in both the Darjeeling hills and Sikkim Himalayas started to be unproductive inflicting huge economic loss showing severe citrus decline. Gradually the malady assumed alarming proportions and threatening the cultivation of the crop. The major causes for the bottleneck of the citrus industry in these areas are due to heavy incidence of insect pests, diseases and nutritional deficiencies.

Operation of Software

This software is a collection information about the diseases, insect-pest and nutritional deficiencies in the form of text, Photographs, Sound clip, Movie clip in an organized way so that any information can be retrieved as and when required. This software has been written in visual basic platform. It is available in two languages English and Nepali. It is user friendly very easy to operate. All usual operations of a database like enter, exit, print, yes, no, has been incorporated in the software.

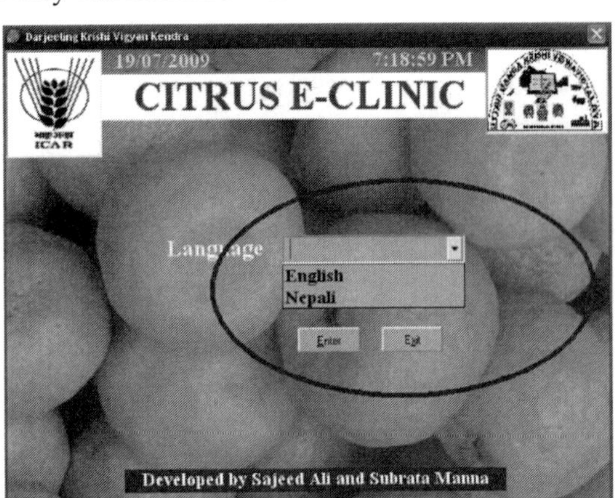

The software can be installed in the hard disc or the user can run it directly from CD. The user has to select the language (English OR Nepali as shown in the picture) on the home page of the software. After selecting the language, user should click enter to get into the software.

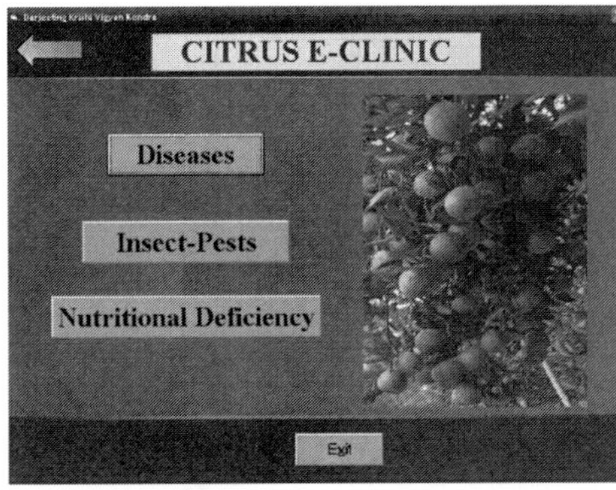

Another page appears in the screen, now the user has to select from the given option like DISEASES or INSECT-PEST or NUTRITIONAL DEFICIENCIES (shown in the picture).

Suppose, if you click insect-pest option a photograph/movie of an insect appears in the screen with a question (Have you seen this type of insect in your orchard ?) and several options like YES, NO, MAY BE, PRINT, EXIT.

And, if you click YES, detailed information about that pest (Entomological name, Symptom of Damage, life cycle and management practices) appears on the screen in the form of text, Photographs, Sound clip and Movie clip. If you click MAY BE then different picture of the same pest appears on the screen but if you click NO then picture of different pest appears in the screen. All the pages are interconnected, if the user do not have idea on differences between insect-pest and diseases, he/she will have to go on by clicking NO, after all diseases, page of insect-pest automatically opens and then page of nutritional deficiencies.

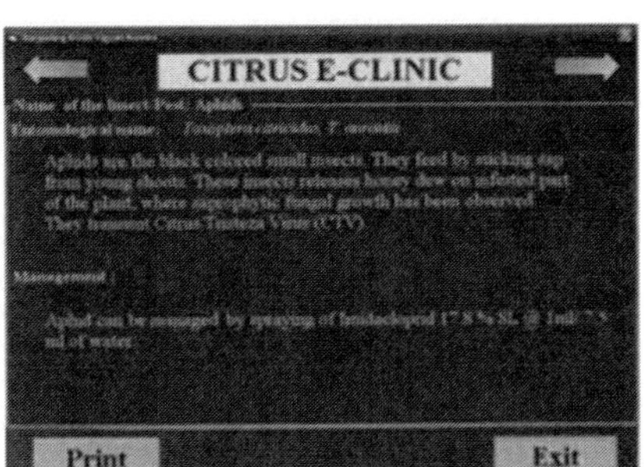

Yes

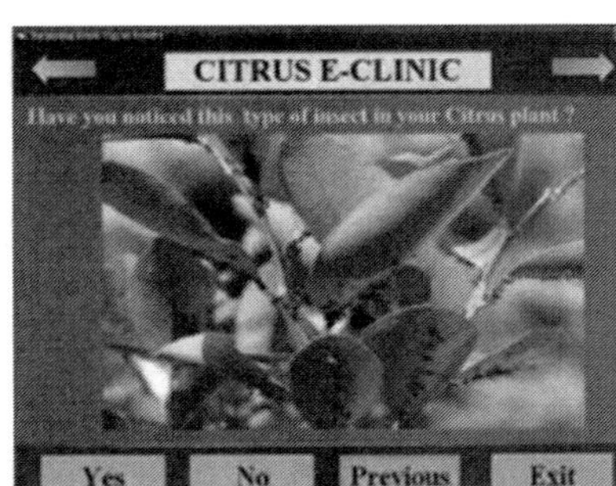

May be

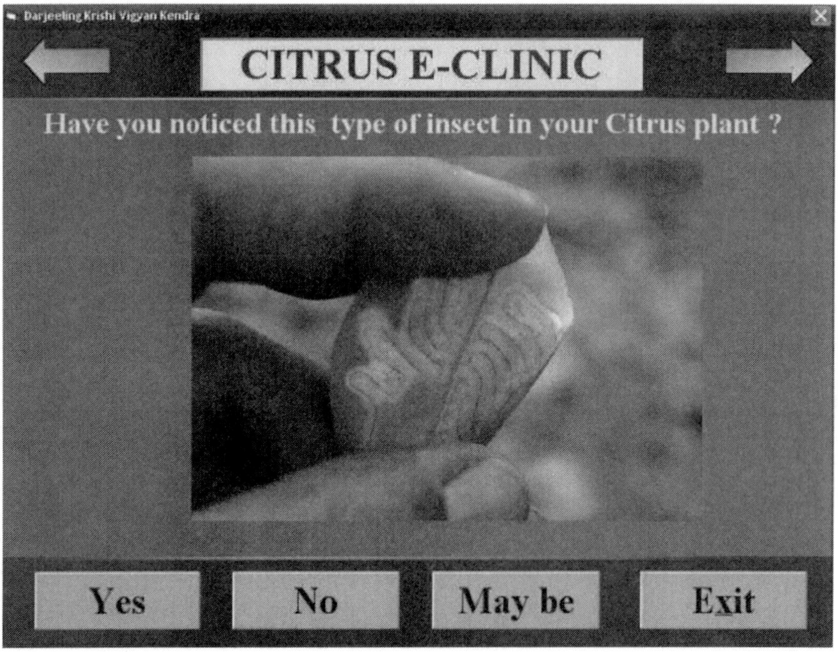
Darjeeling Krishi Vigyan Kendra
CITRUS E-CLINIC
Have you noticed this type of insect in your Citrus plant ?
Yes
No
May be
Exit

No

Darjeeling Krishi Vigyan Kendra
Thank you from Sajeed and Subrata for using this software
OK

Exit

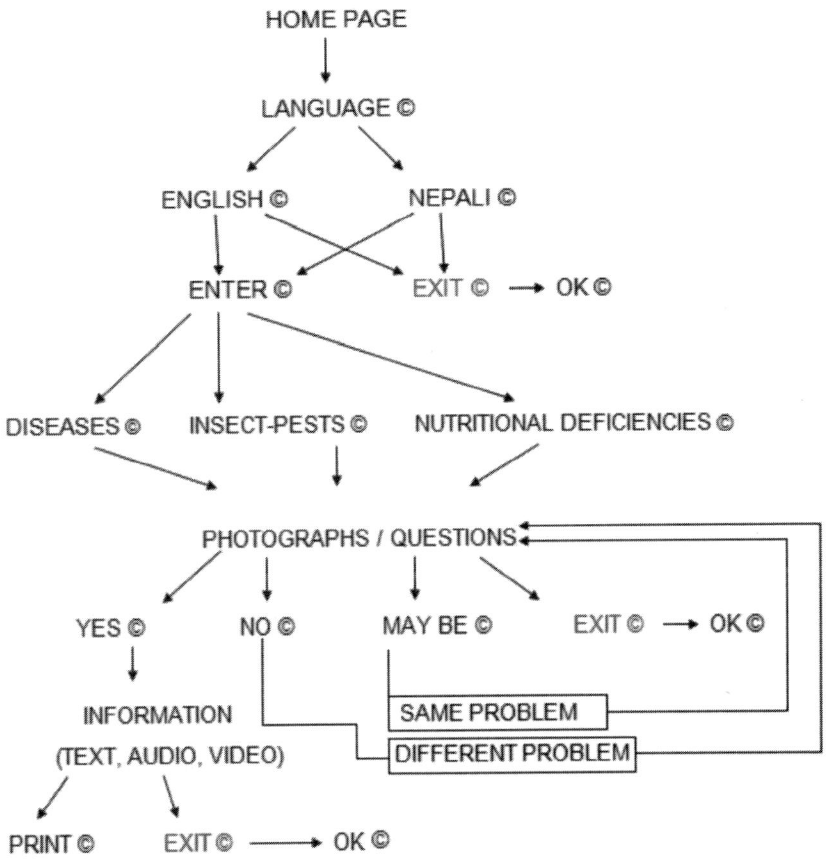

Conclusion

This software has also been performed in touch screen kiosk. We have installed it in our farm information unit at Bunkulung Village of Darjeeling District. Farmers are interested to use similar type of interactive programme on other crops also. Hence, further expansion of this software on other crops and implementation in different villages is under process.

17

Effectiveness of Interactive Multimedia Compact Disc (IMCD) in Diffusing Technologies among Farm Women in Wetland Ecosystem

N. Anandaraja, N. Sriram, Shibi Sebastian and D. Periyar Ramasamy

ABSTRACT

Agricultural growth is a catalyst for broad based economic development in most of the low-income countries. Over the last few decades, the share of agriculture in GDP has steadily declined in India from over 50 per cent in the 1950s to 18.5 per cent in 2006-07. The role of extension personnel in this juncture has become immense and there is a mammoth task ahead in the era of contract farming, organized retailing and market led extension. Information and Communication Technology (ICT) is an umbrella that includes all technologies for the communication of information. It caters for individualized instruction and the contents could be easily updated and expanded. IMCD is a part of the vast ICT arena. This paper describes the experiences of using this IMCD as a transfer of technology tool to farm women. Further, it elaborates the identification of felt needs of the wetland growers, the development of IMCD's and the knowledge gained by the farm women due to their exposure to the IMCD's. It was found that the knowledge gain was significant with variation for the various presentations.

1. Introduction

Very few countries have experienced rapid economic growth without agricultural growth either preceding or accompanying it. Agricultural growth is a catalyst for broad based economic development in most of the low-income countries. Over the last few decades, the share of agriculture in GDP has steadily declined in India from over 50 per cent in the 1950s to 18.5 per cent in 2006-07. In contrast, the share of industry and services has increased to 26.4 and 55.1 per cent respectively due to greater focus on these sectors. But, agriculture still remains the backbone of Indian economy as nearly as 72 per cent of the rural population depends upon agriculture and allied activities for livelihood. The National Agricultural Development Policy (NADP) of India envisages a growth rate of 4 per cent per annum in the agriculture sector so as to achieve a target of over 300 million tones of food grain production by the year 2020 (Survey of Indian Agriculture, 2007).

Rural women form a very significant part of the productive work force in the agriculture sector in India and in most of the developing countries. Women are central to the selection, cultivation, preparation and harvest of food crops. About 80 per cent of all economically active women in the country are engaged in the agriculture sector. Also, about 48 per cent of India's self-employed farmers are women. There are 75 million women engaged in animal husbandry as compared to 1.5 million men (www.digitalopportunity.org). The findings of Chaudhary and Singh (2003) support the above statement. They reported that women played considerable role in various agricultural activities such as pre-sowing, sowing and transplanting operations. Women's role was also reported in the preparation of compost and rising of vegetable nursery. Women's role was found to be maximum in hoeing and weeding practices on intercultural operations. Harvesting and post harvesting operations were reportedly more suited to the physical condition of women. Reaping of crops, storage of food grains, storage of seed and processing of grains were reported to be mostly done by women. It is an acceptable fact that women possess talent which could be harnessed for the productive purpose. But, it is most unfortunate that the role of women in agriculture has been neglected. A holistic development of women requires empowerment in all aspects involving them as managers, planners, scientists, and technical advisories. This is supported by Singh (2005) as "A large part of women's contribution in agriculture is often not recognized and is also unpaid".

The world is in the midst of a knowledge revolution, complemented by opening up entirely new vistas in communication technologies. Recent developments in the fields of Information and Communication Technologies (ICTs) are indeed revolutionary in nature. Hundreds of millions of dollars are being spent on ICTs, reflecting a powerful global belief in the transformatory nature of these technologies. ICT, when used as a broad tool for educating the rural communities especially women, heralds the formation of a new class of society - the Knowledge Society. Knowledge thereby becomes the fundamental resource for all economic and developmental activities. The one resource that liberates people from poverty and empowers them is knowledge. *Possessing knowledge is empowering, while the lack of knowledge is debilitating.* The World Bank organized a forum called "Voices of Poor" during 2000, which got feedback from 60,000 people in 60 countries and concluded that people wanted access to knowledge and opportunities instead of charity to fight conditions leading to poverty. A reorientation of existing strategies is essential to meet the new requirements of farming systems in the country and challenges in the areas of relevance, accountability and sustainability of agricultural extension. Ponnusamy and Gupta (2004) supports with their findings that "The paradigm shift is required to meet the challenges of the 21st century and to bring food, nutritional and environmental security and for serving all sections of the rural community including farmers, farm women and farm youth". "To give visibility to rural women, they must be organized into Self Help Groups (SHGs). Group approach is a viable set up to empower women economically, socially and technologically for improved life", reported Singh and Mathur (2005).

Information and Communication Technology (ICT) is an umbrella that includes all technologies for the communication of information. It encompasses any medium to record information (magnetic disk/tape, optical discs - CD/DVD, flash memory, *etc.*) and also technology for broadcasting information - radio, television; any technology for communication through voice and sound or images - microphone, camera, loud speaker, telephone to cellular phones.

Interactive Multimedia Compact Disc (IMCD) is a modern electronic gadget which demonstrates complex farm technologies in simpler terms using text, graphics, audio and video. It is user friendly encompassing labeled instructions to learn the contents, structure and information using the mouse without depending on the help of a facilitator. It is highly self controlled which could be retained and viewed repeatedly as per the demand of the users. It is highly self controlled, comprehensive, user-friendly, highly reliable, pragmatic, innovative, cost-effective and simple tool which offers an unprecedented potential of providing knowledge to the users. It caters for individualized instruction and the contents could be easily updated and expanded. IMCD is a part of the vast ICT arena. This paper describes the experiences of using this IMCD as a transfer of technology tool to farm women.

1.1. Objectives

1. To study and document the technical know-how required by women farmers in different ecosystem through participatory approach.

2. To develop a user-friendly Interactive Multimedia Compact Disc (IMCD) on identified agricultural technologies.
3. To assess the knowledge gained by the farm women due to the exposure to the developed IMCD.

2. Methodology

The study initially focused to build base line data on farming status of women from Wet land eco system. Several discussions were held with KVK scientists and officials of the department of agriculture and horticulture, village administrative officers on the identification of a village under wetland ecosystem. The research team identified and selected Mettuneerathan village in Vadipatti Taluk of Madurai District in Tamil Nadu as study area under wet land eco-system. In order to elicit the felt technologies from the farm women, it was highly felt that only participatory techniques would help rather than applying conventional data collection methods. Hence, the Participatory Rural Appraisal (PRA) techniques were applied in the above village. Only select PRA tools namely Resource Mapping, Seasonal Calendar and Pair-wise matrix ranking techniques were selected.

2.1 Resource Mapping

This is the construction of a map of the village by the participant village using rangoli powder/chart and marker pens. The map shows the resources, infrastructure facilities available in the village and also depicts the social set up of the village. Further, it gives an idea about the resources like soil, water, forest *etc*. For the above exercise, charts which were pasted together to form a big sheet was given to the participants along with a variety of colour marker pens. The participants were given a briefing on the exercise and its purpose. The participants located few important structures of the village such as Temples, Schools, Water tank, Shops, Phone booths, Public latrines, Primary health centres, Milk society unit, *etc*. They differentiated each structure using different colours like green colour for paddy, yellow colour for settlements, blue colour for water bodies so on and so forth.

2.2 Seasonal Calendar

It shows the seasonal patterns in rural areas related to rainfall, farming practices, employment, *etc*., over the months in a year. The participants were explained about the seasonal calendar and its purpose. A pre-drawn seasonal calendar indicating months on one column was pasted in the wall for reference. The participants were given free space on the chart to enroll themselves on the crops, varieties and operations carried out in a particular season. They also involved themselves fully and brought out a complete seasonal calendar including the major crops being cultivated each season namely rice-rice- pulses. Various agricultural operations starting from sowing, transplanting, weeding, fertilizer application and harvesting of the crops grown and varieties were also collected with the application of the exercise.

2.3 Pair Wise Matrix Ranking

According to Mukherjee (1993), Pair wise ranking uses two items or attributes at a time for ranking to explore people's criteria for choosing one alternative over another. Group of villagers compare one pair at a time and give reasons for the choice made. At the end, the most favored choice is identified if it is a question of choice or the major problem is identified in case of problems being ranked.

Through consultation with farmers, KVK scientists and department officials of the study area, a list of agricultural technologies were identified which suits to the particular farming system. They are; SRI in paddy, Rat control in paddy, Coconut eriophid mite control, Vermicompost production, Fodder cultivation, Mushroom production, Amla package of practice, Identification and uses of medicinal plants, Package of practices for black gram, Land Leveling and Mango cultivation. As the matrix consisted of equal number of rows and columns, the 11 technologies concerned were listed as 1 to 11 for the study of preference. Each matrix was administered separately one after another. First the subject matter matrix was administered. After completing the data for subject matter, the other matrix for training methods was administered. The number of times each technology was preferred was placed in total column. Ranks were assigned to the technologies according to the total number of times they were selected. The technology preferred more times was ranked first, similarly other ranks were assigned. Thus the ranks of felt agricultural technologies were finalized and the result is presented below:

Table 1 : Ranks of the Farmers Felt Agricultural Technologies

Sl.No	Agricultural Technologies	1	2	3	4	5	6	7	8	9	10	11
1.	SRI		1	1	4	5	1	7	8	9	10	11
2.	Land leveling			3	4	5	6	7	8	9	10	11
3.	Fodder cultivation				4	5	6	7	8	9	10	11
4.	Black gram Package of practices					5	4	7	4	9	10	11
5.	Vermicompost p roduction						5	7	5	9	10	11
6.	Mushroom production							7	8	9	10	11
7.	Rat control in paddy								7	7	7	7
8.	Mango Package of practices									8	10	11
9.	Coconut Eriophid mite control										9	10
10.	Amla - Package of p ractices											11
11.	Identification & uses of medicinal plants											

Though the village was predominant for paddy cultivation, they felt that agricultural technologies were not confined to paddy cultivation alone. Rather, they wished to shift their cultivation by raising other remunerative crops as per market demand.

Table 2 : Identified and Prioritized Felt Agricultural Technologies

Sl.No	Agricultural Technologies	Score	Rank
1.	Rat control in paddy	10	I
2.	Identification and uses of medicinal plants	8	II
3.	Amla – Package of practices	8	II
4.	Coconut Eriophid mite control	7	IV
5.	Blackgram -Package of practices	6	V
6.	Vermicompost production	5	VI
7.	Mango – Package of practices	5	VI
8.	System of Rice Intensification (SRI)	3	VIII
9.	Mushroom production	2	IX
10.	Fodder cultivation	1	X
11.	Land leveling	0	XI

Further, the research team and villagers held a deliberation over the finalization technologies through IMCD. By using Focused Group Discussion (FGD), farmers and research team sorted out the following eight techniques to be devised in the form IMCD for the benefit of farm women in wet land eco system. They are as follows: SRI methods, Rat management techniques, Hybrid seed production technologies for Paddy, Eriophyid mite management, Vermi compost technology, Mushroom production technologies, Drip based techniques and IPM in Paddy.

2.4 Development of Interactive Multimedia Compact Discs (IMCD)

Computer multimedia offer learners more complete and individual control over their learning. The main reason behind recommending the use of computer based multimedia system for farm women is to facilitate interactivity and better understanding between individual learners and the subject matter. IMCD is a modern electronic gadget which demonstrates complex farm technologies in simpler terms using text, graphics, audio and video. It is highly self controlled which could be retained and viewed repeatedly as per the demand of the users. The step involved in devising IMCD is given in Figure 1.

The pre-production phase involved all preparatory works including review of literature, content preparation/script writing, collection of video/ pictures/

video capturing *etc*. In the second phase the software selection was given importance. Among the available different software, Microsoft Office 2003 Power Point was selected specifically for the study purpose. This software possesses number of unique features such as simple operation; create highly interactive multimedia presentation *viz*., possibility to add sound, photo, picture, animation *etc*., play movies directly with in the power point, create simple hyperlinks, convert the presentation into web pages, pack and wizard have the advanced facility to compress and save our presentation across the multiple disc *etc*. Further, the research team used Pinnacle Studio 11.00 version for audio, video editing which was well supported by the MS PowerPoint.

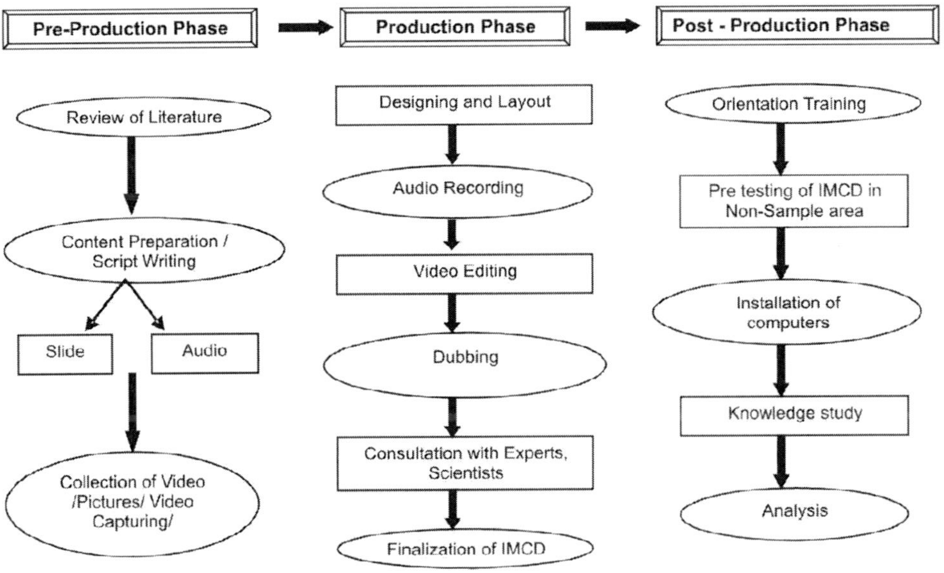

Figure 1 : Steps Involved in Devising IMCD

Each IMCD differed in terms of designing and layout *i.e.*, font size, type, colour, background colour and animation. Audio recording was done in a digital audio recording studio through professional voice artists and research team with good vocal ability. The captured videos were edited into a very short duration movie of one to two minutes. In the third phase developed IMCDs were exposed to user groups (farm women) through orientation training. During the training, farm women groups were allowed individually to handle the computer including the monitor, mouse, CPU and CD so that they get rid of computer phobia. Before handing over the IMCDs to the user groups, pre testing of all IMCDs were done for content validation in non sampling areas and with experts. With necessary incorporations of valuable suggestions obtained from pre-testing, the final IMCDs were handed over to the user groups by installing computers at the respective villages.

2.5 Sample and Sampling Technique

Since the respondents of the study were farm women, the farm women of the entire village were considered as samples. The selected village had 150 households and the entire farming population families were considered as population of the study. Regarding the knowledge gain of developed IMCD module, 70 respondents participated in the pre exposure and post exposure knowledge test. These 70 farm women constituted the sample size of the study.

The main aim of the study was to diffuse the farm technologies to farm women through IMCD and to assess the effectiveness of IMCD in terms of knowledge gain was assessed. The knowledge level of farm women was assessed as pre exposure and post exposure to the felt technologies through the IMCD.

2.6 Measurement of Knowledge Gain

Knowledge occurs when an individual or other decision making unit is exposed to the innovation's existence and gain some understanding of how it functions. Knowledge function is mainly cognitive or knowing. Knowledge seeking is initiated by an individual and is greatly influenced by one's pre dispositions. Testing Knowledge gain, the research team prepared the knowledge question for each module. Each question carrying 20 objective type questions.

The knowledge test developed for the study was used to measure the knowledge level of the felt technologies at two stages namely pre and post exposure. Before viewing the felt technologies in the IMCD, the knowledge questions that were prepared in local language (Tamil) were exposed to farm women. This is called as pre-exposure of the IMCD. After exposure the respondents to IMCD again the farm women were administered the same knowledge test. This is called as post exposure. The difference between the pre and post exposure scores was taken as knowledge gained by an individual respondent on felt technologies. The possible range of knowledge gain score was between zero and twenty.

Data Collection was done with carefully constructed interview schedule by reviewing various literature, journals, theses, books and in consultation with experts and scientists. The knowledge gain was measured using percentage analysis. The Paired-Samples't' test was used to know the significant of knowledge gained by the respondents. The paired–samples 't' test procedure compares the means of two variables that represent the same group at different times (*i.e.*, before and after exposure to IMCD).

3. Results and Discussion

In the wetland ecosystem the farm women were exposed to eight different IMCD's based on the priority set. The percentage of knowledge gain was calculated and the result is given below.

The knowledge test scores collected from the respondents for each module (IMCD) is pooled and the average given in Table 3.

Table 3 : Knowledge Gain for the Eight IMCD Modules

S. No.	IMCD	Before Exposure	Immediately after Exposure	Mean Knowledge Gain	Percent of knowledge Gain	't' Value
1.	SRI Techniques	3.3	17.3	14	70	29.69**
2.	Rat management Techniques	6.1	16.6	10.5	52.5	16.054**
3.	Hybrid seed production technologies for Paddy	2.3	11.2	8.9	44.5	21.873**
4.	Eriophyid mite Management	2.5	14.1	11.6	58	29.000**
5.	Vermi compost Technology	3.6	14.1	10.5	52.5	21.000**
6.	Mushroom Production Technologies	2.8	14.3	11.5	57.5	17.583**
7.	Drip based Techniques	3	16.8	13.8	69	18.976**
8.	IPM in Paddy	8.1	17.1	9	45	11.84**

** Significant at 0.01% level of probability

From Table 3, it is evident that SRI IMCD had brought about a significant knowledge gain among the farm women respondents. The average per cent knowledge gain was 70 per cent which was found to be significant at one per cent level of probability.

The results indicate that there occurred a significant increase in knowledge level at the post exposure stage of System of Rice Intensification IMCD. Hence, it could be concluded that the IMCD on System of Rice Intensification technology, acted as an intervening variable in changing and increasing the cognitive domain of respondents in the wet land eco system. The reason for the highest knowledge gain (70 per cent) on SRI IMCD as expressed by the respondents were newer technology, suitable to their cropping system, relevant and attractive video clippings, pictorial representation , number of relevant video clippings and clear voice.

Among the Eight IMCDs, second highest average knowledge gain (69 per cent) was found among the respondents on Drip based sugar cane technologies. This might be because of the following reasons; newer technology, suitable to overcome water scarcity, higher yield, pictorial representation of field layout and field preparation.

The third highest knowledge gain (58 per cent) was observed in respect on Eriophyid Mite Control. The knowledge gain was found to be significant at one per cent level of probability. This is due to newer to that eco system, suitable as an alternate crop during water scarcity, clear package of practices and relevant video clippings.

From the table, it was found that the average knowledge gain of Mushroom production technology is 57.50 per cent. The higher percentage of knowledge gain be due to following reasons; some of the farm women are interested in producing mushroom in their home and they are keenly observed the different types of mushroom and their production technology, video clippings regarding milky mushroom production, video clippings regarding the oyster mushroom production, several illustrations of different verities of mushroom and clear text message with the voice support.

The results indicated that there is a significant increase in knowledge level at the post exposure stage of Mushroom production technology. Hence it could be concluded that IMCD on Mushroom production technology need to be up scaled in a bigger way which will improve the standard of life and purchasing power of Rural Women. The other IMCD's on Vermicompost Technology, Rat management Techniques, Hybrid seed production in paddy and IPM in paddy had shown improved level of knowledge gain due to the exposure of IMCD.

4. Conclusion

The innovative interactive multimedia modules developed have been successful in delivering significant levels of benefits to the rural farm women. It has been used to transform and revive the traditional agriculture, also it is helpful in better dissemination of information using low-cost technologies and empowering the farm women. Multimedia Module has enormous potential to transform farm activity, create new forms of enterprises among farm women and access to the intervention.

5. The Way Forward

Empowerment of women in the context of knowledge society is understood as building the ability and skills of women to gain insight of the actions and issues in the agriculture and allied activities which influence them, and to build their capacity to get involved and voice their concerns in these activities, to make informed decisions. It entails building up capacities of women to overcome social and institutional barriers and strengthening participation in the economic and political processes for an overall improvement in the quality of their lives.

Access to information can be seen as a central issue concerning empowerment of women. Women in developing countries, however have been traditional excluded from the external information sphere both deliberately and because of factors which inherently work to their disadvantages such as little freedom of movement, low education-levels *etc.* However, IMCD, a component of ICT opens up a direct window for women to the outside world. Information now flows to them without distortion or any form of censoring, and they have access to the same information as their male counterparts. This leads to broadening of perspectives, building up of greater understanding of their current situation and causes of poverty, and initiation of interactive processes for information exchange and in turn empowerment and transformation of rural women community.

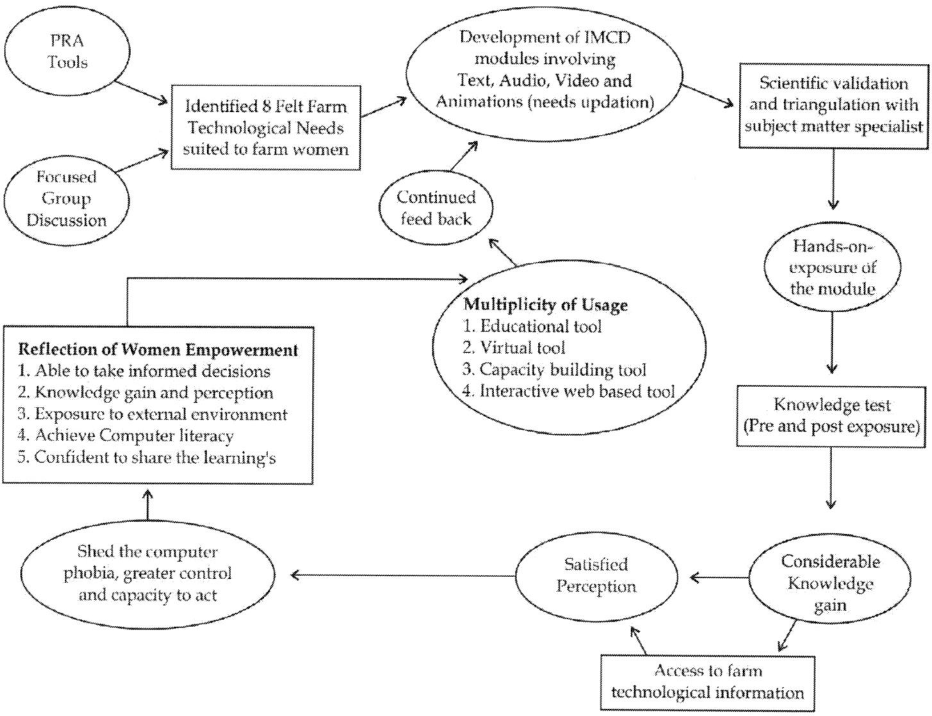

Figure 2 : Suggested Model to Disseminate Farm Technologies through IMCD.

The project as a whole aimed at empowering the farm women in using the new communication technology advancements, helps in developing new education modules and in exploring the feasibility of this new communication tool was established. Moreover, this venture can be used for administrative convenience in a longer run. Educationalist, policy makers and administrators can take a lot from this project in developing new modules in educational, research and extension institutes in the near future.

Acknowledgement

The authors wish to thank the Department for Scientific and Industrial Research, Ministry of Science and Technology, Govt. of India for funding the research work.

References

Chaudhary, H., and Singh, Sangram, 2003. Farm Women in Agriculture Operations. *Agricultural Extension Review*. 15(1): 21-23.

Mukherjee, Neela, 1997. *Participatory Appraisal of Natural Resources*. Concept Publishing Company, New Delhi.

Ponnusamy, K., and Gupta, Jancy, 2004. Agricultural Extension - New Paradigms. *Agricultural Extension Review*. 16(6): 3-6.

Singh, Radha, 2005. Gender issues are central to the development dialogue.www. digitalopportunity.org.

Singh, Suman and Mathur, Puja, 2005. Self Help Group - A Successful Approach for Women Empowerment. *Agricultural Extension Review*. 17(6): 27-30.

The Survey of Indian Agriculture, 2007. The Hindu Publication, Chennai, India

18

Development of Farmers Database for Darjeeling District

Subrata Manna

ABSTRACT

Future is the era of ICT (Information Communication Technology) which is equally applicable in the field of Agriculture. Farmers database will be an important part of ICT which is required for smooth functioning of various agricultural activities like Farmer's Trainings, Demonstrations, Crop monitoring, Support to farmers, implementation of various government schemes *etc.* This software records farmers Name, Phone number, Sex, Category (SC/ST/Gen), Number of dependents, Annual agricultural income, whether fully or partially dependent, Land holding size, Major crops, Village, Block, Sub-division, Land situation (Plains/ Low Hills/ Mid Hills/ High Hills), Soil Reaction, Soil type, (Highly Acidic/Moderately /acidic/Neutral/ Alkaline/Highly Alkaline), NPK status, Irrigation facility and information regarding Cattle. From the above record various reports can be generated according to the needs of the scientific personnel. For example soil scientist will have record of soil pH & NPK status of different farmers across the district. Scientist can filter the database to get farmers of a particular soil type to conduct training; no need to give training on Acidic soil management to the farmers who are having alkaline soil or training on Potato whose major crop is Rice. Similarly, off season vegetable cultivation under poly-house can be implemented with govt. aids and should be encouraged more among small land holding and irrigated farmers. Veterinary personnel can filter those villages where cattle are more for conducting training. For various govt. schemes farmers can be selected from the database on basis of different criteria. Moreover time to time Mobile SMS advice can be sent to selected farmers. Various statistical reports can be generated to analyze socio-economic status of the farmers.

Introduction

The raw facts of information are data, which has been derived from the Latin word *datum*. "Darjeeling district", "reducing", "Alkaline soil, "agricultural productivity", "hilly area" are data. Data can't convey any meaning but when it is arranged in a proper sequence it express a meaning. "Alkaline soil of hilly area of Darjeeling district is reducing agricultural productivity" -is an information.

Database is a computerized filing system that stores the information in an organized way so that any information can be retrieved as and when required. A database is a group of tables which contain data coming from a same domain. For example all tables of farmer's database should contain various information of farmer's only; it can not have record from other domain like market place. A database can contain not only the text material (written document); it contains any form of digital information like Photographs, Sound clip, Movie clip *etc*. A database is a collection of several tables & all tables are connected to each other by a common field called primary key. Table is a logical collection of row and column. Row is called record touple and column is called field or attribute

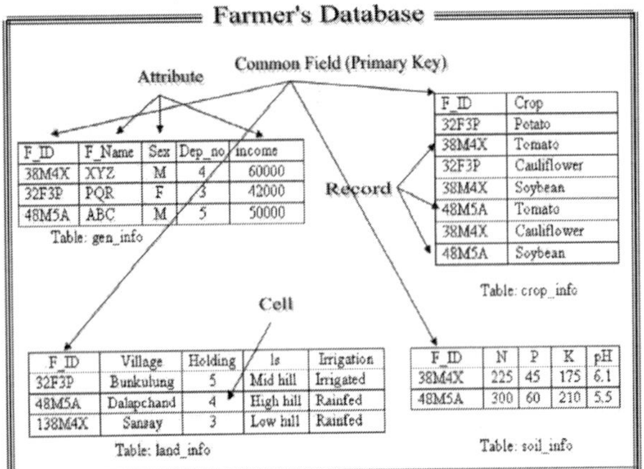

F_ID : Farmer's Identification Code
Sex : Sex of the Farmer
Income : Yearly approxmiate income
Village : Name of the native village
ls : Landing Situation
N : Existinge Nitrogen containt

F_Name : Farmer's Name
Dep_no : Number of dependent member
Crop : Main crops cultivated by the farmer
Holding : Land holding size in acre
Inrrigation : Whether Irrigated or Rainfed land
P & K Existinge P_2O_5 & K_2O_5 containt

The record from the database is fetched by SQL (Structured Query Language) query. For example, to get the name of those farmers whose yearly income is 50,000 or less the query is,

```
Select F_Name
From gen_info
Where income <= 50000
```

To get the name of those farmers who is cultivating soybean the query is,

```
Select F_Name
From gen_info
Where gen_info.F_ID=crop_info.F_ID and crop_info.crop='Soybean'
```

Here the Front end (Forms & Reports) of the software has been designed by Visual Basic and Back end (Database) has been designed in MS Access. MS Access is a better choice for small database with portability feature (can be installed easily anywhere)

All usual operations of a database like Save (to Add a new record in the database), Find (to search any record of a particular farmer), Update (to change the any information of a particular farmer), Delete (to delete any record from the database) has been incorporated with the software. Also print option has been added with the software so that filtered data can be printed for future activity.

The field level data collection process has been initiated in collaboration with the ATMA, Darjeeling. A verification column has been kept in every from to check the authenticity of every information by an Agricultural Development Officer (ADO) or Panchayet member of the respective village. Data is being collected Farm Family wise, not individual farmer wise. A sample of the form is given on page 248.

Agriculture Technology Management Agency (ATMA), Darjeeling

Farmer's Survey Form Serial No.: 1110

Date: / /2009 Sub-Division:

Block: .... Village: ....

Name of the Farmer: ....

Father's / Husband's Name: ....

Phone No.: .... Sex: Male / Female ....

No. of Dependent: .... Category: SC / ST / OBC / General ....

Land Ownership : Land owner / Bargadar / Landless Labour

Annual Income: ....

Involvement in agriculture: Full Time / Part Time

Holding Size: .... Acre Current Fallow .... Permanent Fellow ....

Major Crops (specify area): Cerials / Vegetables / Fruits / Flowers / Spices / Mushrooms / Plantation Crops

Land Situation: High Hills / Mid Hills / Low Hills / Plains

Irrigation: Rainfed / Irrigated

Animal (No. & Type): ....

Any member of the family can operate computer? Yes / No.

Data Collected by .... Data verified by ....

Name .... Name & Designation ....

Signature .... Signature ....

The Areas of Application of the Database

- To keep the individual farmers record of entire district
- To implement any Government scheme (financial help) to the farmer we can filter the database baed on annual income and more number of dependent members, who really need the help.
- We can keep the NPK status of individual land in the database and hence if required a gross assumption of nearby land is possible.
- We can follow up the cropping pattern of individual farmer and if require can suggest to alter cropping pattern for increasing income of the farmers.
- Agricultural advice can be mailed as a SMS to those farmers who have mobile phone.

- Any agricultural activity (training, demonstration *etc.,*) can be informed to individual farmers well in advance through SMS.
- For Women Empowerment those village can be adopted, where more active female farmers are recorded.
- To analyze the Computer Literacy rate among farming community & to analyze the real feasibility to implement various Information Technologies related projects.
- To setup irrigation facility village can be adopted on the basis of various criteria available in the database.
- To compare the socio economic status of the Backward community farmers (SC/ST) in the village.
- To conduct training on specific soil management no need to call all farmers. We can filter the database and call the right person for training.
- To conduct any training on specific crop cultivation no need to call all farmers who are not cultivating the crop, we can call the right farmer from the database for the particular training.

19

Pest Surveillance of Rice Crop Using Satellite Data – A Case Study for Brown Plant Hopper (*N. Lugens*)

Sujay Dutta, I. Choudhury, Y.G. Prasad, M. Prabhakar, G. Katti, K. Vasantbhanu and S. Vennila

ABSTRACT

A methodology for depiction of Brown Plant Hopper (BPH) pest incidence on rabi rice crop using Remote Sensing (RS) data was attempted. The reflectance and geographically registered RS images of Feb.12, Mar.13, and April 30, 2009 as well as Feb.23 and March 14, 2007 were used to classify the Godavari district for generation of rabi rice crop area map. The rice crop map thus generated was used as an overlaying layer of rice crop mask in GIS to derive other RS based indices described below over this crop mask. Land Surface Temperature (LST) data product for rice growing areas was generated using rice crop mask. Normalised Difference Water Index (NDWI), Temperature Vegetation Dryness Index (TVDI) onices derived from MODIS data and NDWI from AWiFS data were correlated with pest count data for two years 2007 and 2009. A grid based approach (5 x 5 km) was attempted to develop RS based input parameters required for this pest prediction model. The RS based indices were used to show variability in crop growing environment like leaf wetness factor, chlorophyll content and soil moisture status susceptible for pest incidence. The anomaly in the indices combined with the weather parameters were used to locate the susceptible areas for pest incidence and were compared with the ground truth location taken with Ground Positioning System (GPS).

Introduction

An attempt has been made to develop a prototype model for detecting distributed location of pest incidence in spatial scale by linking satellite based Remote Sensing (RS) data as input at grid level (5 x 5 km). Spatial modeling using GIS (Geographical Information System) will be able to actually locate the foci of infestation. This detection is crucial to warn about the potential pest affected locations and allow for timely preventive decisions like spraying operations in an optimized way. Willrich *et al.* (2004) have studied the optimum insecticidal control through equations describing the relationship between damage type, and intensity of pest buildup. Pest infestation in crops is highly influenced by agro-meteorological parameters. Many research articles published on pest-weather relationships used pest-monitoring data from light traps (for example yellow stem borer in rice), pheromone traps (for American bollworm) and sticky traps (for whitefly) apart from population counts data. Long-term data is preferable as it better captures the patterns in relationships. The rates at which insects complete their life cycles depend mainly on temperature (Prabhakar *et al.* 2002), so that the times of activity of a given pest insect can vary greatly both from region to region and from year to year. Many models have been developed using temperature data to forecast insect activity. A multi-station study is preferred as it facilitates corroboration of the general surmises and leads to maximization of data in a short period if observations are recorded on crop stands sown at periodic intervals at a number of stations (Venkataraman and Krishnan, 1992). Weather based early warning of pest infestation is being practiced using statistical and dynamic simulation models on point scale. Many empirical pest prediction models have been developed describing its population build up in relation to environmental weather variables during the crop season would guide in formulating strategies for its forecasting and management. Elsewhere, the influence of temperature in determining emergence and development of insect population has been established (Higley *et al.*, 1986; Wilson & Barnett, 1983) and thermal/heat summations or degree-days are used for predicting *P. gossypiella* emergence (Gergis *et al.*, 1990; Chu & Henneberry, 1992; Beasley & Adams, 1996). However, their use in largeareas has been limited because most pest prediction models were developed at the field scales and the performance of the models is not so satisfactory when they are extended from field to regional scales. The most critical reason is that they have been constrained by the lack of an efficient means of incorporating the spatial variability of input parameters. Satellite based inputs and epidemiological models can extend the application to areas with irregular and non-existing ground observations. The main advantage of remotely sensed information is that it provides quantification of the crop-growing environment in the form of indices on a regional scale and shows the spatial variability. While empirical models provide a continuous description of pest growth over a period of crop cycle, remotely sensed observations are discrete time events. Hence, integrating the inputs as RS based indices to the pest prediction models can be a useful technique to predict distributed location of pest incidence at large scale.

In developed countries, dynamic websites that include interactive models, GIS based decision systems, real-time weather, and market information is being rapidly developed and made available on the Internet (*www.effita.net*) for enabling farmers for real time use in crop management. In United States and the Netherlands commercial firms are applying mesoscale modelling techniques to forecast insect development and producing gridded products for regional and on-farm planning and pest management (Strand, 2000). Extrapolation of model results over larger areas is possible through remote sensing techniques (Dutta *et al.* 2008). In epidemiological modeling, survey data are usually collected at sampling sites and then regionalized in Geographical Information Systems (GIS). To enhance the spatial data density, continuous field data such as Land Surface Temperature (LST) and vegetation indices are commonly derived from satellite data. Agro-ecosystems need to be monitored continuously to understand the ecological processes that are in operation. One major event is likely to trigger multiple effects. In the context of pest scenario this has been witnessed very clearly in the case of Bt cotton replacing conventional cotton. To improve the risk mapping at high resolution, other sources of climatic and vegetation data are required. This demand can be fulfilled by using remote sensing data. Complementing field studies, laboratory experiments under controlled environmental conditions will generate data on developmental growth rates for mealy bug and mirid bugs in cotton; Brown Plant Hopper (BPH) and leaf folder in rice. These studies will lead to insect phenology models for estimating the timing of pest attack. Extrapolation of model results over larger areas is possible through remote sensing techniques. The integration of satellite data into epidemiological research enhances the spatio-temporal resolution of climatological data (through RS based sounder data) in regions where weather stations and ground surveys are sparsely available.

Since the Earth's surface temperature influences pest growing environment and vegetation growth, satellite based Land Surface Temperature (LST) values have been used as criteria, in addition to other RS based indices like NDWI, TVDI and NDVI, for evaluating the status and development of pest.

1.1 Study Area, Crop and Pest

The major pest in rice crop grown during the rabi season in East and West Godavari districts of Andhra Pradesh has been attempted. The main pest is *Nilaparvata lugens (Stål)* (Brown Plant Hopper - BPH) which is assumed to be the most destructive pest since 1970s (Heinrichs and Mochida, 1984). *N. lugens* is probably the most serious insect pest of rice in Asia. Its feeding causes plants to wilt and causes a symptom called hopper burn. Heavy infestations produce symptoms of 'hopper burn'. Leaves dry and turn brown after insect feeding, and patches of 'burned' plants are often lodged.

Nitrogen fertilizer, high tillering varieties and/or close spacing and good water management are factors that contribute to its population increase. Double/continuous cropping associated with good water management may also contribute to this gradually increasing pest problem. However, in areas where these contributing factors are present, infestation has not always been serious (IRRI, 1975). During the 1970s, *N. lugens* caused severe damage to the rice crop in many countries in tropical Asia. Several outbreaks have occurred in the Philippines, Solomon Islands, Thailand, Sri Lanka, Vietnam, India (Manjunath *et al*., 1978), Malaysia and Indonesia. In the Philippines, *N. lugens* damaged at least 80,000 ha in 1973-74 (IRRI, 1979).

In Korea Republic, the yield loss of rice infested at the booting and heading stages with *N. lugens* was studied in the field by Kim *et al*. (1984). Losses were greater from infestation at the booting than at the heading stage, but infestation of plants in either stage reduced the weight of 1000 grains and the numbers of filled grains compared with uninfested plants.

1.2 Data Used

IRS-P6 AWiFS data of

i. February 12, 2009
ii. March 13, 2009
iii. April 30, 2009
iv. February 23, 2007
v. March 14, 2007
vi. Weekly 2009 MODIS AQUA data from February to March 2009 (February 2, 10, 18 and 26, 2009 and March 6, 14, 22 and 30, 2009)
vii. Weekly 2007 MODIS AQUA data from 27 February, 7 and 14 March 2007.
viii. Meteorological data from ground based weather stations.

2. Methodology

A holistic approach for pest prediction model using combination of weather and RS based inputs has been attempted in conjunction with finding the endemic areas of perennially growing rice crop area in the region. There are two layers of pest prediction model

i. A model based on weather data to determine the timing of pest incidence.

ii. A model based on RS indices to locate the affected areas within the 5x5 km grid at 1 km resolution which are further analysed at 56 m resolution.

2.1 Weather based Pest Prediction Model

Since historical pest incidence data was not available for this region as the project was initiated from this current year, no empirical relationship could be established with weather data. However, an existing BPH pest prediction model for rice crop developed for this state had been selected through literature survey. An existing BPH pest prediction model was used which was developed for Hyderabad region in Andhra Pradesh using data from 1984 to 1998 (Prabhakar *et al.*, 2002) as stated below:

$$BPH_N = -87.843\ T_m - 0.803\ BPH_{CP} + 1.505\ T_{MAX} + 2.497\ T_{MIN} \qquad (1)$$

$r^2 = 0.63^{**}$

Where, BPH_N = Square root of BPH population of next week

BPH_{CP} = Mean BPH population of current and previous week transformed into Square root values

T_{MIN} = Minimum temperature of two and three previous weeks

T_{MAX} = Maximum temperature of two and three previous weeks

** Significance at 5% level

eq. 1 was tested to predict the BPH pest incidence for 2007 and 2009 rabi season using the ground based weather station data. The pest count obtained using eq. 1 was valid for whole district as there was one input temperature data for whole district. Within the district the spatial level foci of pest infestation were attempted at 5x5 km grid using a model developed based on RS indices.

2.2 RS Index based Pest Incidence Model

Normalised Difference Water Index (NDWI) and Temperature Vegetation Dryness Index (TVDI) and Normalised Difference Vegetation Index (NDVI) were derived from MODIS 8-day composite Surface reflectance products of 1 km spatial resolution. Similarly Land Surface Temperature (LST) products were obtained at weekly interval from February to April for both the years 2007 and 2009. The AWiFS reflectance images at 56 m spatial resolution of 2007 and 2009 were also used to develop Normalized Difference Water Index (NDWI) index of the crop area. Pest count information (BPH) from field observations in the study area was used for development of pest incidence model with these indices.

2.3 Generation of Rice Crop Map and Persistent Rice Growing Areas in Godavari

Spatial distribution of rice area map was prepared using multi-date IRS-P6 AWiFS data (February 12, March 13, April 30, 2009 and February 23, March 14, 2007) with 56 m spatial resolution. Atmospheric correction of the multi-date AWiFS data was carried out using FLAASH in ENVI s/w. Multi-date reflectance image were generated. All images were geo-rectified using 1:250,000 scale toposheet published by Survey of India. Rice area was classified using hierarchical decision rule classifier. The classifier performs multistage classifications by using a series of binary decisions to place pixels into classes. Each decision divides the pixels in a set of images based on defined expression of rules and NDVI threshold values for classification. The rice growing areas in rabi season 2007 and 2009 were delineated using AWiFS data. The two years' rice area were compared pixel by pixel and common rice growing areas in the two years were separated by taking the difference of both the classified images. Thus persistent rice growing areas over the two years could identify the endemic areas of pest occurance.

2.4 Generation of RS-derived Indices as an Input for Pest Prediction Model

Multi-date reflectance images of AWiFS and MODIS sensor were used to derive various remote sensing based indices *viz.* TVDI (Temperature Vegetation Dryness Index), NDVI (Normalized Difference Vegetation Index) and NDWI (Normalized Difference Water Index) from the rice-growing pixels of the study area. Vegetation has a low level of reflectance in red region due to its sensitivity in chlorophyll contents of the leaves, whereas it shows high reflectance in the NIR region is due to the internal structure of the leaves (Figure 1). The value of this index ranges from -1 to 1. The common range for green vegetation is 0.2 to 0.8. NDVI is defined by the following equation:

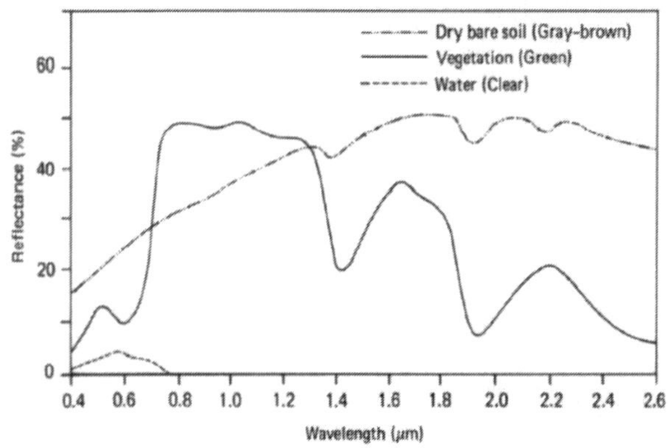

Figure 1 : Reflectance Curve of Vegetation

$$NDVI = \frac{NIR - Red}{NIR + Red}$$

NDWI is sensitive to changes in vegetation canopy water content. It uses two reflectance bands, one is centered at 857 nm and other at 1241 nm (Figure 1). Since, both the bands are located in the high reflectance plateau, therefore, the vegetation scattering properties for the two reflectance channels are expected to be about the same but has slightly different liquid water absorption properties. Two liquid water absorption bands are centered, respectively, at 980 nm and 1200 nm in the green vegetation can be comparable and vegetation canopies enhance the weak liquid water absorption at 1241 nm as this band is sensitive to liquid water changes.

The value of this index ranges from -1 to 1. The value of NDWI for the dry vegetation is spectrum is negative (-0.056). In general, NDWI value for green vegetation is positive due to the weak liquid water absorption near 1241 nm. The common range for green vegetation is -0.1 to 0.4. NDWI is defined by the following equation:

$$NDWI = \frac{r_{857} - r_{1241/1600}}{r_{857} + r_{1241/1600}}$$

Corresponding to the acquisition of MODIS LST dates, MODIS 8-day composite surface reflectance images were acquired for the same year and same duration. Weekly NDVI and NDWI images were generated from the surface reflectance data products using the formula as stated above.

TVDI is a simplified land surface dryness index (Temperature–Vegetation Dryness Index, TVDI) based on the relationship between surface temperature (T_s) and vegetation index (NDVI) as depicted in Figure 2. Theoretically, the scatter plot formed by vegetation index and land surface temperature is like a triangle. The upper edge of the triangle is defined as dry edge, while the lower one is called wet edge. Pixels close to the dry edge are comparatively drier while those close to the wet edge are wetter. The basic approach is to interpret the so-called T_s/NDVI space in terms of surface soil moisture status. The spatial variation in TVDI reflects the variation in moisture on a finer scale. Theoretically, TVDI should be between 0 and 1 (Yang *et al.*, 2008). The larger the TVDI, the drier the pixel is. Thus, TVDI should be negatively correlated with soil moisture. TVDI was generated for the study area using the following formula:

$$TVDI = \frac{T_S - T_{S_{min}}}{a + bNDVI - T_{S_{min}}}$$

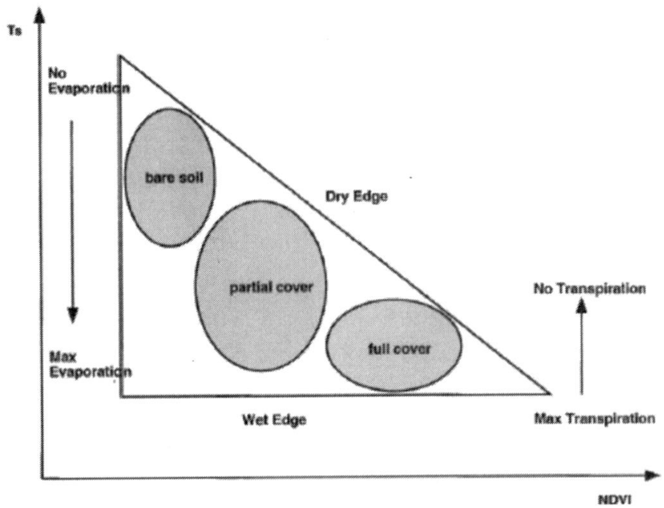

Figure 2 : T_s/NDVI space

Where, T_{smin} = Minimum surface temperature,

T_s = Surface temperature for given pixel, *a* and *b* are parameters resulting from a linear fit to data: $T_s = a + b$*NDVI (Sandholt *et al.*, 2002). TVDI gives the variability of soil moisture over the agricultural area.

Weekly TVDI were developed from LST and NDVI images using the formula as stated above. Weekly interval of NDVI, NDWI and TVDI images were an important input for vegetation status and leaf wetness condition over the agricultural area. Analysis was carried out at 1 km spatial resolution aggregated at weekly interval throughout the rabi rice growing season. Remote sensing derived indices were correlated with the BPH pest count data obtained from field observation on the corresponding date. The pest-infected areas were showing discrepancy from the healthy areas. Infected areas were then further scrutinized using remote sensing based indices derived from AWiFS data at 56 m spatial resolution to locate the pest affected area in a spatial distribution form.

3. Results and Discussion

3.1. Temperature based foci of Infestation at District Level

Using the ground based temperature model (equation 1) developed by Prabhakar *et al.* (2002) the pest incidence timing was estimated and validated for two pest incidence level in march 14, 2007 and April 8, 2009. The estimated v/s observed pest count data for the two years is shown in figure 3 and 4.

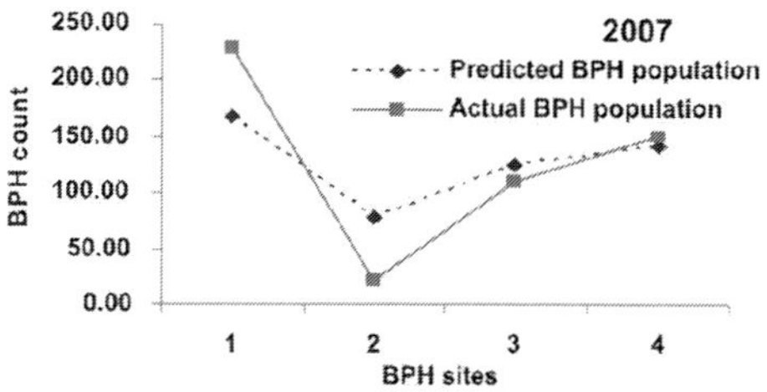

Fig.3

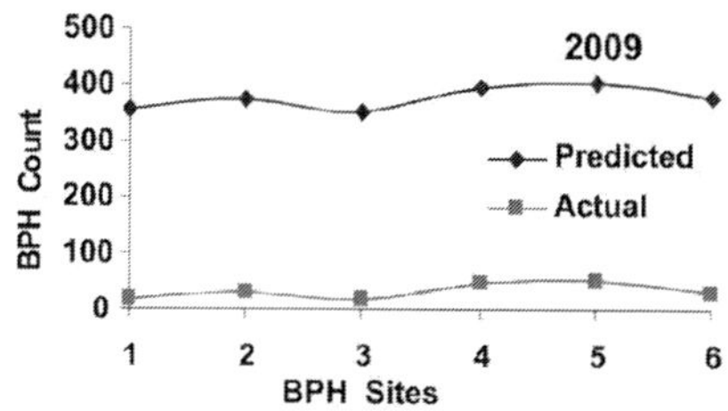

Fig.4

Figure 3&4 : Comparison of Model v/s Observed Pest Incidence Level in 2007 and 2009.

Although, the estimated count in 2007 was at par to observed data in 2007 but an over estimate of 300 pest count was observed in 2009. This showed that the model worked in high infestation time but does not work at low infestations. This model suggests to work better in epidemic years compared to low infestation years. As the temperature used was obtained from ground based weather station, the model depicts uniform situation for all endemic rice growing areas within the district. To generate the location of pest infestation in a more precise form RS based model was developed.

3.2. Development of RS based Model at 1 km Resolution

Using MODIS derived LST, which signifies the surface level composite temperature of the ground and canopy, was taken into account along with RS indices TVDI, NDWI and NDVI. The TVDI reflects the surface level soil moisture conditions. BPH finds conducive environment in hot and humid environment at the base of crop canopy so TVDI would be effective in depicting the humid environment. A model was developed using multiple regression of two years pest count data for 2007 and 2009 is as follows.

$$Y = -10991.6 + 737.1244 * NDVI - 236.408*NDWI + 34.903*LST - 93.092*TVDI \quad r^2 = 0.84 \qquad Adj.\ r^2 = 0.72 \qquad SEOE = 37.39 \tag{2}$$

Where, Y = BPH pest count levels in numbers

This model was developed using MODIS data for classified rice pixels only at 1 km resolution. Within 1 km the variation of pest incidence level can be known at 56 m resolution using NDWI derived from AWiFS data

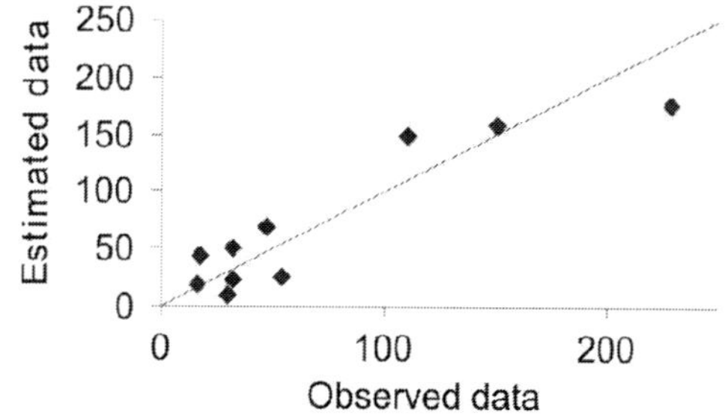

Figure 5 : Model Estimated Pest Count Data v/s Observed Data with 1:1 Line

The multiple regression model of eq. 2 was used to estimate the pest count data over the two years data points (figure 5). This shows that the RS indices have good correlation with the pest severity level in the infested crop.

3.3. Development of RS based Model at 56 m Resolution

The variation of NDWI index with pest infestation level showed strong negative correlation. The relationship shows the hopper burn patches produced by high BPH concentration causes low NDWI values due to low leaf wetness characteristics in hopper burn areas.

The variation of NDWI index with pest infestation

$$Y = -1376.88 + 720.42 * NDWI \quad (3)$$

$r^2 = 0.81$ Adj. $r^2 = 0.77$ SEOE = 37.30

where, Y = BPH population count in numbers

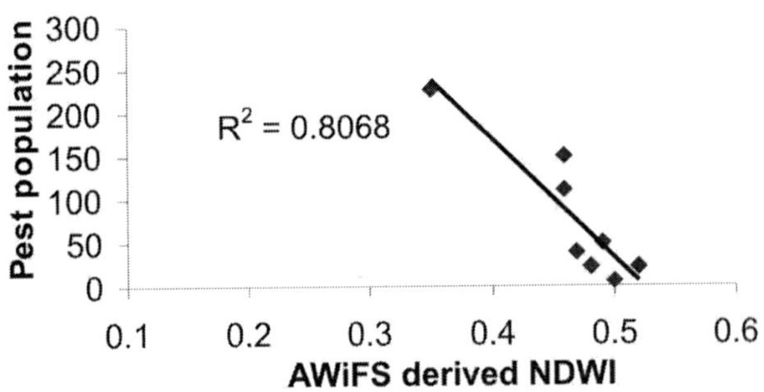

Figure 6 : Estimated BPH Population Count with AWiFS Derived NDWI Index

3.4. Identification of Pest Infected zones using AWiFS RS Data

The rice crop classified images of 2007 and 2009 were superimposed to get the common rice growing areas which are considered to be endemic BPH infested areas. In 2009 data the multidate classification showed that the infestation occurred at sites where the late maturing rice were dominant. Although all rice were transplanted at the same time the late maturing rice growing areas were affected by BPH this year as reported by ground truth GPS locations (Fig. 7). These areas could be identified using multi-date AWiFS data by decision rule based classification. The NDWI based reflectance data was used to depict the pest affected sites at 56 m resolution. Fig.6 shows the classification of early and late maturing rice areas. Figure. 8 shows the endemic BPH infested areas where rice was grown persistently in 2007 and 2009.

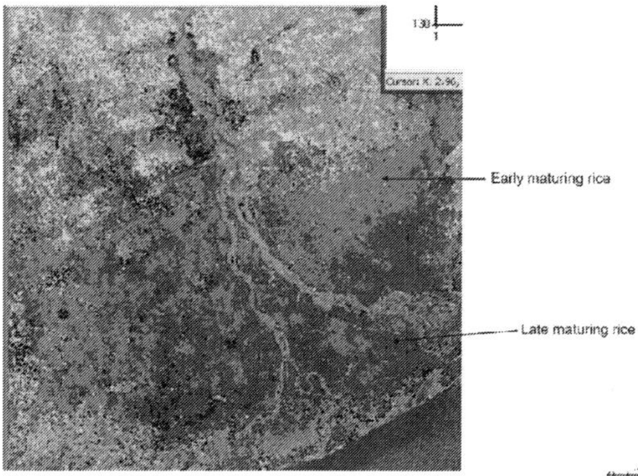

Figure 7 : Classified Early and Late Maturing Rice in 2009.

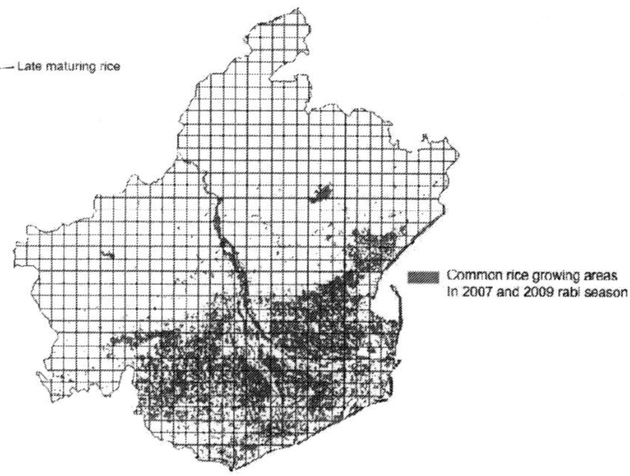

Figure 8 : Common Rice Growing Areas in 2007 and 2009.

3.5 A Prototype for Spatial Level Depiction of Pest Incidence Using RS and GIS

A grid (5 x 5 km) was prepared and all weather and RS derived parameters were incorporated after overlaying the grid. The minimum and maximum temperature and RH derived from AIRS sensor was obtained at 50 km resolution. The ground based BPH incidence model described in eq. 1 will be used and replace ground based temperature to satellite derived temperature to depict the pest infestation at 50km grid interval. Assuming that the temperature and RH remained same for every 10 grids, the spatial variability of other RS based indices has been depicted at each 5 x 5 km grid at weekly interval using GIS (Fig. 9). Using MODIS data the 5 km grid would be attempted along with the model developed in eq. 2. This approach would show the spatial variability of crop growing environment like leaf wetness factor, chlorophyll content and soil moisture status of the pest infected areas of the study region. Further at 56 m resolution the AWiFS derived model given in eq. 3 would be used to specifically locate the foci of infestation within the endemic rice growing areas.

This model was used to demonstrate a procedure for spatial level implementation of pest prediction model and validation of the model prepared at a grid level as discussed above.

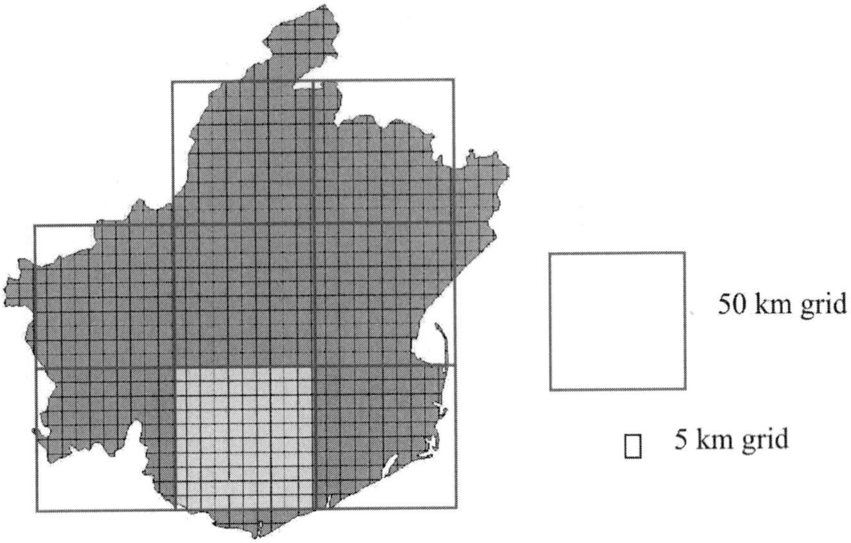

Figure 9 : Spatial Depiction of Input Parameters Using RS and GIS for Godavari Region

4. Conclusion

An attempt to locate foci of a major pest (BPH) infestation in rice crop was made in three stages from district level to field level. At first, district level weather based model was used to develop pest infestation severity level using two years' data of 2007 and 2009. Second, a pest model was developed at one km resolution using pest and RS derived indices for those two years. Lastly, at 56 m resolution a pest and NDWI index based model was developed to locate the foci of infestation. A prototype of spatial distribution of pest infestation will be applied at grid level using combination of RS and GIS which will be further enhanced using satellite derived meteorological parameters to give regional level likely pest infestation zones.

Acknowledgements

Authors are grateful to Dr. R. R. Navalgund, Director, SAC for his constant encouragement and support for this project. Guidance and support provided by Dr. J.S. Parihar, Deputy Director, SAC, Dr. S. Panigrahy, Group Director, SAC and Dr. N.K. Patel, Head, CMD Division, SAC is thankfully acknowledged. The facilities provided by Dr., Dr. S. Venkateswarulu, Director, CRIDA and Dr. S. Bandopadhyay, National Coordinator, National Agricultural Innovation Project (NAIP), the funding received for the investigation from the Indian Council of Agricultural Research under the World Bank Funded NAIP and the MODIS data made available through internet are gratefully acknowledged.

References

Beasley C. A., and Adams C. J., 1996. Field-based, Degree-Day Model for Pink Bollworm (Lepidoptera: Gelechiidae) Development. *Journal of Economic Entomology,* 89: 881–890.

Chu C. C, and Henneberry T. J., 1992. Pink Bollworm (Lepidoptera: Gelechiidae) Male Moth Trap Catches in Gossyplure-Baited Traps in Relation to Accumulated Heat Units. *The Southwestern Entomologist,* **17**: 79–87.

Dutta S., Bhttacharya B.K., Rajak D.R., Chattopadhyay C., Dadhwal V.K., Patel N.K., Parihar J.S. and Verma R.S., 2008. Modeling Regional Level Spatial Distribution of Aphid (*Lipaphis erysimi*) Growth in Indian Mustard Using Satellite -Based Remote Sensing Data. *International Journal of Pest Management,* 54(1): 51-62

Gergis M. F, Moftah E A, Soliman M. A, and Khidr A. A., 1990. Temperature-Dependent Development and Functional Responses of Pink Bollworm, *Pectinophora gossypiella. Assiut Journal of Agricultural Science,* 21: 119–128.

Heinrichs E.A., and Mochida O., 1984. From Secondary to Major Pest Status: The Case of Insecticide-Induced Rice Brown Planthopper, *Nilaparvata lugens,* Resurgence. *Protection Ecology,* 7(2/3): 201-218.

Higley L. G., Pedigo L. P., and Ostlie K. R., 1986. DEGDAY - A Program for Calculating Degree-Days and Assumptions Behind Degree-Day Approach. *Environmental Entomology,* 15: 999-1016.

IRRI, 1975. Brown Planthopper Infestation in Various Countries. *The Entomology Newsletter,* International Rice Research Institute, Los Banos, Laguna, Philippines. 3(3).

IRRI, 1979. Brown planthopper: Threat to Rice Production in Asia. International Rice Research Institute, Los Banos, Laguna, Philippines. [6+], 369 pp.

Kim J.D., Kim H.J., Rho S.P., and Bae S.H., 1984. Analysis of Damage on Rice by Brown Planthopper (Nilaparvata lugens StÅl). Effects of Infested Stages of Rice and Appearance Days of Hopper-Burn on Yield Loss. *Korean Journal of Plant Protection,* 23(3): 153-157.

Manjunath T. M., Rai P.S, and Gowda G., 1978. Parasites and predators of *Nilaparvata lugens* in India. *PANS,* 24(3): 265-269, 387

Prabhakar M., Ramakrishna Y.S., Prasad, Y.G., Reddy C.S., Krishnaiah N.V., Krishna Reddy P., Masood Khan M.A., NagaLakshmi T., and Yella Reddy, D., 2002. Weather based forewarnoing models for insect pests and disease of Rice. *Proceedings of the National Symposim on Crop Weather Models in Agriculture,* UAS, Dharwad, Dec.19, 2002.

Sandholta I., Rasmussen K., and Andersen J., 2002. A Simple Interpretation of the Surface Temperature/Vegetation Index Space for Assessment of Surface Moisture Status. *Remote Sensing of Environment,*79 : 213-224.

Strand, J.F. 2000. Some Agrometeorological Aspects of Pest and Disease Management for the 21st Century. *Agricultural and Forest Meteorology,* 103: 73-82.

Venkataraman, S. and Krishnan, A., 1992. Weather in the Incidence and Control of Pests and Diseases. In: Venkataraman, S. and Krishnan, A., (Eds.) *Crops and Weather*, Publications and Information Division, Indian Council of Agricultural Research: 259-302.

Wilson L. T., and Barnett W. W., 1983. Degree Days: An Aid in Crop Pest Management. *California Agriculture*, 37: 4-6.

Willrich, M. M., Leonard, B. R., Gable, R. H., and LaMotte, L. R. , 2004. Boll Injury and Yield Losses in Cotton Associated with Brown Stink Bug (Heteroptera: Pentatomidae) during flowering. *Journal of Economic Entomology*, 97: 1928-1934.

Yang X., Wu, J. J., Shi, P. J. and Yan, F., 2008. Modified Triangle Method to Estimate Soil Moisture Status with Moderate Resolution Imaging Spectro-Radiometer (MODIS) Products. *The International Archives of the Photogrammetry, Remote Sensing and Spatial Information Sciences*. Vol. XXXVII. Part B8. Beijing 2008: 555-560.

20

Techmode Approach for Distance Learning Courses for Field Veterinarians

P.S. Lonkar, A.S.Ninawe and Ajit Maru

ABSTRACT

Veterinarians, appointed in field, rarely get the opportunity to update their knowledge and skills, through trainings or courses, due to their working conditions. More over, they feel isolated and neglected from colleges due to poor interactions. Looking to this need, the courses on Post Mortem Techniques and Diagnosis of Parasitic Diseases in Livestock were developed, based on the Need Assessment survey. The courses were developed on interactive CD ROM with audio, video clips and photographs along with some reading material. The courses were of one credit each and had minimum one face-to-face session. The response by the veterinarians was tremendous and around 210 veterinarians offered for the courses by paying fees. Participants and stakeholders expressed the need for additional courses and accordingly two more courses on surgical techniques in large and small animals were launched which also received good response.

Introduction

Importance of Animal wealth in Indian economy is well established. Animal holders, usually small and marginal farmers and agrarians depend on field veterinarians, mostly in Government or cooperative sector for solutions to their animals' production problems.

Their linkages with the College or University are also meager. Updating their knowledge and skills, though essential, is not possible due to paucity of resources, financial crunch and their inability to remain away from their assignments for a longer period to attend face-to-face sessions.

Distance learning, through information and communication technology, was thought to be a useful tool for the same. The Maharashtra Animal and Fisheries Science University (MAFSU) initiated a Continuing Veterinary Education Program for Field Veterinarians in 2002.

Methodology

To understand the needs of field veterinarians, a survey was conducted where in 288 veterinarians responded. The analysis revealed the need for such courses. The people though had poor computer facilities, appreciated the CD based courses or even online. The compulsion of MS CIT and accordingly two courses namely.

1. Necropsy Techniques and
2. Diagnosis of Parasitic Disease were developed

Content Development

The courses were developed on Interactive CD ROM with some write up in the form on small booklet. The main concept kept in the mind was learning rather than teaching. In modern era, learning is considered as a life long process.

The course was divided in to chapters and each chapter was further subdivided in to modules. The contents in each module comprised of maximum 5-6 lines. Hyperlinks were provided in each module for the important terms/ subjects/topics and details of the same were provided in the CD.

Guidelines for operating the CD were provided to the participants, though the CDs were prepared with *AUTORUN*. The benefit of the CD is its Pedagogy, where in one can select and shift to the chapter he wishes. He can also select the topics/ modules/ illustrations and videos.

Standard formats for recording the postmortems and for submission of the materials to various laboratories are also provided. Model questions, with options are also provided. The examination is developed keeping in view the online programmes.

Implementation

The courses were offered to the qualified veterinarians with registrations by the Veterinary Council of India (VCI). The courses, though prepared with an objective of online availability, initially were launched at two places with a total of 210 participants, on payment of fees of Rs. 500/- for each course. Most of the participants were able to open the CD and try to understand about the compatibility of the computer programmes. All the difficulties and queries were answered through phone or e-mail. Two face-to-face sessions were conducted for the participants where they asked the questions, even not related to the course. Out of 210, 140 participants appeared for the examinations, mid term and final, with total 100 marks of objective type. After the initial success and assessment, two additional courses, namely:

1. Surgical techniques in Large Animals and
2. Surgical techniques in Small Animals were developed.

They were also appreciated by the stake holders.

An assessment of the program for Commonwealth of Learning (COL) was made in 2004 by Dr. Ajit Maru. The assessment report, in addition to other issues, indicated:

Follow-up and "hand-holding" will be needed for development of at least four more courses to provide a deeper experience to a larger group of faculty to make courseware development fully self-sustained in MAFSU. The experience has to spread from the Bombay Veterinary College to include faculty from the seven other colleges of the University.

MAFSU has very poor capacities in enabling effective ICT use and information and communication management (ICM) in University administration and management and its libraries and information centers. Without strengthening this area, especially in student records management systems, decentralized student administration and inter-institutional collaboration sustaining more Continuing Veterinary Education (CVE) activities will be very difficult. MAFSU requires an ICT use and ICM audit urgently and concerted efforts to bridge the gaps and weaknesses.

There is a need to develop a "learning objects" repository at MAFSU and in collaboration with other similar Institutions. Preliminary efforts, through collection of digital objects of pathological conditions and parasites have been very successful but needs systematic organization.

After the report, the University has developed four more courses in Surgical Techniques in Large and in Small Animals, Herd Health Management and Poultry Disease Diagnosis. All these courses have been developed only at Bombay Veterinary College where more faculty were trained in use of ICTs in ODL by Dr. Ajit Maru during August-September 2004. Participation of faculty in course development and production from the other six colleges of MAFSU has been limited by lack of training of faculty outside Mumbai.

The University is constrained by funds, necessary skills and bureaucratic processes to implement all recommended areas of improvements in its ICT infrastructure and ICT use.

The University lacks skills in managing its learning objects. As experienced in the development of the ODL courses, this not only limits rapid development of new courses and increases costs but also limits collaboration across Departments in designing new courses which cover several disciplines and are of more practical use for field veterinarians.

The initial momentum of the CVE program at MAFSU that uses the TECHMODE approach as being evolved under guidance by COL was gained through kick start funds provided by COL. This went a long way to convince the faculty and the management of the University of the potential of ODL for CVE and the hidden capacity of the University for providing ODL. Further, the outputs and outcomes of the COL supported CVE project has gained the attention both of the Indian Council of Agricultural Research and the Veterinary Council of India as being pioneering. Many State Agricultural and Animal Sciences Universities have now started inquiring about initiating ODL in their own Institutions.

This momentum can be continued and can be further accelerated to offer CVE to more field veterinarians and in a larger area, which include not only in Maharashtra but also Western India covering the states of Gujarat, Southern Rajasthan, Western Madhya Pradesh, North Karnataka, Western Andhra Pradesh and Goa. The area has about 12000 field veterinarians. Further, the experiences of MAFSU, if shared as per the initial project proposal for initiating CVE at MAFSU and TANUVAS in 2002, could bring TechMODE applications across all state agricultural and animal sciences universities and Institutions in India and elsewhere in the world.

At the moment there are three major constraints for the enlargement and expansion of the CVE program for field veterinarian and its further institutionalization at MAFSU.

1. Lack of trained faculty so that the repertoire of courses available can be increased in number and quality.

2. Lack of a computerized method, preferably on-line through use of the Internet, for enrollment and evaluation of participants in the CVE which will allow more enrollment, registration and evaluation across a larger area than is possible by the manual administrative and evaluation methods followed at the moment by MAFSU from Bombay Veterinary College.
3. Lack of a Learning Objects Repository (LOR) and the skills to manage it so as to lower the costs of course ware up-gradation and development.

The constraints for MAFSU to initiate action in this area are bureaucracy within the University, shortage of funds and the necessary skills and experience. COL, through its technical support and funds to "kick start" the process, as it did to initiate the 1st phase of the CVE project, can help MAFSU further institutionalize TechMODE, build greater capacity and confidence and be self sufficient in providing CVE.

The following proposal includes three projects for technical support and part funding by Commonwealth of Learning:

1. To train more faculty at MAFSU.
2. Establish an on-line enrollment, registration and evaluation system for CVE, if possible in collaboration with Maharashtra Knowledge Corporation Limited (MKCL).
3. Initiating development of an LOR in MAFSU.

Experiences Gained

1. Development of the faculty suitable for developing such courses.
2. Awareness amongst the staff and participants regarding the use of ICT.
3. Confidence amongst the staff and participants for having access to communicate with each other for a common cause and medium.
4. Feeling of being a part of the system, inspite of leaving the college, by the participants.
5. Development of linkages and clients-all students-which may further help in numerous collaborative projects.
6. Participants expressed the views for new courses and suggested even the topics.

Weakness

1. Non-availability of computer at some places.
2. Paucity and unsuitable time for using office computers.

3. Phobia for computers, especially among the senior participants.
4. Mentality for asking benefits.
5. Apathy by some, thinking of no loss if not participated- Why to do?
6. Inability to express the needs by some people-shy of expressing the difficulties *etc.* mostly due to age, experience and authority.
7. Face to face session is not possible if sufficient number of participants is not available.
8. Chances of missing the sessions or examinations due to transfer, family or personal problems and official assignments.
9. Modified courses will have to be provided to the participants regularly, as they are not online.

Suggested Solutions and Plans

1. To have the courses and admissions online so that the participants may register at any time and they may appear for examination after stipulated period (12 weeks).
2. The courses when kept online-modifications would be easy.
3. Contact point for face-to-face sessions can be many, at all the veterinary colleges of the Maharashtra Animal and Fishery Sciences University and students may come there as per need with prior intimation and appointment.
4. Examination would be online-may be at various centers.
5. As observed, many of the field veterinarians rarely get a chance for upgrading to advancement of his knowledge and skills. They may get this opportunity. Incentive can be in the form of certificate and participants completing stipulated number of courses may be awarded suitable diploma *etc.*
6. Administration should encourage the acquired additional qualification/ certificates/skills and desire to learn and acknowledge and reward them by some incentives.
7. Basic computer knowledge-certificate by an authorized agency- Maharashtra Knowledge Corporation Limited- is compulsory for the Government staff, same pattern maybe used for the field veterinarians also.

Major concern may be the quality of the course materials. Continuous monitoring, up gradation and evaluation of the same is essential, where in very few experts are available.

Acknowledgements

Authors are thankful to the Commonwealth of Learning for financial and technical help and authorities of Bombay Veterinary College and Maharashtra Animal & Fishery Science University for encouragement and the facilities.

21

Information Retrieval System for Buffalo Reproduction

Y. P. Singh, S. K. Singh, S.K. Agrawal, S. K. Ghosh, G.K. Das, G. S. Bisht and Sanjay Kumar

ABSTRACT

The domestic buffalo occupies an important niche in agricultural systems by providing milk, meat and draught power. Indian buffalo population which is about 1/3 of the cattle population contributes more than half of the total milk production of the country. Several biological and socio-economic problems restricted the efficiency of buffalo production which needs careful and intensive research. Studies pertaining to reproduction are one of the important areas in buffalo research. The reproductive process in buffalo is complex in nature, which differs on many aspects with the cattle. There is a need to undertake separate strategies to understand buffalo reproduction to disseminate the information to the livestock owners and researchers. Information Communication Technology (ICT) initiatives can provide a strong base for effective coverage and dissemination of livestock outreach information to the farming community. An information system was developed on buffalo reproduction in the present study. Information system covers different aspects of buffalo reproduction which includes information on male and female reproductive system as broad subjects and dealt individually. This information retrieval system was designed using Adobe Flex3.0 with Action Script 3.0 language. System will assist veterinary professionals and persons engaged in buffalo management to learn/ understand the buffalo reproduction in self teaching mode at their own speed. It will also assist to understand the complex phenomena of buffalo reproduction and diagnosis of reproductive disorders in the easy and better mode.

Introduction

The domestic buffalo occupies an important niche in Indian livestock industry, providing milk, meat and draught power. Indian buffalo population, which is about 1/3 of the cattle population, contributes more than half of the total milk production of the country. Over last few decades, this species has attracted considerable scientific attention in many parts of the world especially where buffalo forms an integral part of farming system. Food and Agricultural Organization (FAO) has also emphasized the potential value of the water buffalo (*Bubalus bubalis*) and endeavored to stimulate interest to carry out research in buffalo to reduce the gaps in knowledge and to establish an up-to-date documentation on buffalo reproduction.

Several biological and socio-economic problems restricted the efficiency of buffalo production which needs careful and intensive research. Studies pertaining to reproduction are one of the important areas in buffalo research. During the year 1978, FAO organized a seminar on "Reproduction and Artificial Insemination of Buffaloes" of which one of the important objectives was to establish an up-to-date documentation on buffalo reproduction. After a lapse of about two and half decades, many facets in buffalo reproduction are still poorly understood. Besides, the documentation in the form of book, bulletin, and manual *etc.* on buffalo reproduction as a whole is very meager or not available (Agarwal and Tomer, 2003) and as such no electronic form is available.

The reproductive process in buffalo is complex in nature which differs with cattle. Despite of the above facts, most of the work attempted on buffaloes are by simulating cattle resulted into limited success or underwent unsuccessful attempts. Hence, there is an immediate need to undertake separate strategies to understand buffalo reproduction for exploiting maximum productive potential from the species. In India, the veterinary services in the field are limited. The challenge to reach millions of livestock owners spread in length and breadth of the country remains insurmountable. Thus, it requires developing an alternate mode of communication for transferring the knowledge as well as latest technologies evolved in buffalo reproduction to the ultimate stakeholders rearing the species. Information Communication Technology (ICT) initiatives provide a strong case for effective coverage and dissemination of livestock outreach information to the farming community (Reddy and Reddy, 2009).

The ICT is a combination of computer and communication technologies, mainly used for processing, storing and retrieving of information in multimedia format (text, graphics, video & sound). Thus, it is the base for Information Systems. India is experimenting with Information and Communication

Technology (ICT) projects in different sectors and circumstances. ICTs are being used in agricultural development to provide market information, extension advice, information about rural development programmes and other information from government and private sources to end users. The past research experiences suggested that there are benefits in incorporation of ICTs in all endeavors related to livestock development.

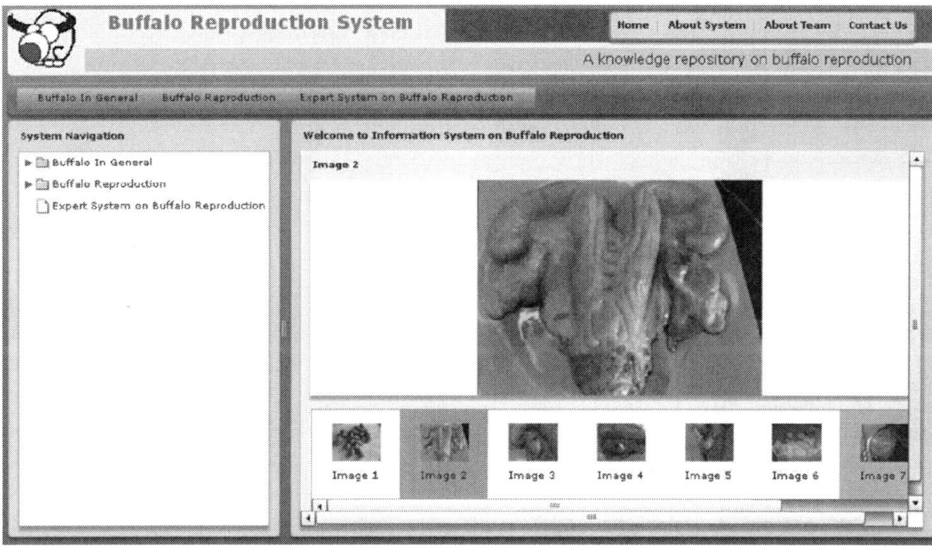

Keeping in view, a study was carried out to develop a web based information retrieval system on different aspect of buffalo reproduction. The practical utility of this system is that it will assist veterinary professionals and persons engaged in buffalo management to learn/ understand the buffalo reproduction in self-teaching-learning mode at their own speed. It will also assist to understand the complex phenomena of buffalo reproduction and diagnosis of reproductive disorders in buffalo in the easier mode. The ultimate goal of the proposed expert system is to improve the fertility and productivity by acquiring and maintaining expertise on different aspects of reproduction and providing expertise in an integrated package to work as a permanent consultant for the professionals.

Design Methodology

A typical web based information system design methodology has the following system architecture.

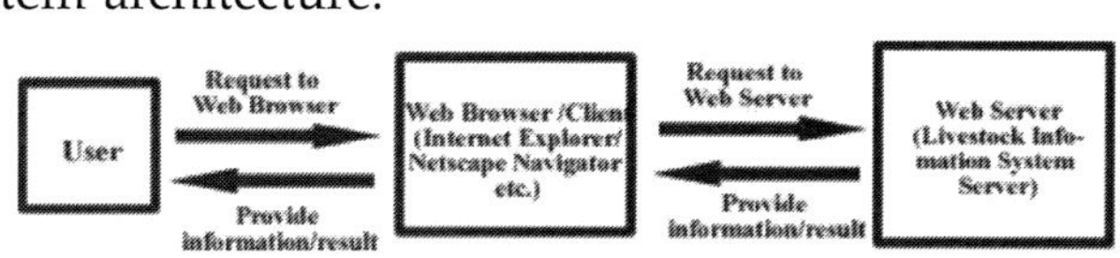

The system will be a Rich Internet Application (RIA), which can be used in internet as well as a stand alone system. Rich Internet applications are fast becoming an important differentiator in the area of web presence. RIAs combine the reach of the Internet with a rich and compelling user interface that provides greater level of satisfaction and success in a user's web-based interactions. Rich Internet applications can provide a significantly improved user experience for many different types of user tasks. There was a option to develop this information system using any of the prevailing tools such as AJAX, JAVA or FLASH technology but looking at the requirement as we need animations *etc.,* hence, it was decided to develop this application using FLASH technology (Hammond *et al,* 2006).

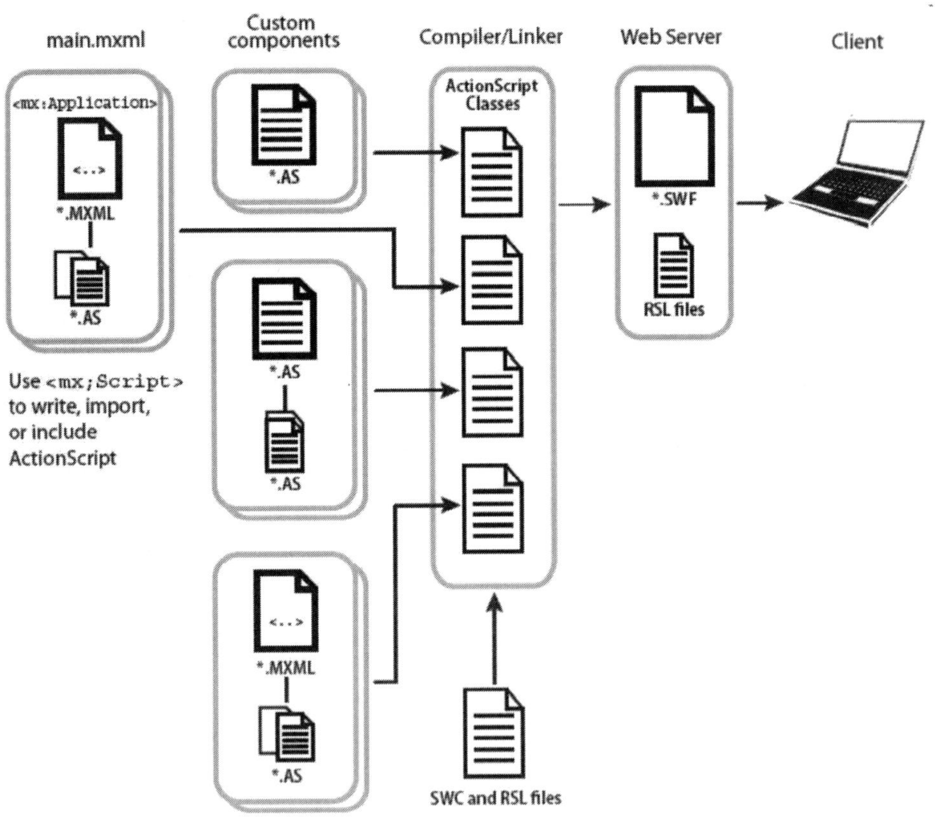

Flex applications are created by using a combination of the declarative markup language, MXML and the scripting language ActionScript. Both MXML and ActionScript were used together, to create the project. AJAX applications use HTML as the declarative markup language to lay out the user interface elements of application, and then use JavaScript to add interactivity. MXML and ActionScript work much the same way, with MXML often serving to create the visual layout of application and ActionScript adding powerful programming capabilities. By compiling the application, it converts application files and assets into a single SWF file.

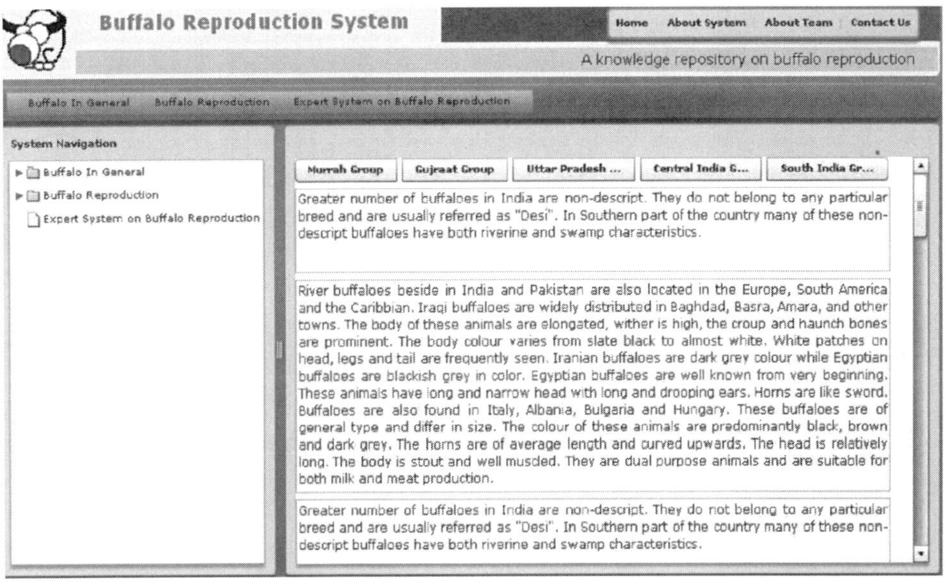

Results

Information system on Buffalo Reproduction is designed in three main parts *viz*. General, Buffalo Reproduction and Expert System for Reproductive Diseases/ disorders.

'Buffalo in General link' cover the buffalo reproduction in general, breeds, its population and production trend, importance in livestock production system, landmarks in buffalo reproduction and measurement of reproductive efficiency.

The female reproduction system interface deliver the information on female reproductive organs, embryological development, morphology and their functions, endocrinology of female reproduction, puberty and sexual maturity, estrus and estrus cycle, gynecological examination and reproductive disorders.

Female reproductive system chapter delivers information on female sexual organs and their functions like ovaries includes folliculogenesis and corpus luteum development, oviducts, uterus, cervix, vagina and vulva. It also includes information on puberty and sexual maturity, gestation length, parturition, calving interval and seasonality in behavior.

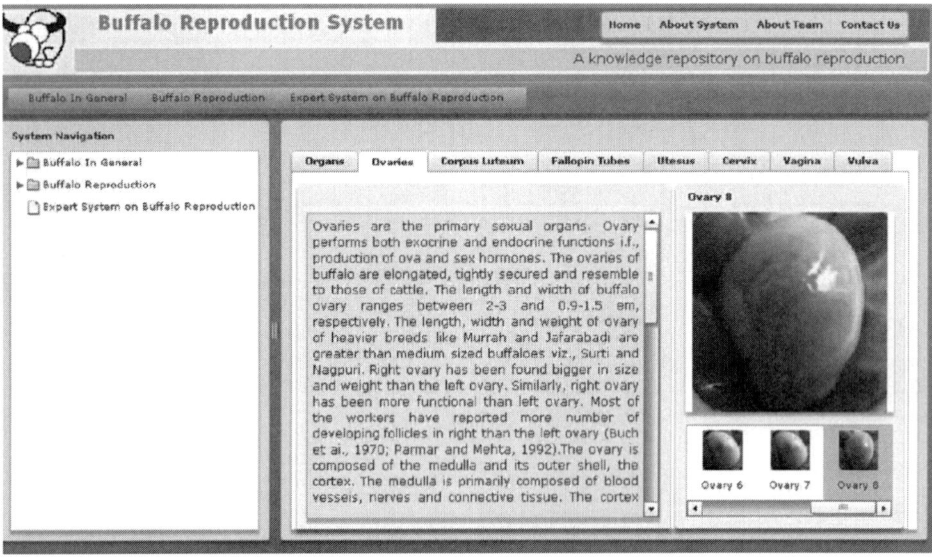

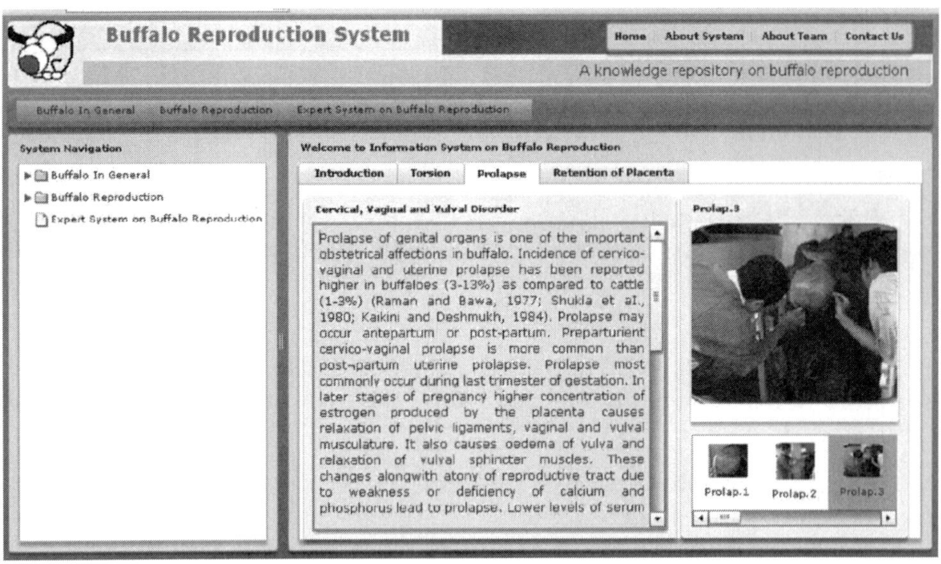

Estrus and estrus cycle includes information on duration and time, cycle length, endocrinology of estrus, symptoms of estrus, techniques for estrus detection and synchronization of estrus. Pregnancy diagnosis link shows information on placentation and methods of pregnancy diagnosis during first, second and third trimesters.

Reproductive disorder depicts information of various type of reproductive diseases prevailing in buffalo.

The male reproduction system interface deliver the information on male reproductive organs, puberty and sexual maturity, endocrinology of male reproduction, spermatogenesis, andrological examination, management, selection and training of bulls, examination of semen, semen collection and processing, semen preservation and handling, factors affecting Semen production, advance techniques for evaluation, artificial insemination and reproductive disorders, breeding soundness of bulls, reproductive behavior of bulls *etc.*

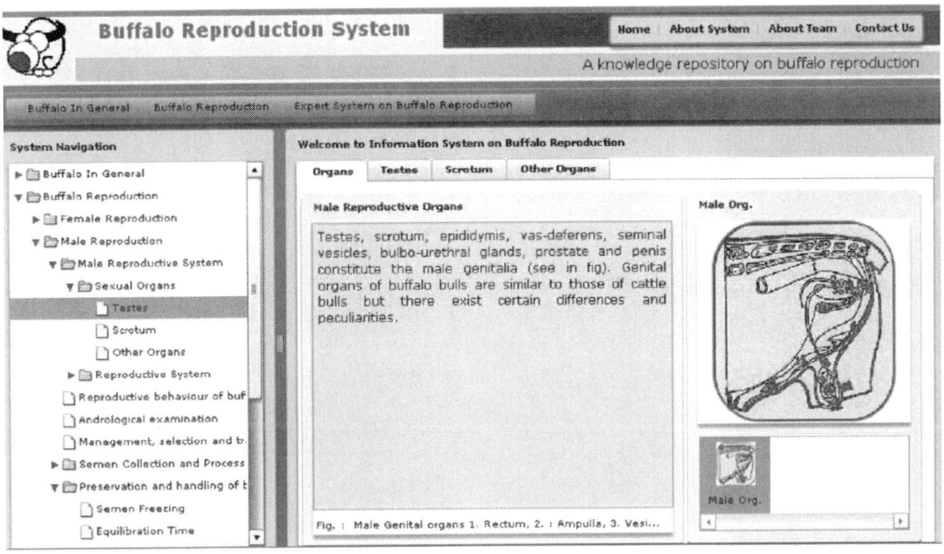

References

Agarwal, S.K. and Tomer, O. S., 2003. Reproductive Technologies in Buffalo- A Monograph. Published by Indian Veterinary Research Institute, Izatnagar.

Reddy, Sudarshan Aligireddy and Reddy, Krishna. 2009. The Application of ICT in Indian Agriculture: The Case of eSagu Model of Web-based Agricultural Expert Advice Dissemination System, http://t06.cgpublisher.com/proposals/343/index_html (Accessed on 8th August, 2009).

Hammond, Jeffrey S., Carey Schwaber, Ron Rogowski and Jacqueline Stone, 2006.- Ajax Or Flex?:How To Select RIA Technologies, http://www.forrester.com/Research/Document/Excerpt/0,7211,40989,00.html (Accessed on 8th August, 2009).

22

Useful Databases and Websites on Agroforestry and Farm Forestry

P. Ratha Krishnan and R. Indumathy

ABSTRACT

Probably since the beginning of agriculture, farmers have recognized the value of rows of trees/ shrubs planted in the farm (called as Agroforestry) which offers environmental protection, crop and livestock production and reduce the soil health hazards. The role of agroforestry is immense in the undulating terrains like North East India, rain fed area like Central India or the problem soils in general. Even though research and extension work is on the way to increase the trees out side the forest, the information and knowledge on best tree + crop combination is either scarce or limited to regions in the form of research articles, books or journals. The recent scope of Information and Communication Technology (ICT) and its applicability due to its easy, fast and any where accessibility have opened the window of opportunity towards the scientific agroforestry and farm forestry developments. One such means of ICT is through the availability of useful databases and websites on agroforestry. This article highlights the range of commercial (searchable for a fee) and non-commercial (freely searchable by anyone) databases and websites related to agroforestry and farm forestry. It gives clear and introductory information about the site address (URL), contents, and uses. In this aspect, the online databases and publications *viz.*, Agroforestry systems (www.kluweronline.com/issn/0167-4366/contents), Elsevier publishing (www.elsevier.com), Joint venture agroforestry Programme (www.rirdc.gov.au/programs/aft), CSIRO Journals (www.publish.csiro.au.journals), Infoscan databases (www.infoscan.com.au), World Agroforestry Centre (www.worldagroforestrycentre.org/level1a) and agroforestry BASE (www.mirror.iasri.res.in/net/nrcaf) were presented in this article. The list of international and national websites *viz.*, agroforestry.net and agroforester.com (www.agroforestry.net; www.agroforester.com), Subtropical Farm Forestry Association (www.turboweb.net.au), Agroforestry Research Trust (www.agroforestry.co.uk), Association for Temperate Agroforestry (www.missouri.edu/~ifta), Glossary for Agroforestry (www.bugwood.org/glossary/), USDA National Agroforestry Centre (www.unl.edu/nac/), World Agroforestry Centre (www.worldagroforestrycentre.org/level1a) and National Research Centre for Agroforestry (www.nrcaf.ernet.in) and short introduction of centers are also given in this article. This lists are not comprehensive, but the sites listed have links to a far wider range of international agroforestry and farm forestry sites, and hence, these are useful start points for looking for information. Using such online information hub to access the needed information is an art and need of the hour too. This kind of ICT and its role in agriculture/ agroforestry made the world as an e-village. But the applicability of information mainly depends on the local source availability, ecological status and the interest of the farmers/ tree growers.

Introduction

Agroforestry is practiced traditionally since time immemorial to meet diverse needs of households and presently much emphasis placed due to its potential of climate mitigation. Three decades of organized agroforestry research in India and the considerable research has been undertaken by many countries were not always easily accessible to the public. This can make it difficult to know what research has been undertaken, and what new research is still needed. Application of information technology in this anciently originated field is the needed strategy of this hour. Computer made the calculations easier, predicts accurately and also used for the information management and support decisions. Using computer/ information technology by the way of net work and databases for the dissemination and extension of technology in agroforestry can optimize the production while conserve the resources (Rathakrishnan and Indumathy, 2009). This article is an attempt to provide a list of useful databases and online indexes of publications and websites which provide research and related information on agroforestry and farm forestry.

Database

It is the collection of information or data which is possible to access in any mode and answerable for any query. It may be of a personal system supported files or available in the internet. It covers the information from many sources. It organizes and facilitates the management and querying of large quantities of data and information.

Importance of Database in Agroforestry

Three decades of agroforestry research in our country has led to development of several agroforestry systems that are based on people's need and site specific characteristics. Presently, the expectations from agroforestry are quite high in this country both in rural and urban areas. In recent years, sizeable increase in tree cover outside the forest areas has taken place in many states (FSI, 2005). However, it is realized that there is a tremendous scope of dissemination of knowledge to land owners, farmers and extension agents about not only the availability and suitability of various agroforestry options but also their management in such a complex environment. In order to achieve this, an appropriate dissemination mode in national and international level is the prerequisite.

The following are some of the useful databases range from commercial (searchable for a fee) to non-commercial (freely searchable by anyone) databases and websites related to agroforestry and farm forestry.

Table 1 : Following Table Gives the Type of Database and their Important Features

Sl.No.	Database	Important Features
1	AFSI (Agrofore stry Systems Inventory Database) (Oduol *et al.* 1988)	It describing geographic location , bio-physical, socio - economic and species characteristics.
2	MPTS (Multipurpose Tree & Shrub Database) (Schroder and Jaenicke, 1994)	Multipurpose tree and shrub database used for tree selection and species information
3	Agroforestry Tree Database (Salim *et al.* 1998)	Internet and CD -ROM application for referenc e & Selection guide of agroforestry trees.
4	Forestry Compendium (CABI 1998; Tree CDs)	Compilation of knowledge on forestry, Agroforestry and plantations and information on trees for management decision -making and species selection.
5	Southeastern Agrof orestry Decision Support System (SEADSS) (Ellis *et al.*, 2003)	Landscape and site -scale AF planning and species selection, DST for landowners and extension agents of Southeast US that integrates GIS, tree and shrub database and expert knowledge.
6	Agrofor estry BASE (Agroforestry Information System) (Ajit *et al.*, 2005)	An internet based agroforestry information quarry and data retrieval system.

Agroforestry Databases

Agroforestry Systems Inventory Databases (AFSI) has been developed by the World Agroforestry Centre (WAC), Nairobi, Kenya. It was done by a global collection of data/ information on agroforestry systems using a questionnaire as the survey instrument. Data and information collected and entered into the database included general description, geographical location, biophysical characteristics, socioeconomic aspects, system evaluation, components of the system and their uses and identification of research gaps. With AFSI, the user is able to query the database, extracting information such as geographical locations of different agroforestry systems and the species found within these systems in different locations (Nair, 1987). AFSI was apparently developed as a research and information tool for researchers. But, AFSI was failed to be maintained and upgraded and therefore, faded with time and presently no more.

Another database *viz.*, Multipurpose Tree and Shrub Database (MPTS) was developed by ICRAF in 1991 (Von Carlowitz *et al.*, 1991). It was meant for the researchers and extension agents as an aid to select the right tree or shrub species for agroforestry practices. This database was primarily meant for the tropics and subtropics containing information on 1093 species on the aspects related to site-specificity, morphological and phenological descriptions, management characteristics and environmental responses (Schroder and Jaenicke, 1994). A simple climate model was included to predict climatic conditions based on the input of geographical coordinates. Tree and shrub species were selected *via* a database search that matched the descriptors that the user selected. The descriptors included 19 different criteria covering aspects of location, climate and soil conditions, products, environmental services, management, cultivation and references on selected agroforestry species (Schroder and Jenicke, 1994). The current and revised version of MPTS is now the Agroforestry Tree Database (AFTD). It is a database management system intended for use by researchers and fieldworkers to select agroforestry trees that are being deliberately grown and managed for more than one output and expected to make significant economic and/or ecological impacts (Word Agroforestry Centre, 2003).

In India, the Web based application "Agroforestry BASE" was designed and developed by NRCAF, Jhansi (Ajit *et al.*, 2005). It includes six databases *viz.*, Research Project Database, Database on Economic Analysis, Database on MPTs, Agroforestry Interventions and Poplar and Eucalyptus based Bibliographic Systems. This data base is upgraded and serves the need of the researchers till date.

A. Online agroforestry related publications

1. *Agroforestry Systems*: http://www.kluweronline.com/issn/0167-4366/contents

   This is the primary international journal of agroforestry research. Subscriptions and articles are available to purchase online.

2. *CABI Agroforestry Abstracts*: http://www.cabi-publishing.org/journals/Abstract /AFA /Index.asp

   This is an online subscription service. Agroforestry Abstracts is a database of agroforestry papers - including journals, conference papers, reports and books - published internationally. It includes citations for Australian publications, and includes abstracts. The database has 13 years of citations and adds around 1500 citations annually. The subscription rates can be found on the website - an annual subscription must be taken out to use the database, or it can be accessed in libraries that subscribe to it. It is possible to have a free trial use of the database before subscribing.

3. *CSIRO Journals*: http://www.publish.csiro.au.journals/

   From this website you can access CSIRO journals with articles about agroforestry. The site provides subscription and pay-for-view services. Journals published by CSIRO with particular relevance to aspects of agroforestry include Australian Plant Pathology, Australian Journal of Agricultural Research, Australian Journal of Botany, Australian Journal of Experimental Agriculture, Australian Journal of Soil Research and Wildlife Research.

4. *Elsevier Publishing*: http://www.elsevier.com

   Elsevier publish several journals and many other publications with relevance to agroforestry. Journals of particular relevance include Agricultural Water Management, Forest Ecology and Management, and Agricultural Systems. All Elsevier publications can be searched for from the central website and most can be ordered online. There are subscription and pay-for-view services for journals.

5. *Infoscan Databases*: http://www.infoscan.com.au/

   This is one of the best online resources for finding citations of agroforestry, farm forestry, forestry and agricultural management research in Australia. You can search several databases online, and can also browse the database by topic - agroforestry is one of the topics that can be browsed. A considerable amount of research is cited in the databases with detailed descriptions of the citations given. You can do a 'Power Search' that searches all the Infoscan databases, or search any of the following databases separately:

- *ABOA*: Contains citations for published information on Australian agriculture, including abstracts for many of the citations.
- *ARRIP*: Contains citations for current research projects being undertaken on Australian agriculture and natural resources, including detailed descriptions of project goals, methods and progress.
- *ARRIP Archive*: Contains citations for completed research projects on Australian agriculture and natural resources, including detailed descriptions of project goals, methods and progress.
- *Streamline*: Contains citations for published information on Australia's natural resources, including abstracts for many of the citations.
- *Geosearch*: Allows the user to search for information related to a particular agro-ecological region or drainage division of Australia by clicking on a map interface.
- *NamesNet*: Allows users to search for contact details of researchers by name, organisation, or State/Country. It is then possible to link to details of research projects listed in ARRIP and ARRIP Archive by that researcher.

6. *USDA National Agroforestry Center* : http://www.unl.edu/nac/insideagroforestry

   "Inside Agroforestry" is published three times per year by the USDA National Agroforestry Center (NAC). It began as a Forest Service Research and State and Private Forestry effort in 1992 and expanded into a partnership with the Natural Resources Conservation Service (NRCS) in 1995. The USDA National Agroforestry Center (NAC) is a partnership of the Forest Service (Research and Development and State and Private Forestry) and the Natural Resources Conservation Service. It is administered by the Forest Service, Southern Research Station; and its program manager and headquarters are located in Huntsville, AL, on the campus of Alabama A&M University, while NAC's staff are located at the University of Nebraska, Lincoln, NE; University of Idaho, Moscow, ID; and in Blacksburg, VA. NAC's purpose is to accelerate the development and application of agroforestry technologies to attain more economically, environmentally, and socially sustainable land use systems. To accomplish its mission, NAC interacts with a national network of partners and cooperators to conduct research, develop technologies and tools, establish demonstrations, and provide useful information to natural resource professionals.

B. Agroforestry Websites (Jacki Schirmer, 2001)

1. *Joint Venture Agroforestry Program*: http://www.rirdc.gov.au/programs/aft.html

   This is the homepage of the Joint Venture Agroforestry Program of the Rural Industries Research and Development Corporation. Many research reports on agroforestry and farm forestry can be downloaded free (only for Australian) from this site; others can be ordered online.

2. *World Agroforestry Centre*: http://www.worldagroforestrycentre.org/level1a.htm

   This website provides access to a wide range of resources on agroforestry & farm forestry, including databases. The following databases can be accessed from the site:

   - *The Agroforestree Database*: Provides information on the management, use and ecology of a large number of tree species that can be used in agroforestry systems
   - *The Tree Seed Suppliers Directory*: Provides a list of suppliers of seed and microsymbiont for nearly 6000 tree species, including 26 Australian suppliers.
   - *The Botanical Nomenclature Database*: Provides information on taxonomy of trees including synonyms and common names.

3. *Subtropical Farm Forestry Association*: http://www3.turboweb.net.au/~sffa/

   The Subtropical Farm Forestry Association is an association for agroforesters and farm foresters in the subtropics providing a range of information and services.

4. Agroforestry.net and agroforester.com: http://www.agroforestry.net/ & http://www.agroforester.com/

   These two sites provide online information and resources for tropical agroforestry and forestry, including a range of publications, online information, and links to other sites. You can subscribe to a free email newsletter on agroforestry, *The Overstory*, and access all back issues (over 100) of the newsletter online.

5. *Agroforestry Research Trust*: http://www.agroforestry.co.uk/

   The ART is a UK based organisation and focuses on temperate agroforestry systems. The information on the website is primarily focussed on agroforestry in the British and European context. The ART produces *Agroforestry News*, a quarterly journal (not to be confused with Australia's *Agroforestry News*, for which articles are cited in the *Farm Forestry and Agroforestry Reference Library* database), and other publications are available online.

6. *Association for Temperate Agroforestry*: http://www.missouri.edu/~afta/

   This organization is based in the USA and most of its resources were developed with the USA in mind. The website is not large, but does contain links to online publications. Some publications can be ordered online.

7. *Glossary for Agroforestry*: http://www.bugwood.org/glossary/

   Useful glossary of agroforestry terms.

8. *USDA National Agroforestry Center*: http://www.unl.edu/nac/

   This website has a range of information on agroforestry, centered on information relevant to the United States. It contains a number of useful links to other sites. A number of publications are available for download from the site.

9. *World Agroforestry Centre*: http://www.worldagroforestrycentre.org/level1a.htm

   This website contains a large range of information on agroforestry and farm forestry, targeted primarily at agroforestry development in developing countries. The World Agroforestry Centre was previously called the International Centre for Research in Agroforestry. The site contains considerable information on a wide range of aspects of agroforestry, links to many other sites, and access to three databases (The Agroforestree Database, The Tree Seed Suppliers Directory, and The Botanical Nomenclature Database, as listed above in the guide to useful databases).

10. *FAO Sites Related to Agroforestry:*

a. *Forestry Department* : Genetic resources, Forest Resource Assessment and Non- Wood Forest Products (FORM : Forest Resources Development Service). http://www.fao.org/forestry/foris/webview

b. Agriculture Department

   1. Livestock, Environment And Development (LEAD) : Initiative by AGAL (Animal Production and Health, Livestock Information, Sector Analysis and Policy Branch).

      http://www.virtualcentre.org/selector

2. Dry lands forestry and agroforestry: Initiative by AGLL (Land and Water Development Division, Land and Plant Nutrition Management Service).

   http://www.fao.org/ag/agl/agll/drylands/forestry

3. Grassland and pasture crops website: Initative by AGPC (Plant Production and Protection Division, Crop and Grasslands Service).

   http://www.fao.org/ag/AGP/AGPC/doc/pasture/pasture

C. *Regional site*: Asia-Pacific region, Forestry department APANews : Asia-Pacific Agroforestry Newsletter.

   http://www.fao.org/forestry/foris/webview/forestry2

11. Non-FAO Sites Related to Agroforestry

Agroforestry.net	http://www.agroforestry.net/
CATIE: Tropical Agricultural Research and Higher Education Center	http://www.catie.ac.cr/Magazin.asp
CIGAR: Consultative Group on International Agricultural Research	http://www.cgiar.org/centers/index.html
Alternative to Slash-and-Burn Programme	http://www.asb.cgiar.org/home.html
CIFOR: Center for International Forestry Research	http://www.cifor.cgiar.org/docs/_ref/research_tools/flores/
CIAT: International Center for Tropical Agriculture : Research and Development	http://www.ciat.cgiar.org/asia/rfd_series.htm
ICRAF : World Agroforestry Centre	http://www.worldagroforestry.org/
IFOAM Organic Agriculture	http://www.ifoam.org/
ILEIA: Centre for Information on Low	
External Input and Sustainable Agriculture	http://www.leisa.info/
Regional Land Management Unit	http://www.relma.org/
CTA: Technical Center for Agricultural and Rural Cooperation	http://www.cta.int/index.htm

Conclusion

The progress made by agriculture in the last four decades has been one of the biggest success stories of free India. Agriculture and its allied activities are the means of livelihood of about 2/3 of the work force in the country. But, the present climatic uncertainties, increased cost of cultivation and scarcity of labor made the farmers to opt for the alternate land use technologies like agroforestry. The use of computer-based technologies in agroforestry and natural resource management has been impressive to tackle the threats. Even though providing solution to the agroforestry queries are region/ farm level specific the farmer's query can be collected and stored in the database and maintained in the proper Relational Database Management Systsm (RDBMS) with corresponding data dictionaries. According to the queries answers can be provided by researchers, expert in that field or by refereeing the above databases and websites. Already such kind of networking is going on as gyandoot, echoupal, ikisan, n-logue, *etc.* This will serve the answers with in short duration as well as helps to formulate the future strategies of research in agroforestry. Since this serves information to any where, any time and any person, it assures the availability of information for the development of rural areas.

References

Ajit, Rai, P., and Handa, A.K., 2005. Structure Design and Utility of Agroforestry Information System. *Agroforestry BASE.* National Research Centre for Agroforestry, Jhansi.

CABI (Commonwealth Agricultural Bureau International), 1998. *The Forestry Compendium : A Silvicultural Reference.* CABI, Oxford, U. K. CD-ROM.

Ellis E.A., Nair P.K.R. and Jeswani S.D., 2003. The Southeastern Agroforestry Decision Support System (SEADSS): A Web-Based Application for Agroforestry Planning and Tree Selection. In: Vacik H., Lexer M.J., Rauscher M.H., Reynolds, K.M. and Brooks R.T. (Eds). *Decision Support for Multiple Purpose Forestry.* Pp.1-12.

FSI, 2005. State of Forest report- 2005. Forest survey of India, Dehradun, P. 171

Jacki Schirmer. 2001. A report for the RIRDC/Land & Water Australia/ FWPRDC/MDBC Joint Venture Agroforestry Program by RIRDC Publication No. 04/061

Nair, P.K.R.,1987. Agroforestry System Inventory. *Agroforestry System*, 5 : 319-338.

Oduol, P.A., Mauraya, P., Fernandes, E.C.M. and Nair, P.K.R., 1988. The Agroforestry Systems Database at ICRAF. *Agroforestry System*, 6 : 253-270. 420.

Ratha Krishnan, P. and Indumathy, R., 2009. Decision Support System and Data Bases in Agroforestry. In: O.P. Chaturvedi, A. Venkatesh, R.S. Yadav, Badre Alam, R.P. Dwivedi, Ramesh Singh and S.K. Dhyani (Eds.) *Agroforestry Natural Resource Sustainability, Livelihood and Climate Moderation*, Satish Serial Publishing House, Delhi, 617 - 628 p.

Salim A.S., Simons A.J., Waruhiu A., Orwa C. and Anyango C., 1998. Agroforestry Database : A Tree Reference and Selection Guide. *Version 1.0.* ICRAF, Nairobi, Kenya. CD-ROM.

Schroder, J.M. and Jaenicke, H., 1994. A computerized Database as Decision Support Tool for the Selection of Agroforestry Tree Species. *Agroforestry System*. 26 : 65-70.

Von Carlowitz P.G., Wolf G.V. and Kemperman R.E.M., 1991. Multipurpose Tree and Shrub Database : An Information and Decision Support System. ICRAF, Nairobi, Kenya & GTZ, Eschborn, Germany. 104 pp.

World Agroforestry Centre 2003. Agroforestry Database. http://www.worldagroforestrycentre.org/Sites/TreeDBS/AFT/AFT.htm

23

Watershed Modelling for Development of Sustainable Livelihood through Convergence of National Rural Employment Guarantee Act (NREGA) and Integrated Watershed Management Programme (IWMP) using GIS and Remote Sensing

K. Binoy and Yogiraj V. Shete

ABSTRACT

The major part of the agriculture in India is Rainfed. The increasing pressure of population on land has resulted into over exploitation of land, water and other ecological resources, resulting in water scarcity, land degradation and rapid depletion of ground water-table. Government of India has adopted watershed management strategy to address these problems, but all existing watershed management programmes are mainly fragmented and governed by different departments. This has resulted in formation of providing better service delivery and governance. In this study an attempt is made to identify the scope of convergence between National Rural Employment Guarantee Act (NREGA) with Integrated Watershed Management Programme (IWMP) for generation of social capital and self-employment in the command area of a watershed using geospatial technology. The study was been conducted in Nagavasan micro-watershed of

Saraswati basin, which is situated in Siddhpur block, Patan District of Gujarat. A comprehensive assessment of Nagavasan micro-watershed was taken up to assess the impact of watershed development programme and NREGA programme on the livelihood of the people with a study area of five Kilometer radius around the watershed. The slope analysis, soil loss and run-off was calculated using DEM (Digital Elevation model).Various theme map details about the present land use/ land cover map, soil map, Drainage map were layered to obtain a proper Decision Support Model. The overall outcome of the study was to formulate a model which helps to: Identify the gap and constraint of the watershed development programme which will result in better planning of new programme, to identify the possibilities of coupling NREGA and IWMP with particular emphasis on season and location and assessment of benefits generated in terms of employment and physical improvement in quality life in a sustainable way of the user groups.

Introduction

Watershed is an area that drains to a common point. Watershed development includes a lot of activities that deal with the management of the natural resources with a key emphasis on increasing productivity of land, which in turn supports rural livelihood to alleviate poverty. Increasing population has created an enormous pressure on land and water resulting in degradation and over exploitation of both, ultimately reducing forest cover and water-table which in turn has created a shortage of drinking water, fodder and fuel. Watershed development has developed as a new paradigm of development with the aim of:

a) Optimal utilization of natural resources in order to mitigate the adverse effect of drought.

b) Employment generation and development of human and economic resources of watershed.

c) Restoration of ecological balance in watershed through sustained community action (Government of India, 2001).

Experience of watershed development project is quite varied. One of the major problems of the watershed development is lack of sustainability. The major reason is the lack of participation of the stakeholders and the degree of ownership, which leads to lesser interest in operating and maintaining the asset. The New Common Guidelines, 2008 has increased the level of investment from rupees six thousand per ha to rupees twelve thousand per ha and simultaneously included the concept of convergence which will result in coordination and harmonization of different development programmes. The high level of flexibility in the guidelines has opened the participation of various agencies, departments and organizations. The coupling of new technology with the Indigenous Technical Knowledge (ITK) will give a broad framework for the next generation watershed programme (Government of India, 2008).

The study is fundamentally done to provide strong platform for convergence of various government supported programmes in a collective manner with the help of Geographic Information System (GIS). The Primary emphasis is made on the convergence of Integrated Watershed Management Programme (IWMP) with National Rural Employment Guarantee Act (NREGA). The main objective of convergence is to enhance the livelihood security of the rural people by providing at least 100 days guaranteed wage employment in a financial year to every household whose adult member volunteer to do unskilled manual labour (NREGA, 2005). The department of Land Resources has integrated the existing scheme of Drought Prone Area Programme (DPAP), Desert Development Programme (DDP) and Integrated Wasteland Development Programme (IWDP) into Integrated Watershed Management Programme (IWMP), since more than fifty per cent of NREGA work are related to soil and water conservation/ development and these works are highly labour intensive, the convergence with IWMP of NREGA will provide additional resource to fill the critical gaps in treatment of watershed with detailed technical planning.

Objectives

This paper explains in detail the practicable aspect of various parameters of convergence of NREGA with IWMP. The paper concentrates on the role of operational term of watershed activities with NREGA and uses of geo-spatial techniques at different levels of institution, planning, implementation and monitoring. The major objectives of study are:

- Identify the contours of convergence between NREGA works with the watershed development program using GIS and RS.
- Usages of GIS and RS for technically identifying the area for coupling NREGA and IWMP for generation of social capital and self employment in the command area of watershed.
- Mapping the command area of watershed and understanding the demographic profile and economic status of the stakeholders.
- Usage of the model as a platform for capacity building of the local community, Project Implementation Agency (PIA), Self Help Group and Watershed Committee.
- Identifying and prioritization of the watershed action plan according to the model for development, conservation, and generation of natural

resources with integration of various economic activities like livestock and dairy marketing for better sustainable livelihood.

Study Area

Study was conducted in Nagavasan watershed, Siddhpur Taluka, Patan District of Gujarat. The watershed falls under the Government designated Integrated Wasteland Development Programme. The watershed covers an area of 1050 ha. The watershed includes complete Nagavasan village parts of Sandesari, Ganwada and Lukhasan. There are an average 423 household with more than 90 per cent dependent on agriculture for their livelihood. The watershed is located between 23^0 57′26.17″ N land 72^0 26′1.76″ E to 23^0 55′1.45″ N and 72^0 78′37.08″ E. The watershed has 110.2 ha of wasteland which is characterized by undulating and Ravinous landforms. The drainage density is very high due to this topography. The average rainfall of the area is about 700 mm. The rainfall is influenced by South-West monsoon from Arabian Sea. The majority of waste-land is degraded and devoid of vegetation due to illicit felling, open grazing and is prone to high Rill and Sheet erosion ultimately leading to Gully erosion. The total population of the watershed is 2580 people with 423 household. The literacy rate of the people is about 61 per cent. The major castes are Chaudhary, Thakoor, and Prajapati. There are about 120 schedule caste household with 12 Schedule tribe household of Bhils. The average land holding of the people is about 2.02 ha. with a majority of the farmers are marginal or small. The extreme left portion of the watershed is highly flood prone which causes high amount of migration towards the upland due to sudden flash floods in Saraswati River. The other occupation of the people is dairying but the major breeds available are indigenous with very less milk productivity.

Problem Analysis

1. The Nagvasan watershed falls under rainfed area, the soil is highly productive due to deep alluvial in nature. Water availability for irrigation can increase the livelihood of the people of which majority is small and marginal farmers.

2. Due to lack of vegetation and ravenous nature of wasteland, there is a high amount runoff leading to excessive soil erosion every year which has to be controlled thereby decreasing the silting of the water harvesting structure like check dam.

3. The status of ground water is moderate due to inadequate recharging combined with erratic, uneven and excessive rainfall.

4. There is enough availability for NREGA Card holders in Sandesari village, but the lack of professionalism and accountability of PRI and ground staff has resulted in lack of proper asset creation for sustainability.
5. There is sufficient livestock population in project watershed. However, due to poor quality of livestock and lack of fodder availability has resulted in the low productivity of milk in the area. Lack of private fodder cultivation has resulted in excess pressure on village common land resulting in degradation of it.
6. There is an ample scope of organic horticulture production in the area due to high availability of livestock population which can be used for vermicompost and vermi-wash production.
7. The human capital available there has to be motivated and encouraged to develop their own local institutions for the management of these assets created during the project period.
8. There is need of proper scientific approach towards a sustainable planning through proper technical and management aspect to converge different Government schemes resulting in proper coordination of line departments by sharing of information, fund and materials.

Methodology

Data Collection

The study is conducted in Nagavasan watershed, Siddhpur Block of Patan District for modelling. A transect walk for identifying various soil and water conservation structures was done with the Watershed Development Team (WDT) and Sarpanch of the village. As the records are maintained by the watershed committee and village panchayats, the secondary data were collected from them. Additional information regarding the watershed activities was collected by Focused Group Discussion with farmers; Self Help Groups and beneficiaries of NREGA using unstructured closed end questionnaires. Study of various geo-spatial data was done in Bhasakaracharya Institute of Space and Geo-Informatics (BISAG) for GIS modelling of the watershed and creating various theme maps and treatment maps.

Strategy to be Adopted

Generation of Various Thematic Maps

a. *Land Use and Land Cover Map*: Satellite imagery of IRS 1C LISS III FCC was used to identify the land use/ land Cover of the available area. The satellite imagery of October, December and April were used to classify the land accordingly to agricultural land (on the basis of Kharif, Rabi and Fallow) and wasteland (with and without scrub), residential land and river area.

b. *Drainage Map*: Nagavasan watershed forms a part of drainage of Saraswati River. The streams are being demarcated using satellite image of the area.

c. *Soil Map*: The soil map was adapted from the map available at Bhasakaracharya Institute of Space and Geo-Informatics (BISAG) for Patan District of Gujarat.

d. *Soil and Water Conservation Structure Map*: The available soil and water conservation structures are mapped using the ground data with the help of cadastral map overlaid on the satellite image. Various structures like Check Dam, Sand bag check dam, plantations *Gram- vatika* (Village Garden), Self Help Groups, Vermicompost units are mapped accordingly.

e. *Slope Map*: The slope map is prepared from DEM simultaneously classified according to 0 – 1%, 1- 3%, 3 - 5%, 5 - 10%, 10- 15% and above.

f. *Demographic Map*: A demographic map is prepared showing the population, economic status, number of Dairy Units, Milking Animals and NREGA Card holders in the watershed area.

Prioritization and Analysis

After creation of various theme layers the task is to develop a sustainable model which decreases soil erosion, silting and degradation of present watershed structures, simultaneously resulting in regeneration of ground-water. The major factors which results in soil erosion are soil type, soil depth, rain-fall, slope, slope- length and the vegetation. Soil erosion was estimated using Universal Soil Loss Equation (USLE) (Wischemier and Smith, 1978).

A (Soil Loss) = RKLSCP

Where,

'K' is the erodibility factor which depends on soil characteristics such as percentage of sand, silt, organic matter content. K- Factor map for soil is prepared accordingly.

'LS' is been calculated from slope map which is mainly dependent on per-cent slope and slope-length of each category of slope

'C' factor is dependent on crop and the local vegetation there. This can be mapped from the land use and land cover map

'P' factor is dependent on conservation practices applied or ought to be applied in watershed under NREGA.

By integrating the three maps we will get the areas in the watershed which will be highly erosion prone and where the soil and Water conservation activities (P- factor) have to be performed in the watershed by converging it with NREGA.

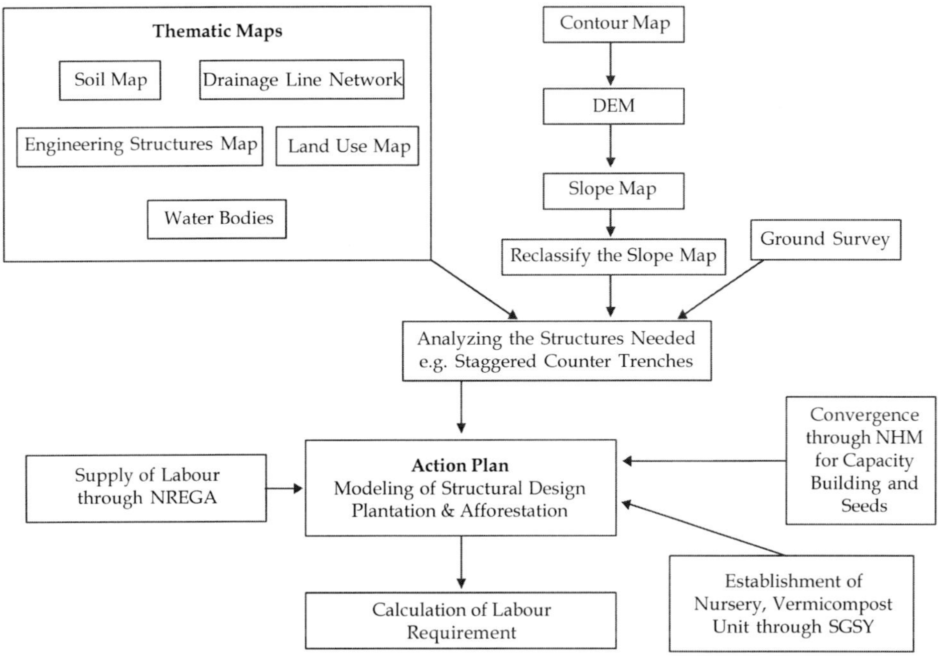

Figure 1 : Model for Convergence of IWMP with NREGA.

Analysis of Action Plan under NREGA

Nagavasan watershed area is implemented under IWDP-II, therefore, additional watershed work has to be prepared and incorporated in the perspective plan and annual work plan of NREGS and will be executed by watershed committee. Proper framework for the execution of the project has to be worked out. Total number of NREGA card holders has to be analyzed in the watershed area. The work should be labour intensive with no machinery and contractors with the involvement of communities in planning. The work should help in generating income, reducing temporary employment dependency of the people and ultimately leading to poverty alleviation.

Generation of NREGA Planning Map

Under NREGA almost all activities of soil and moisture conservation required for watershed development are permitted, but the guidelines of NREGA stipulate the formulation of a perspective plan for a advance planning and to provide development perspective for a district (NREGA, 2005) (Joint Convergence Guideline, 2008). The planning analysis involves the need of the labour and gap in the district so there is dire need of production of annual work plan (AWP) map for NREGA activities. The map has to be passed by the Gramsabha. The map will facilitate the better understanding of activities by both literate and illiterate (major beneficiaries). This will result in proper transference of work. The map also help in deriving a platform for mutual coordination between NREGA and IWMP. This being in digital form can be easily shared between the line departments which results in better fund flow and act as a monitoring and evaluation Fund.

Result and Discussion

Generation of Action Plan Maps

The various data collected from the watershed regarding demographic, agriculture land, infrastructure details established during the project period including the expenditure and installments were analyzed and were mapped accordingly, which gives complete details of the watershed in a pictorial way (Map-1). The priority map for convergence of IWMP and NREGA was

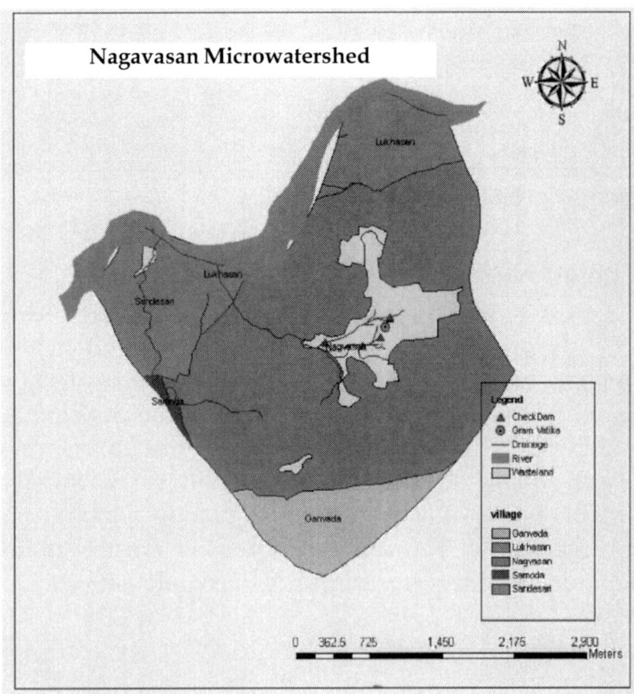

WaterShed Details

Demographic Details	
No. of Household	423
Population	2586
Male Population	1331
Female Population	1255
Literacy rate(%)	61
Below Poverty household (BPL)	20
Agricultural Details (in Ha)	
Cultivated Land (Soil Type)	873.6
i) Irrigated	500
ii) Rainfed	206.9
Waste land	110.2
Area not available for cultivation	56.4
Average Land holding	2.02
Cattle population	2010
Average milk collection(per day in lit)	5500
NREGA Details	
Nagavasan	115
Sandesari	100
Ganvada	142
Lukhasan	162

Infra-Structure Details

Infrastructure	Number	Expenditure (in Lakh)
Check Dam	4	4.21
Sand bag	28	0.75
Vermiculture Unit	14	1.37
Gram-vatika	1	1
Afforestation	20Ha	2.24

Map 1 : Details of Demographic, Agriculture Land, NREGA and Infrastructure in Nagavasan Microwatershed, Patan district, Gujarat

generated using Soil Map, Slope Map, Land use/ Land cover Map, giving weightage to each of its attributes accordingly then it was overlaid on the cadastral map of the village which will give the exact area of action in the watershed. (Map-2). After the identification of the area, map was generated for NREGA work. Generally, the manual work in the watershed is specified to the area which has the slope percentage less than 25 per cent and above it generally vegetative activities are preferred.

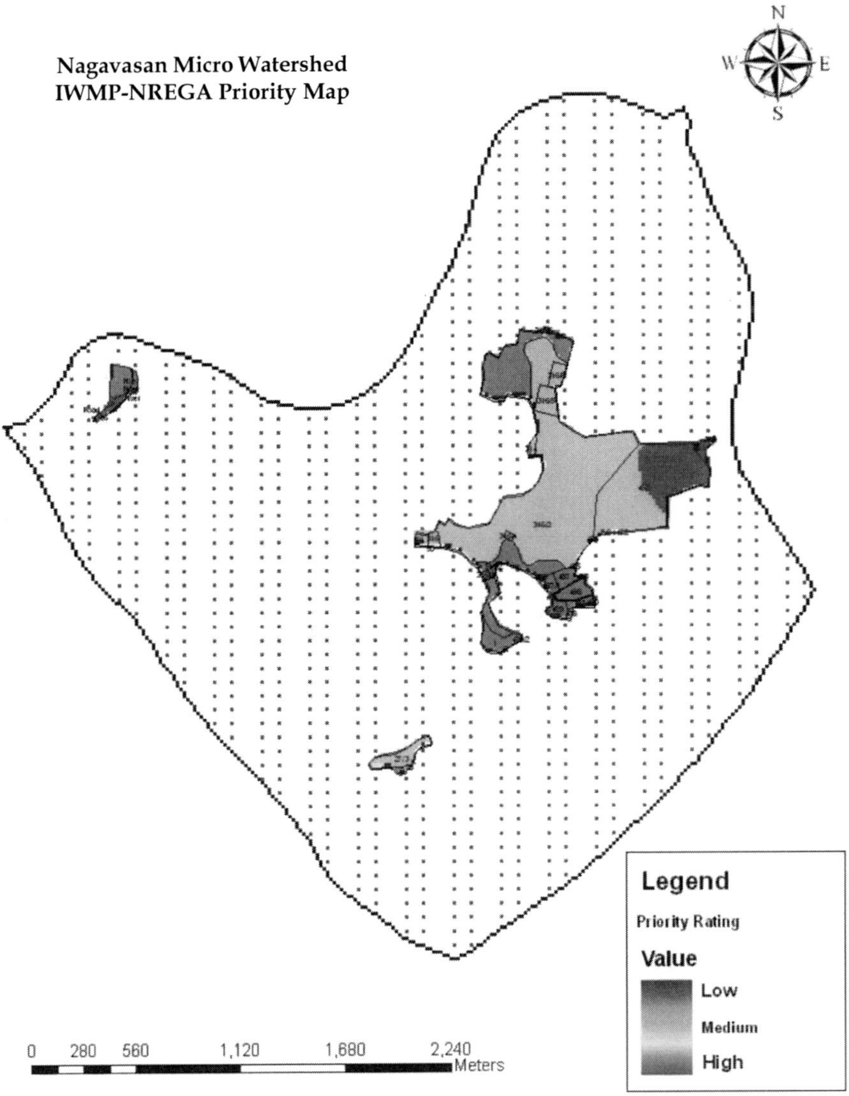

Map 2 : IWMP-NREGA Convergence work priority area map with Cadastral Details

NREGA Work

6.2.1 Manual Work

During the survey and discussion with the WDT members and User group, it was found that the major problem of the area is excessive soil loss and runoff of water. Since the soil depth is above 100 cm to 150 cm, it is ideal for trench activities, which will help in soil and moisture conservation.

Details of Staggered Contour Trenches

Trenches are any form of depression or micro pit constructed over the land surface in order to prevent soil erosion and to absorb rainwater in non arable lands. Trenches are constructed along the contours (called contour trenches) on hill slopes above 15 per cent with vegetative supports for forestry and horticulture land uses.

Contour trenches are used both on hill slopes as well as on degraded and barren waste lands for soil and moisture conservation and afforestation purposes. The trenches break the slope and reduce the velocity of surface runoff. It can be used in all slopes irrespective of rainfall conditions (*i.e.* in both high and low rainfall conditions) and varying soil properties. For vegetative supports forage grasses as well as economically important trees can be planted.

The trenches were working as silt trap earlier. Periodical maintenance by way of unearthing of the trenches and depositing on the downhill side ought to be given due attention by the beneficiaries themselves.

In order to design the Staggered Contour trenches based on hydrologic and hydraulics consideration, following stepwise procedure as described for graded bund was adopted.

1. Vertical interval between trenches (VI)

   $VI= 0.305(S/3+2)$

   Where,

   VI = Vertical Interval, m

   S = Slope of land in %

2. Horizontal Interval (HI) between trenches

   $HI = (VI/S)*100$

   Where,

   HI = Horizontal Interval *i.e.* Horizontal Distance

   S = Slope of land in %

   For the practical purpose, the HI is considered as half of the calculated values,

   Practical HI = HI/2

Table 1 : Numbers of staggered counters per unit hectare

S.No.	Slope %	Distance between trenches	No. of Staggered Contour trenches per Ha.
1	0-6	10 to 12 meters	130

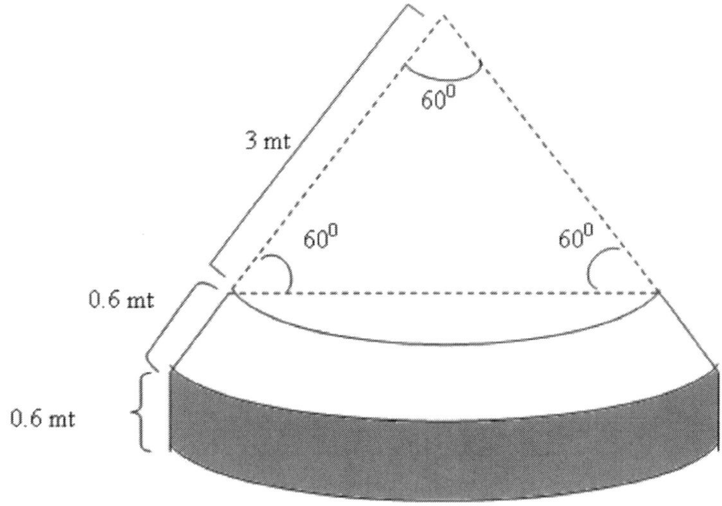

Figure 2 : Cross Sectional View of Staggered Contour Trench

Vegetative Cover

Since the soil is fine sandy loam and the waste land is highly degraded, there is need of plantation and vegetative ground cover. It is also required to stop the grazing activities until the establishment of the forage grasses. It has to be protected by bio-fencing using local available species like Jatropha. There is significant need of arranging participation of people for the activities. The stakeholders should take the responsibility and accountability of these tasks. Need for planting of local varieties of the trees should be advocated and proper nursery is to be maintained by the user group. This can be initiated with the help of Self Help Groups under Swarnjayanti Gram Swarojagar Yojana (SGSY) Scheme of Government of India. There is also a need of increasing the organic content of the waste land to make it arable. This can be done by utilizing the available Vermi-Compost units in that area.

Calculation of NREGA Man Days

The area where manual work is to be done is about 97 hectare out of which 30 per cent of land is under check dam and Bori (Sand-Bag) Bund *i.e.*

67.9 Hectare of land has to be treated. The total numbers of staggered contour trenches are about 130 per hectare (from above). So, 827 trenches has to excavated in the treatment area and the amount of manual labour required is calculated using the NREGA schedule of rates (SOR) of Rs.44.93 per Cubic meter and this is converted into man days. Simultaneously, the NREGA man days required for plantation pits, desilting of check dams, repairing and formation of Bori-bunds is being calculated and the total required is projected in NREGA annual planning map.

Table 2 : Details of NREGA work to be done in Convergence with IWMP in Nagavasan Watershed

Sl. No	Detail of NREGA work	Specification	Unit	Net Cost (in Rs.)	Man Days
1.	Staggered Contour Trenches,	1.24 m3	8827	491780.4	4918
2.	Plantation pits	0.045 m^3	17654	35693.74	357
3.	Desilting of Check- Dam	220 m^3	4	39560	395
4.	Repairing of Bori-Bund	28000	280		

Recommendations

1) Growing of trees like *Simarouba glauca, Pongamia Pinnata, Colophospermum mopane, Acacia senegal, Acacia tortilis.* Growing of soil erosion tolerant species of grasses like *Chrysopogon zizanioides* (Vetiver) will prevent the soil loss and help in moisture conservation.

2) Growing of forage crops like *Lasiurus sindicus, Cenchrus ciliaris;* Legumes: *Stylosanthes guianensis, Vigna trilobata, Aptenia cordifolia* and selling it to the communities will help in development of Watershed Development Fund (WDF) and also decreases the overgrazing of community land .

3) Uses of Vermi-Compost will increase the Organic content of the soil and help to revitalize the degraded wasteland and simultaneously will help in generating non-farm based income activities. It will also help in value addition of the available cow-dung.

4) Need of the capacity building activities like the nursery cultivation, Vermi-composting. Dairying and Fodder Cultivation methods has to be done in the area.

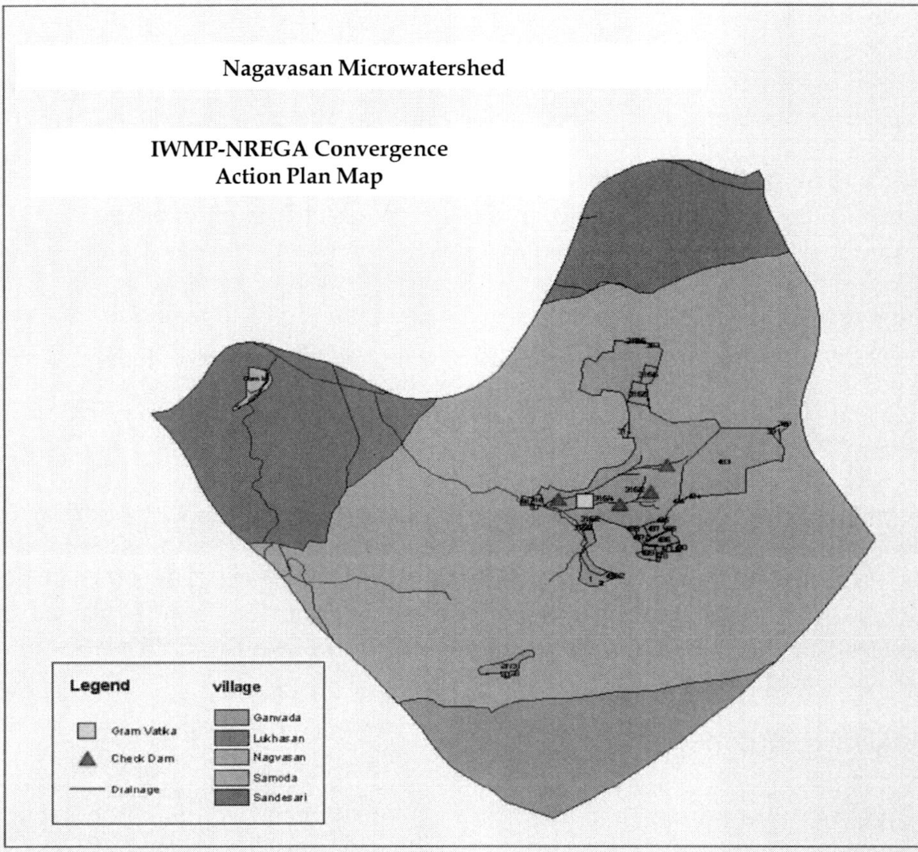

NREGA WORK

S.No	NREGA Work	Man Day	Expense (in lakh)	Survey-No.
1	Staggered contour trench	4918	4.92	316/2,391,390,483, 316/5,316/6,90,91
2	Plantation Pits	357	0.35	316/2,391,390,483, 316/5,316/6,90,91
3	Desilting of check dams	400	0.4	316/2
4	Repairing and formation of new sand	280	0.28	316/2,316/4
	Total	5955	5.95	

Other Activities

S. No	Other activities	Govt.Scheme
1	Nursery Cultivation	SYSG,NHM
2	Vermicompost	SYSG,IWMP
3	Fodder Cultivation	IWMP

S.No.	Capacity building activities
1	Nursery Technology
2	Diary Technology
3	Fodder Generation
4	Vermi-Composting
5	Organic-Horticulture

Map 3 : IWMP-NREGA Joint convergence action plan map for Nagavasan Micro-watershed, Patan District, Gujarat

Conclusion

Decentralization of watershed activities and convergence of it with other government Scheme has to be induced by participatory approach of the local stake-holders and Panchayat Raj Institution (PRI). The notion of real convergence can only be achieved by having proper mechanism of information, fund, and material sharing between the collaborating departments, institutions, and organizations. The sustainability of a watershed project or any other developmental programme is achieved by creating basically five capitals namely Human, Social, Physical, Financial and ultimately Natural Capital. An exploration of these can explicitly explain us the concept of decentralization, coordination and harmonization as the sole pillars of the development. With the development of information as tools of modern era, achieving these pillars has become quite easy. Geographic Information System (GIS) has a vital role in designing, planning, implementing, monitoring and evaluating watershed project. GIS having the capability of storing both spatial and non-spatial information, which inturn act as a eminent Decision Support System. This can be further developed into an expert system by providing it with timely and precise information and with proper ground truthing. But, there is a need of addressing some of the issues like the capacity building of the ground staff for using the software and information collection. Secondly the need of development of sound database for storing the relevant thematic layers and proper sharing of it between users. Thirdly, there is a need of addressing the infrastructure problem in ground like the GPS reader and software availability.

Acknowledgement

This work is done under Gujarat Watershed Management Agency (GWSMA), Commissionarate of Rural Development and in collaboration with Bhasakaracharya Institute of Space and Geo-informatics (BISAG), Gandhinagar which is an apex organization working in projects related to Remote Sensing and GIS in the State.

References

Government of India, 2001. Guidelines for Watershed Management, Department of Land Resources, Ministry of Rural Development.

Government of India, 2008. New Common Guidelines for Watershed Development Projects, Department of Land Resources, Ministry of Rural Development.

NREGA, 2005. NREGA Perspective Plan Guideline, Department of Rural Development, Ministry of Rural Development, Government of India.

Joint Convergence Guideline, 2008. Report of the Task Force on Convergence, Department of Rural Development, Ministry of Rural Development, Government of India.

Wischemier, W.H and Smith, D.D, 1978.Predicting Rainfall Erosion Losses. A Guide to Conservation Planning. *USDA Agricultural handbook No.537*

24

Effectiveness of ICT Enabled Village Information Centres among Dairy Farmers in Tamil Nadu

S.Senthilkumar and Mahesh Chander

ABSTRACT

The study was undertaken to assess the effectiveness of ICT-Enabled Village Information Centres (VICs) with special reference to dairy farming in Tamil Nadu. Ex-post-facto research design was followed, wherein 180 users and 90 non-users of VICs constituting 270 dairy farmers covering three ICT-VIC initiatives *viz.*, Tamil Nadu Veterinary and Animal Sciences University (TANUVAS), M.S.Swaminathan Research Foundation (MSSRF) and n-Logue communication in Tamil Nadu were selected using random sampling technique. Data collected through personal interviews were analysed using appropriate statistical tools. TANUVAS, MSSRF and n-Logue Communication beneficiaries had total knowledge score of 0.80, 0.56 and 0.70, respectively and overall total knowledge score of 0.69 was gained by dairy farmers from three initiatives, of that, 0.59 from VICs and 0.10 from other sources. Knowledge gained through VICs is the indicator for assessing the effectiveness of Village Information Centres and results of this study revealed the above fact. The respondents attained high total knowledge score on dairy farming activities only by the way of utilising VICs not from any other sources. Among thirteen independent variables, time of using VICs, mass media exposure, extension agency contact and credibility were found to be positively correlated with knowledge level of dairy farmers for three VIC initiatives. Gender was negatively correlated but distance covered was positively and significantly ($P<0.05$) correlated with knowledge. For the overall VIC usage level, only two variables *i.e.* land holding and the knowledge level of the farmers were significantly associated with VIC usage level at 0.01 level of probability and the regression coefficient for these variables was positively related to the VIC usage level. General problem was considered as foremost constraint with highest mean score of 5.11 and was ranked first followed by socio-personal constraint (4.12), content problem (3.15), physical facilities (0.93) and socio-cultural constraint (0.05) and these were ranked second, third, fourth and fifth, respectively. The respondents and service providers suggested that, battery back-up may be needed especially in summer, in order to reduce connectivity problem, broad band connection may be required. Content should be updated regularly based on the need of the people. Infomediaries play a vital role in increasing the knowledge level of the farmers, hence, they should be appointed based on their educational background. Preferably they should be from veterinary, agriculture and allied disciplines. Impart training to infomediaries for updating their knowledge by use of line departments' *i.e.* veterinary university, agricultural department, *etc.*

Introduction

The present era is rightly termed as an "Information age". People want adequate and authentic information as early as possible. In recent years, there is a visible shift from the old ways to the modern ways of information delivery systems. With the advent of global information society, information and communication technologies are increasingly being adopted as effective tool for reaching rural communities. Yet, the benefits of the information revolution are still much debated, particularly, in the case of developing countries like India. There is very much concern about the gap between the information "haves and have not's", which will continue to grow unless the developing countries acquire the infrastructure and resources to access these new technologies. Many in the rural community do not have individual access to these technologies but Village Information Centres or Telecentre offer shared access.

A telecentre is "a public place where people can get a variety of communication services, and major part of the operators' purpose is to benefit the community" (Roman and Colle, 2002). Even the term 'telecentres' is arguably out-dated, because of its implicitly narrow focus, and a broader term like 'knowledge centres' or 'information centres' may be more appropriate. Other related terms in the Indian literature, often used interchangeably with 'telecentres', are 'information kiosks' and 'cyber dhabas' (Conroy, 2006).

Tamil Nadu is one of the leading states of India in implementing ICT projects and three sectors are being involved in this process. Major government, non-governmental and private initiators of ICT-VICs in Tamil Nadu State are Tamil Nadu Veterinary and Animal Sciences University (TANUVAS), M.S. Swaminathan Research Foundation and n-Logue communication, respectively. With the revolution in the IT sector, ICT is being integrated with animal husbandry, agriculture and rural development. In spite of number of projects oriented towards the transfer of technologies through ICTs, still the acceptability and utilization of ICTs is very low in the rural areas. This created the need of investigating knowledge gained through VICs and constraints in utilization of VICs by the end users particularly dairy farmers. So far, very limited studies have been attempted to evaluate the effectiveness of these initiatives that too deal with broader aspects, not addressing specific issues like ICT in relation to transfer of technology for dairy development. Keeping in view these concerns, the study was undertaken with the main objective of studying the Effectiveness of ICT-Enabled Village Information Centres among Dairy Farmers in Tamil Nadu.

Research Methodology

Following Ex-post-facto and survey research design, the present study was carried out to cover three types of ICT-VIC initiatives. Accordingly, major Village Information Centres established by Tamil Nadu Veterinary and Animal Sciences University (TANUVAS), M.S.Swaminathan Research Foundation (MSSRF) and n-Logue communication for government, non-government and private initiators, respectively were selected purposively for the study. From each initiator, randomly two Village Information Centers were selected for this study, thus, six Village Information Centres constituted the sample.

In order to find out the effectiveness of Village Information Centres, dairy farmers who are not using these centres were also selected by random sampling method and kept as control group to compare with those who were users of the VICs. As such, 30 dairy farmers who were users and 15 non-user dairy farmers from each centre were chosen. Thus, a total of 270 respondents comprising of 180 users and 90 non-users constituted the sample for the study (Fig. 1). Village Information Centres' operators were also interviewed for their suggestions to improve the effectiveness of VICs.

A. Measuring Knowledge Level

For measuring the knowledge level of dairy farmers, Knowledge test developed by Rao and Kherde (1990) and further modified by Sharma (2004) was followed with suitable modification made, based on the content provided through Village Information Centres.

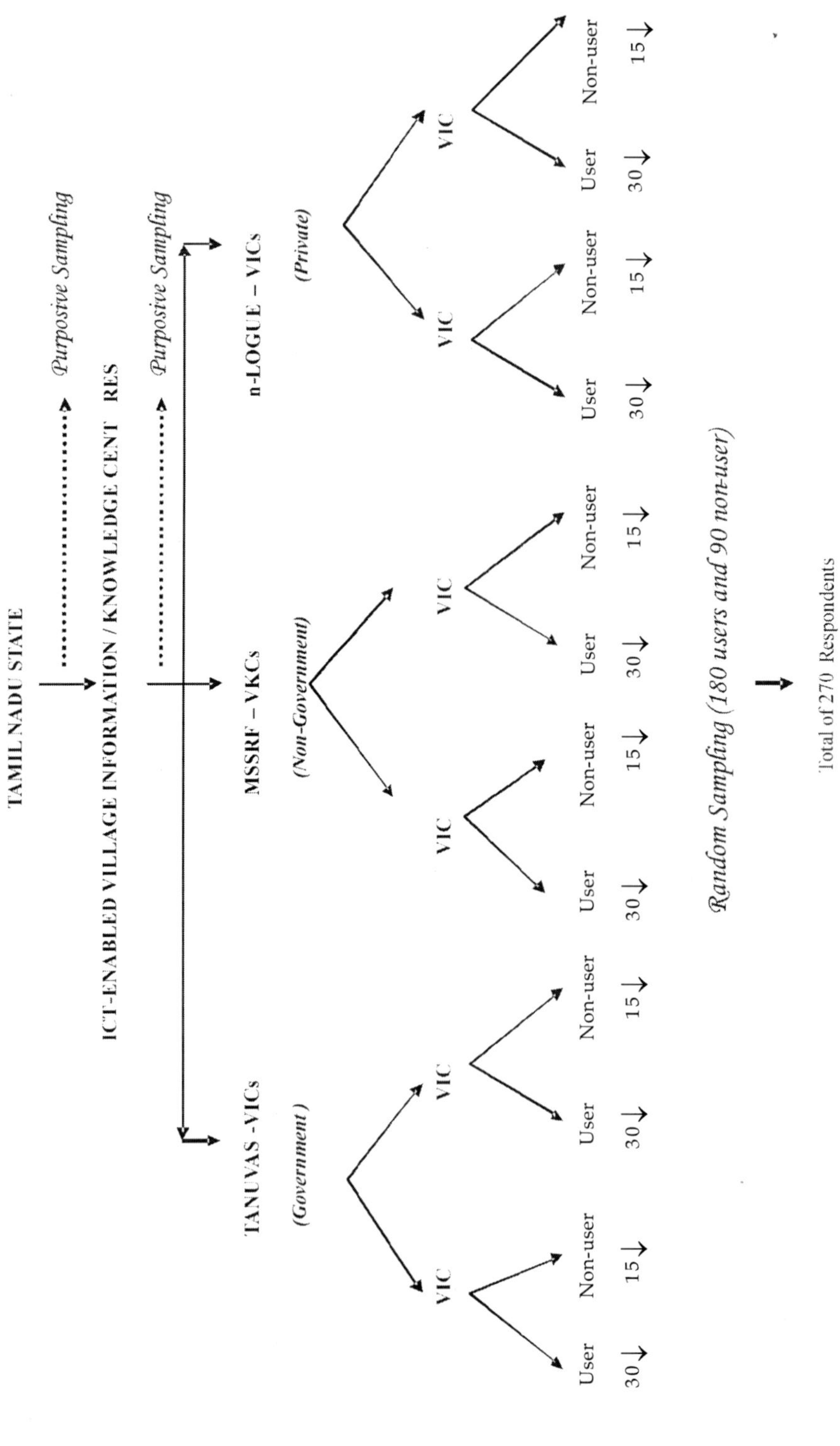

Figure 1 : Sampling Procedure Adopted for the Study

Scoring : The knowledge test consisted of 43 items was administered on the dairy farmers. The test items were of objective type meant to test the knowledge level of dairy farmers about dairying. The response was recorded either as 'Correct' or 'Incorrect' and a score of '1' and '0' were given, respectively.

$$\text{Knowledge score (VIC)} = \frac{\text{Total score obtained through VIC}}{\text{Maximum correct score}}$$

$$\text{Knowledge score (Other source)} = \frac{\text{Total score obtained through other source}}{\text{Maximum correct score}}$$

Total Knowledge score = Knowledge score (VIC) + Knowledge score (other sources)

The antecedent variables namely age, gender, socio-economic status, land holding, herd size, income, frequency of use, time of use, distance covered, innovativeness, credibility, mass media exposure and extension agency contact were selected in consultation with ICT experts.

B. Constraints Analysis

On the basis of available literature, survey of the localities and discussion with different groups of ICT experts, the possible constraints were prepared. The constraints were categorized under the major areas namely socio-personal, socio-cultural, content problem, physical facilities and general problem. The identified constraints were administered through four point continuum as most serious constraint, constraint, less serious constraint and not a constraint with a score of 3, 2, 1 and 0, respectively. The scores for each constraint were added and the mean constraints score was obtained and inference drawn.

Statistical and Analytical Procedures

After data collection, it was compiled, tabulated and analysed by using SPSS (Statistical Package for Social Sciences) keeping in view the objectives of the study. Descriptive statistics (frequency, mean, standard deviation, and percentage), Pearson correlation and Multiple Linear Regression analysis were worked out for analysing and interpreting the data using suitable formulae.

Result and Discussion

I. Extent of Knowledge about Dairy Farming Gained through VICs

The total knowledge score obtained by the respondents after gaining knowledge through VICs and other sources about feeding, fodder production, breeding, management, health care, milk products preparation and marketing of dairy products were presented under the following subhead;

A. Knowledge about feeding aspects

The overall data for feeding aspects were presented in Table 1 revealed that the total knowledge score of 0.68 was obtained by the respondents of three types of VICs initiatives and majority (92.65 per cent) of the knowledge score were gained through VICs and 7.35 per cent gained from other sources *i.e.* extension agency contact, mass media exposure *etc.* The similar trend was observed for each VIC's respondents.

B. Knowledge about fodder production

From the Table 1, it could be inferred that the total knowledge score of 0.55 was gained by the respondents irrespective of the types of VIC initiators. Cent percent of knowledge on fodder production were gained through MSSRF-VICs, this may be due to the fact that the project co-coordinator was from agriculture background. They might have conducted need assessment survey among the dairy farmers on fodder production and fulfilled their needs by providing the content on fodder production through MSSRF-VICs.

C. Knowledge about breeding aspects

The total knowledge score of 0.68 was gained by the respondents irrespective of the types of VIC initiators. Of that, around 80 per cent were gained from VIC source and remaining from other sources.

D. Knowledge about management

The overall data showed that the beneficiaries of three initiatives had the total knowledge score of 0.63 and it was contributed by 87.30 per cent and 12.70 per cent from VICs and other sources, respectively.

E. Knowledge about health care

Table 1 revealed that the total knowledge score of 0.76 was gained by the respondents of three VIC initiatives. Above 90 per cent was contributed by VICs source and around 10 per cent from other sources. This may be due to the fact that, all VICs were providing updated health care information for the needy farmers. By using latest technologies *i.e.*, Videoconferencing facilities in all these centres, face to face interaction was made in between Veterinary experts with dairy farmers and in turn gaining more knowledge.

Table 1 : Dairy farmers' knowledge in accessing ICT-Village Information Centres

S. No.	Subject matter on dairy farming	TANUVAS (n=60)			MSSRF (n=60)			n-Logue (n=60)			Overall (N=180)		
		Total knowledge score	Through		Total knowledge score	Through		Total knowledge score	Through		Total knowledge score	Through	
			VIC	Other		VIC	Other		VIC	Other		VIC	Other
1.	Feeding	0.79 (100.00)	0.71 (89.87)	0.08 (10.13)	0.51 (100.00)	0.50 (98.04)	0.01 (1.96)	0.74 (100.00)	0.69 (93.24)	0.05 (6.76)	0.68 (100.00)	0.63 (92.65)	0.05 (7.35)
2.	Fodder production	0.64 (100.00)	0.56 (87.50)	0.08 (12.50)	0.35 (100.00)	0.35 (100.00)	0.00 (0.00)	0.68 (100.00)	0.63 (92.65)	0.05 (7.35)	0.55 (100.00)	0.51 (92.73)	0.04 (7.27)
3.	Breeding	0.84 (100.00)	0.69 (82.14)	0.15 (17.86)	0.53 (100.00)	0.38 (71.70)	0.15 (28.30)	0.68 (100.00)	0.54 (79.41)	0.14 (20.59)	0.68 (100.00)	0.54 (79.41)	0.14 (20.59)
4.	Management	0.79 (100.00)	0.67 (84.81)	0.12 (15.19)	0.47 (100.00)	0.44 (93.62)	0.03 (6.38)	0.62 (100.00)	0.54 (87.10)	0.08 (12.90)	0.63 (100.00)	0.55 (87.30)	0.08 (12.70)
5.	Health	0.89 (100.00)	0.78 (87.77)	0.11 (12.23)	0.62 (100.00)	0.55 (88.33)	0.07 (11.67)	0.77 (100.00)	0.75 (97.40)	0.02 (2.60)	0.76 (100.00)	0.69 (90.79)	0.07 (9.21)
6.	Milk products preparation	0.87 (100.00)	0.74 (85.06)	0.13 (14.94)	0.50 (100.00)	0.45 (90.00)	0.05 (10.00)	0.81 (100.00)	0.78 (96.30)	0.03 (3.70)	0.73 (100.00)	0.66 (90.41)	0.07 (9.59)
7.	Marketing aspects	0.78 (100.00)	0.68 (87.18)	0.10 (12.82)	0.94 (100.00)	0.47 (50.00)	0.47 (50.00)	0.63 (100.00)	0.54 (85.71)	0.09 (14.29)	0.78 (100.00)	0.56 (71.79)	0.22 (28.21)
	Total	**0.80 (100.00)**	**0.69 (86.25)**	**0.11 (13.75)**	**0.56 (100.00)**	**0.45 (80.36)**	**0.11 (19.64)**	**0.70 (100.00)**	**0.64 (91.43)**	**0.06 (8.57)**	**0.69 (100.00)**	**0.59 (85.51)**	**0.10 (14.49)**

Figures in parentheses indicate percentage

F. Knowledge about milk products preparation

The overall data revealed that, total knowledge score of 0.73 were gained by the respondents. Of that, 90.41 per cent knowledge score on milk products preparation were from VICs and only 10 per cent from other sources. This is because, female farmers accessed these centre more frequently than male and they were very much interested to know, how to prepare milk products like panner, rose milk, channa *etc*. Based on the needs of dairy farmers, information on milk product preparation was provided through VICs. Hands on training also imparted to them by Veterinary University personnel in these aspects and so they had more knowledge in this aspect.

G. Overall knowledge level of dairy farmers

The overall knowledge score of 0.80, 0.56 and 0.70 on dairy farming was gained through TANUVAS, MSSRF, n-Logue communication, respectively. These scores were obtained mainly by the respondents who used these VICs *i.e.*, TANUVAS (86.25 per cent), MSSRF (80.36 per cent) and n-Logue (91.43 per cent), respectively.

The pooled data revealed that the knowledge score of 0.59 from VIC source and 0.10 from other sources contributed to the overall total knowledge score of 0.69 of dairy farmers from three initiatives. These findings were in agreement with findings of Sharma (2004), who mentioned that, all livestock owners accessing internet at Community Information Centres possessed very fair knowledge pertaining to technical, marketing and general aspects of livestock production.

II. Relationship between Knowledge and Antecedent Variables of Dairy Farmers

Thirteen antecedent variables were included for establishing relationship with knowledge on dairy farming among three types of VICs initiatives beneficiaries. The values of correlation coefficient (r-values) for three sectors respondents and overall have been presented in Table 2. It points out that out of 13 antecedent variables, five variables namely time of using VICs, distance covered, credibility, mass media contact and extension agency contact had significant ($P<0.01$) and positive relationship with knowledge level of dairy farmers who had accessed VICs. While, gender of the respondents was found to have significant ($P<0.01$) but, negative correlation with knowledge on dairy farming. It implies that female farmers had more knowledge on dairy farming as compared to male respondents. This was due to the fact that generally more female farmers are accessing VIC and ultimately gained more knowledge. This reflects the involvement of women in most of livestock farming activities. This result was in accordance with the findings of Deepak and Singh (2006).

Table 2 : Correlation Coefficients ('r' values) between Knowledge Level and Antecedent Variables

S. No.	Independent variable	Correlation coefficient with knowledge of dairy farming TANUVAS	MSSRF	n-LOGUE	Overall
1.	Age (X1)	0.085	-0.043	-0.056	0.059
2.	Gender (X2)	0.046	0.044	-0.266	-0.207**
3.	Socio-economic status (X3)	-0.276	-0.101	0.077	-0.121
4.	Land holding (X4)	0.110	-0.118	-0.009	0.027
5.	Herd size (X5)	-0.056	-0.192	-0.088	-0.056
6.	Income (X6)	0.128	-0.286*	-0.111	-0.008
7.	Frequency of use (X7)	0.154	-0.176	-0.005	0.104
8.	Time of use (X8)	-0.049	-0.030	0.211	0.215**
9.	Distance covered (X9)	-0.159	0.112	-0.006	0.168*
10.	Innovativeness (X10)	0.050	-0.033	0.048	0.037
11.	Credibility (X11)	0.256*	-0.022	-0.025	0.197**
12.	Mass media exposure (X12)	0.365**	-0.067	0.140	0.198**
13.	Extension agency contact (X13)	-0.088	0.057	0.218	0.230**

** Correlation is significant at the 0.01 level * Correlation is significant at the 0.05 level

The analysis reveals that those farmers who spent more time in VIC had gained more knowledge due to the fact that they tend to had more exposure to updated dairy farming content provided through VICs. Farmers who are nearer to VICs are easily utilizing the centre and gained more knowledge. To improve knowledge level of rural communities, this type of innovative initiatives or projects should be expanded to other areas as well. Farmers who perceived VICs as more credible source of information gained more knowledge. This might be due to the trustworthiness, latest and complete availability of information and other services that respond to the most pressing needs of the poor. These results were in accordance with the findings of West (2000) and Sharma (2004).

Mass media exposure and extension agency contact are the most important factor to enhance the knowledge level of dairy farmers. The correlation analysis reveals that the dairy farmers having more score on these variables tend to have more knowledge on dairy farming. This could be due to frequent exposure of dairy farmers with different mass media and frequent contact with extension personnel. These observations were in consonance with the findings of Richardson (1996), Cecchini and Scott (2003) and Sharma (2004).

III. Factors Associated with VIC Usage Level of the Sample Farmers in the Study Area

The overall results of the regression analysis to identify the factors associated with the Village Information Centre usage levels in the study area *i.e.* TANUVAS, MSSRF and n-Logue Communication VIC beneficiaries' area are given in the table 3.

Table 3 : Factors Associated with VIC Usage Levels of the Sample Farmers in the Study Area

	Variables (Xi)	Regression Coefficients	Standard Error	t-value	Significant level
	Constant (α)	0.163	0.176	0.926	0.355
X_1	Age	0.004	0.024	0.164	0.870
X_2	Gender	-0.071*	0.032	-2.196	0.029
X_3	Socio-economic Status	0.030	0.028	1.092	0.276
X_4	Land holding	0.084**	0.023	3.586	0.000
X_5	Herd size	0.112*	0.047	2.393	0.017
X_6	Income	-0.111**	0.016	-7.151	0.000
X_7	Mass media exposure	-0.050	0.033	-1.495	0.136
X_8	Extension agency contact	-0.088*	0.036	-2.435	0.016
X_9	Knowledge level	0.027**	0.002	16.627	0.000
	R^2		0.756		
	Adjusted R^2		0.748		
	F		89.694**		
	N		270		

Dependent variable = VIC usage level
** = Significant at one per cent level ($p<0.01$)
* = Significant at five per cent level ($0.05>p>0.01$)

An examination of the table 3 reveals that the adjusted R^2 value for the fitted regression equation was 0.748 which indicates that 74.8 per cent of the variation in the usage level is explained by the nine independent variables included in the analysis. The table also shows that out of nine independent variables taken as antecedents to the usage levels of VIC, only two variables *i.e.* land holding and the knowledge level of the farmers was significantly associated with VIC usage level at 0.01 level of probability and the regression coefficient for these variables was positively related to the VIC usage level which indicates the direct relationship between land holding and knowledge of beneficiaries with the usage level. Herd size was significantly associated with VIC usage level at 0.05 level of probability and the regression coefficient for this variable is positively related to the VIC usage level.

The dairy farmers with higher land holding size might have bigger herd sizes which necessitate good knowledge on dairy farming and so exposure to VICs is more than that of the other beneficiaries with lesser land holding. Since, higher usage level of VIC information contribute to knowledge gain, higher the usages of VIC results in high knowledge level which further shows the effectiveness of VICs.

IV. Constraints in Accessing Village Information Centre

The pooled data (Table 4) revealed that among the constraints perceived by dairy farmers in accessing VICs, general problem was considered as foremost constraint with highest mean score of 5.11 and was ranked first. General problem includes power failure, connectivity problem, inadequacy of computer *etc*. This may be due to the fact that in rural areas most of the time there was no power especially in summer days and so respondents perceived it serious constraints. This finding was in accordance with findings of Benjamin (2000) and Sharma (2004). This was followed by socio-personal constraint (4.12), content problem (3.15), physical facilities (0.93) and socio-cultural constraint (0.05) and these were ranked second, third, fourth and fifth, respectively.

V. Suggestions by Dairy Farmers to Improve the Effectiveness of VICs

It is clear from the Table 5 that majority (61.67 per cent) of the respondents suggested that battery back-up may be needed to over come the power failure especially during summer season. Majority (58.89 per cent) of the respondents suggested that broad band connection may be required to reduce the connectivity problem. Around 49.44 per cent of the respondents sought complete information about dairy farming under one website.

VI. Suggestions by Service Providers to Improve the Effectiveness of VICs

All knowledge workers (6 numbers) suggested that battery back-up may be needed during power failure especially during summer season. They are in need of refresher training programme on livestock farming for updating their knowledge. In order to increase the effectiveness of VICs, awareness campaign and group discussion may be carried out in adjacent villages of VICs.

Conclusion

Knowledge gaining through VICs is the indicator for assessing the effectiveness of Village Information Centres and results of this study revealed the above fact. The respondents attained high total knowledge score on dairy farming activities only by the way of utilising VICs not from any other sources.

Table 4 : Rank order of constraints perceived by dairy farmers in accessing Village Information Center

Sl. No.	Constraints	TANUVAS (n= 60)		MSSRF (n=60)		n-logue (n=60)		Overall (N=180)		
		Total score	Mean score	Total score	Mean score	Total score	Mean score	Total score	Mean score	Rank order
1.	General problem	238	3.97	310	5.17	371	6.18	919	5.11	I
2.	Socio-personal	229	3.82	270	4.5	242	4.03	741	4.12	II
3.	Content problem	188	3.13	191	3.18	188	3.13	567	3.15	III
4.	Physical facilities	47	0.78	60	1.00	60	1.00	167	0.93	IV
5.	Socio-cultural	4	0.07	0	0.00	5	0.08	9	0.05	V

Table 5 : Distribution of respondents according to their suggestions

Sl. No.	Suggestions	TANUVAS (n=60)	MSSRF (n=60)	n-Logue (n=60)	Overall (N=180)
1.	Battery back -up may be needed	42 (70.00)	18 (30.00)	51 (85.00)	111 (61.67)
2.	Broad band connection needed	47 (78.33)	34 (56.67)	25 (41.67)	106 (58.89)
3.	Complete i nformation on livestock farming must be available under one website or package	32 (53.33)	15 (25.00)	42 (70.00)	89 (49.44)

The total knowledge score for MSSRF respondents are comparatively lesser than that of TANUVAS and n-Logue communication respondents. This was due to the fact that MSSRF-VICs were established during 2004-05 in Pudukottai district of Tamil Nadu. They were not using latest technology *i.e.* videoconferencing that would make better face to face interaction between experts and farmers. Kiosk operators at TANUVAS from Veterinary graduates during establishment period and they had analysed the information needs on dairy farming very well as compared to the operators of other service providers. The qualification of the operators is one important factor in running the ICT-Enabled VICs. These operators should be oriented and trained in the area of dairy farming to render better service to the dairy farmers.

More time spent at VICs and lesser distance in accessing VICs, more exposure to mass media and extension agency contact will increase the knowledge level of dairy farmers. This implies that, if dairy farmers were having more exposure to demand driven information provided in VICs, then, even lower socio-economic status, old aged and low income group of dairy farmers can be reached to transfer more knowledge on dairy farming in much less time.

Analysis of the data on constraints perceived by the dairy farmers in accessing VICs indicates that, general problem *i.e.* power failure and connectivity problem was considered as foremost problem. This could be overcome by providing battery back up during summer season. In order to reduce connectivity problem, broad band connection may be required. Content should be updated regularly based on the need of people.

References

Benjamin, P., 2000. African Experience with Telecenters. E-OTI (On the Internet), Nov/Dec,2000. http://www.isoc.org/oti/articles/1000/benjamin.html

Cecchini, S. and Scott, C., 2003. Can Information and Communications Technology Applications Contribute to Poverty Reduction? Lessons from Rural India. *Information Technology for Development*, 10: 73-84.

Conroy, C., 2006. Telecentre Initiatives in Rural India: Failed, Fad or the way Forward? Natural Resources Institute, University of Greenwich, UK. Accessed from www.telesupport.org

Deepak and Singh, S.P., 2006. Knowledge Level of Rural Women about Improved Breeding and Reproductive Practices of Buffalo in Haryana. *Indian Journal of Dairy Science*, 59 (5): 317-321.

Rao B.S. and Kherde, R.L., 1990. A Standardized Test to Measure the Knowledge of the Dairy Farmers on Dairy Production Technologies. *Asian Journal of Dairy Research*. 9 (4) : 224-228.

Richardson, D., 1996. The Internet and Rural Development: Recommendations for Strategy and Activity. *FAO final report*, Rome, Italy.

Roman, R. and Colle, R., 2002. Themes and Issues in Telecentre Sustainability. *Development Informatics Working Paper No. 10.* IDPM: Manchester, UK.

Sharma, G.R.K., 2004. Internet Communication among Livestock Farmers in Chittor District of Andhra Pradesh. *Ph.D. Thesis*, Division of Extension Education, Indian Veterinary Research Institute, Izatnagar, Bareilly, U.P., India.

West, M.D., 2000. Defining and Measuring Credibility of Internet Sources: Developing Scale. *Journalism and Mass Communication Quarterly*, 66: 277-284.

25

Farm Technologies through Interactive Multimedia : A Multi-Stake Holders' Perspective Analysis

N. Anandaraja, N. Sriram, Shibi Sebastian and D. Periyar Ramasamy

ABSTRACT

Communication is now one of the central issues in developing societies. Knowledge expansion, information technology and technology dissemination are the three very important dimensions of any development system be it agriculture, industry and whatever. Revolutionary changes in communication policies, techniques and strategies are needed in order to make information is a fundamental right to the poor rather than making communication a privilege enjoyed by the rich and powerful. Transfer of agricultural technology generally aims at speedy diffusion of novel agricultural information from the place of invention to the place of adoption. In the current Information and Communication Technology (ICT) era, human dependency on computer and allied modern electronic gadgets for executing day to day activities is inevitable. All fields are being influenced by the interventions of ICT and agriculture is in no way an exception to it. The IMCD aims at fastening the diffusion of Good Agricultural Practices (GAP) and novel farm technologies namely System of Rice Intensification, Vermi-compost production, mushroom cultivation, Integrated Pest Management practices and drip based sugarcane cultivation aspects through active involvement and participation of farmers with the help of computer based user friendly Interactive Multimedia Compact Disc (IMCD). The feedback of the farmers were taken to assess the IMCD as a transfer of technology tool. Among the total respondents, 326 respondents were from the farming community, 41 respondents were staff from various agricultural institutes, and the remaining 66 respondents include PG and Ph.D. scholars.

Introduction

Communication is now one of the central issues in developing societies. Knowledge expansion, information technology and technology dissemination are the three very important dimensions of any development system be it agriculture, industry and whatever. Revolutionary changes in communication policies, techniques and strategies are needed in order to make it a fundamental right practiced by the poor rather than making communication a privilege enjoyed by the rich and powerful. It is a mediation tool which brings different social groups together to discuss their interest and need to reach a consensus for action. Communication technology and media are useful tools in this process but should not be considered as an end in them. Sustainable people-oriented development can only be realized if information and knowledge are shared and rural people are involved and motivated (FAO, 1987).

Farming community of India is normally endowed with fragmented land holdings containing marginal and small sectors to the maximum. The affordability, accessibility and possibility to travel a long distance and hunt the information is seldom possible. Moreover, the required message at appropriate time and season is a major bottleneck to the farmers in the rural areas. Access to information and improved communication is a crucial requirement for sustainable agricultural development. At this stage, modern communication technologies when applied to conditions in rural areas can help improve communication, increase participation, disseminate information and share knowledge and skills. It is essential that information availability is demand driven rather than supply driven. The challenges are not only to improve the accessibility of communication technology to the rural population but also to improve socio-economic status and its relevance to local development (Balit *et al.*, 1997). Bringing success and promotion of any technologies could be possible with the formulation of appropriate strategies and blue prints. This might include the purchasing capacity, affordability, financial support, need of subject matter specialists to deliver the modules, adequate training *etc.*

Transfer of agricultural technology generally aims at speedy diffusion of novel agricultural information from the place of invention to the place of adoption. We are in the Information and Communication Technology (ICT) era, dependency on computer and allied modern electronic gadgets for executing our day to day activities is inevitable. All fields are being influenced by the interventions of ICT and agriculture is in no way an exception to it. The scenario warranted the use of ICT to build a knowledge enriched rural society and to ensure reaching the hitherto un-reached. This paves way for enhancing agricultural production and to achieve faster growth of the sector. It is fast becoming the most successful tool for technology transfer and imparting education in agriculture.

Educating Farmers through Computer

In the era of globalization, Indian agriculture gained a new look as an agri business venture which focused on developing profitable and sustainable agricultural enterprises to meet the increasing internal food demands and export. This approach calls for building farmers capacity to absorb and integrate new scientific knowledge and practices with their indigenous wisdom to create and maintain sustainable enterprises (Swaminathan, 2000). The extension system should reorient its roles and responsibilities and equip its human resources to meet emerging challenges. Training is a crucial and continuous input for any efficient and effective transfer of technology system.

The Indian agricultural extension system, being the largest of its kind in the world, is facing innumerable challenges in training its vast human resource base. On the other hand, the emerging communication technologies offer capabilities that redefine the modes of human communication thus improving efficiency and effectiveness of the communication endeavor. It reflects a powerful global belief in the transformatory nature of these technologies. The technology to be transferred should be innovative, remunerative, need based and more beneficial to farmers. The present scenario of agricultural extension in our country is highly conventional with high commands and control over the clientele.

Delivering the technologies in a highly theoretical way has led to collapse many of the public agricultural extension programmes. The extension strategy should not only depend on conventional systems like demonstrations, trainings, radio and television broadcasts with close mindedness. Rather it should be highly decentralized, constructive, open mindset, practical field based and more participatory way of disseminating the technologies. While using computers to disseminate technologies, it allows users to explore information in whatever order suits their needs.

Computer literacy is the need of today and voice of tomorrow to survive in the fast changing world. It has become a dire necessity of life. Nowadays, it is difficult to imagine life without it. Computer-communicated data, information and decision support systems are essential aids to assist farmers in their decision-making. When information is given at right time, it improves production, working and living conditions of farming community. Hence, computer based tutorials will offer learners more complete and individual control over their learning. The main reason behind recommending the use of computer based module for farmers is to facilitate interactivity and better understanding between individual learners and the subject matter. Computer based tools are ideally suited to demonstrate complex and dynamic process that cannot be explained easily with conventional media and methods. It can go a step further by providing the individual solutions in the local language. Farmers obtain information from numerous sources, but they prefer to observe new practices under local situations. Under such circumstances computer based self learning modules help the farmers to learn self to an understanding and thorough comprehension about the subject formulated in the module.

Total access to information is provided because users rely on more than one sense to access the materials. Modern communication like cyber extension, expert system, decision support systems, multimedia, village information kiosks play a vital role in effective transfer of technology. These methods are timely and inevitable in today's rapid era of Information and Communication Technology. They act as very successful methods of extension in combination with traditional methods. Extensionists should create a sense of enthusiasm to develop skills on the use of modern communication gadgets. The application of modern communication methods enhances speedy extension by the officials from their place of work and facilitating farmers to access information at their places concerned. In line with this, a thesis entitled "Developing farmer-friendly interactive multimedia compact disc and testing its effectiveness in transfer of farm technology" was done by the first author of this article, as a part of his doctoral degree from Department of Agricultural Extension and Rural Sociology, Tamilnadu Agricultural University during 2002. Development of IMCD involves preparatory works including review of literature, content preparation, script writing, collection of video/pictures/ video capturing, designing and lay out of IMCD, audio recording, mixing of voice, music, video and finally cross checked with the experts for their valuable suggestions.

IMCD as Self Learning Module to Farmers

Interactive Multimedia Compact Disc is a modern electronic gadget which demonstrates complex farm technologies in simpler terms using text, graphics, audio and video. It is highly self controlled, comprehensive, user-friendly, highly reliable, pragmatic, innovative, cost-effective and simple tool which offers an unprecedented potential of providing knowledge to the users. IMCD is a combination of computer based instruction plus media tools and packages. In simple terms the user can hear the messages recorded through audio, visual and graphics, *etc.*, and the user can test whether their learning is 100 per cent or not. In case their learning is not perfect, they can try again and clear their doubts. Thus the user keeps on using the IMCD till their learning attains perfection. The IMCD aims at fastening the diffusion of Good Agricultural Practices (GAP) and novel farm technologies namely System of Rice Intensification, Vermicompost production, mushroom cultivation, Integrated Pest Management practices and drip based sugarcane cultivation aspects through active involvement and participation of farmers with the help of computer based user friendly Interactive Multimedia Compact Disc (IMCD).

IMCD is a device encompassing labeled instructions helping the user to learn the contents, structure and information using the mouse without depending on the help of a facilitator. IMCD facilitate an interaction asynchronously because both sender and receiver have to be in touch at exactly the same time in order to communicate. It is a motivational cum inspirational tool playing a significant role in sweeping changes in socio-economic development in the era of globalization. It also aims to bridge the digital divide commonly prevailing among rural farming community by orienting them on the usage of computer and IMCD. This self learning module enhances the farmer's ability to learn, acquire knowledge and skill in terms of both modern gadgets and latest need-based farm technologies tailored to their needs and their valuable time schedule in mind. This paper critically analyses various feedback of different stakeholders on using IMCD as a Transfer of Technology tool. The methodology adopted for the study is given below. This paper attempts to highlight the IMCD module - Feedback which was obtained from different stakeholders like farmers, students and scientists.

Methodology

The new millennium era, dawns with the evolving of the information technology, which has been steadfast in its growth and applications. India being endowed with enormous natural resources and a basically agricultural country, it is imperative for the farmers to effectively utilize these natural resources. As the traditional agriculture is transforming into hi-tech agriculture, the need for up to date information is also essential for agricultural production and productivity, thus helping the farmers to know the latest advances in agricultural technologies and preparing them towards progressive farming. At this predicament, the computer based IMCD module can be resorted for clearing the queries and doubts related to subject matter in agriculture. This module can be the better answer to solve the current problems that arise here and there. Here the salient feedback of different stakeholders namely farmers, agricultural graduates, extension workers, staff from various agricultural colleges were discussed in detail. Further, it discusses Strength,

Weaknesses, Opportunity and Challenges of IMCD in the diffusion of farm technologies.

Feed back (Action - reaction interdependence in communication is referred as feedback. Feedback makes persons truly interacting members of a communication system and tells us how our messages are being interpreted) on IMCD (Eriophyid mite management) and its usefulness in imparting knowledge, awareness and skill have been collected from the farmers who had visited the Farmers Mela (It is an annual event organized by the Tamil Nadu Agricultural University in collaboration with various stakeholders in the field to exhibit latest know-how and do-how and technological packages to farmers. Farmers in the region will visit the stalls and have an interactive learning with the Scientist/ Expert), which was organized by TamilNadu Agricultural University during 2005. The mela was conducted for three days. The farmers are allowed to operate the computer system themselves. They were self instructed by the computer to go through the IMCD on various agricultural technologies which are considered to be the felt needs of the farming community. Similarly, in 2006 IMCDs on Vermi-compost production and System of Rice Intensification (SRI) were exposed to farmers. The feedback of the farmers were taken to assess the IMCD as a transfer of technology tool. Among the total respondents, 326 respondents were from the farming community, 41 respondents were staff from various agricultural institutes, and the remaining 66 respondents include PG and Ph.D scholars.

Farmers Feedback on IMCD

1. IMCD as a source for effective information and also it is a new innovation which simplifies the work of extension functionaries.
2. The contents represented in local and linguistic language helps for easy understanding. Use of simple vocabulary helps in clarity about the information.
3. IMCD module helps to create awareness on existing technologies and new innovations.
4. Self explanatory aspect helps to develop individual personality.
5. Farmers' needs are much more diversified.
6. Farmers are in need of constructive, open mindset, practical field based and more participatory way of disseminating the technologies. IMCD allows the users to explore information in whatever order suits their needs.
7. IMCD helps to get convinced and deep understanding about a subject. For any technology it should first satisfy the beneficiary and then leads to conviction and motivation.

8. The theoretical aspects given in attractive manner which includes graphical representations, movies and photo copies helps the illiterate farmers to learn easily and it has enhanced the retention rate of farmers.
9. The video clippings carrying the demonstrations related to the subjects have imparted skill to the farmers.
10. IMCD helps to measure the advantages and disadvantages of a new innovation which helps in adoption or rejection.
11. Majority of the farmers felt that the contents explained in simple and local vocabulary helps in easy understanding. Especially on operation of IMCD their feedback were strong. Because, it is simple to operate mouse with right and left click. The action button provides clear instructions how to proceed.
12. The developed module explains the importance of technology and main heading. Its users option where they want to learn and which one is needy to them and important. Moreover background audio (text over audio) supports the illiterates to navigate the information in the module.
13. The message components such as completeness of the information, understandability of the message, logical presentation, explanation for complex ideas, clarity of the message, appropriateness of the message, summarizing the subject and reinforcement of key ideas are clearly blended in IMCD.
14. In both the farmers mela, there were 423 respondents including farmers, scholars and scientists visited the IMCD stalls. There they had the facility to access, browse, navigate and use of the IMCD through LCD screen. Stake holders feedback received through the feedback register on the learning point of IMCD's.

Agricultural Scientists' Opinion

Majority of the staff from agricultural institutes stated that IMCD is increasingly attracting the attention of farmers as well as the extension functionaries as a way of much needed speedy Transfer of Technology in agriculture. It is a good source of effective information and innovative. The contents developed were self-explanatory which creates interest among the learners and easy learning for illiterates. They also suggest the following aspects as benefits of using IMCD tool.

1. Reduction in training time compared to conventional methods of training.
2. Learner acceptability.

3. Immediate feed back.
4. Superior quality of capacity building, particularly in teaching skills.
5. It caters for individualized instruction.
6. The content is easily adaptable- *eg.* locality and experience.
7. The material may be easily updated and expanded.
8. It is reliable.
9. Cost-effectiveness, particularly in the long run.
10. Easy retrieval and storage.
11. Creative extension and excellence venture.
12. Complete package of practices on selected technologies for better management of illiterate farmers also.

Students' Opinion

The scholars who visited the IMCD stall had given the following points on the usage of Interactive Multimedia Compact Disc as Transfer of Technology tool.

1. Majority of the students expressed that, IMCD acts as an effective teaching material.
2. It offers increased control over the learning process, simulation of laboratory experiences, access to the visual information, vital to scientific studies and more effective tutorial support. It is also user-friendly, invaluable and powerful tool in decision making.
3. New ideas and models are constantly supported which are intended to dramatically improve education. One of these new ideas involves learning through IMCD.
4. Learning through IMCD has a holistic outcome. It is the result of the personal growth, which has elements of knowledge, skill and attitude transition. Competency is another term, which is some times used to describe this growth.
5. Good qualities of graduates with competencies are required for future agriculture is the primary concern of the agricultural education through IMCD.
6. IMCD offers a way of education which is "explicitly consistent" with the sentiment of this statement rather than having it as an "implicit outcome" of educational experience.

SWOC Analysis

This paper attempts to study the SWOC analysis that is strengths, weaknesses, opportunities and challenges of IMCD as a Transfer of Technology tool. The study shows that there are many strong aspects to use IMCD as a tool to disseminate farm technologies. Even though several aspects considered as strength, it also possess some constraints in using IMCD as Transfer of Technology tool. These SWOC factors are given below:

Strengths

- Self paced learning and interactive
- Highly user control
- Long lasting
- Effective information and reliable
- Creates impression on the mind
- Clarifies complex ideas
- Adaptable and flexible teaching aid
- Provides scope for immediate clarification of doubts
- Can be used in different stages of adoption
- Possible of enhancing/ updating new contents/ findings

Weaknesses

- Majority of the farming community residing at remote areas.
- Lack of exposure to the electronic gadgets
- Lack of hands on exposure training
- Lack of affordability and accessibility by the rural farmers
- Initial hesitation on usage of computers even for a literate person
- Needs more active extension functionaries

Opportunities

- Can be used as a virtual tool for distance education
- Contents of IMCD can be explored for online for universal accessibility
- Educate illiterate farmers
- Enhances the capacity building for students, scientists, women and rural youth

Challenges

- Consumes time to convince the farming community to accept IMCD as TOT tool
- Need specialized skill to use IMCD
- Rapid advancement of science
- Deficiency of technical expertise to maintain the system
- High cost of production and distribution of educational multimedia packages

- Greater time needed for preparation of courseware development of instructional packages requires considerably more time and expertise than an equivalent package required for traditional training

Implications of the Study

Based on the farmers, scientists and scholars' feedback and SWOC analysis, the following implications were drawn for policy maker's consideration.

1. Agricultural graduates can be given opportunity in producing IMCD packages on improved and innovative technologies and made available to farming community.
2. Farmers' needs are more diversified and the knowledge required to address them is beyond the capacity of various media. Hence, commercializing the IMCD package to the farming community in a cost effective manner can be undertaken by government and private sectors.
3. Trainings organized by the agriculture and allied departments for the farming community may be entrusted with the task of utilizing IMCD as training module which will lead to large level awareness on technologies as well as IMCD module.
4. Specialized training can be organized to the extension functionaries on updating and expansion of novel technologies.
5. As a part of Mission 2007, each panchayat could get Village Knowledge Centre (VKC) before end of the XI plan period. In this context Government should take a policy initiative and to circulate the Tamil Nadu Agricultural University developed IMCDs to all the Village Knowledge Centers.
6. It can be used in the Common Service Centres

Conclusion

Advance technologies which are used in this new communication system are extremely rapid. However, as we look to the future, IMCD has wide scope for application in agricultural extension activities. This electronic device with its unique interactive and motivation capabilities can offer quality transfer of technology over development, which other conventional medium seldom offers. The use of IMCD in transfer of Technology reduces unproductive travel time, money and energy to the farmer, the actual person who can directly bring about an improvement in efficiency and productivity in agriculture.

References

Anandaraja, N., 2002. Developing Farmer Friendly Interactive Multimedia Compact Disc and Testing its Effectiveness for Transfer of Farm Technology. *Ph.D., Thesis* (Unpub.). AC&RI, TNAU, Coimbatore.

Balit, S., Calvelo Rios, M., and Masias,L., 1997. Communication for Development for Latin America: A Regional Experience. FAO, Rome Italy.

FAO.1987. Development Support Communication Expert Consultation, *Report,* Rome, 8-12 June 1987.

Swaminathan, M.S., 2000. For an Evergreen Revolution. THE HINDU Survey of Indian Agriculture 2000. P.9-15.

http://www.iicd.org/stories

http://newmedia.com/newmedia/8_09.html

26

e-Readiness and Participation Level of AKSHAYA and KISSAN Kerala Beneficiaries

M.S. Nataraju, A. Arun Babu and B.S. Lakshmana Reddy

ABSTRACT

Information is basic element in any development activity. It is useful only if it is available, if the users have access to it, in the appropriate form and language. Information in the field of agriculture, to benefit farmers, has to be tailored to local agro-ecological and socio-economic conditions backed up by relevant input supply services and public policies. The impact of information revolution with its powerful information and communication technologies are making the whole universe into a small global village enabling easy access and faster sharing of knowledge among the world communities. Information Communication Technologies have great prospect and wide scope for disseminating information to the people in remote areas with great accuracy and with no loss or distortion of any information. ICTs can empower rural communities and give them 'a voice' that permits them to contribute to the development process. A silent revolution is taking place in the communication systems in rural India. It is said that in the years to come digital divide will be one important factor in separating people in a society like caste and income. A huge sum of money is being pumped into this sector to reduce the so called digital divide. In this regard, a study was conducted in Kerala state to analyse the e-readiness, perception and extent of participation of beneficiaries in

Information Communication Technology (ICT) projects besides knowing their information needs and problems in using ICTs. The data were gathered from a randomly selected sample of 80 respondents from two projects namely AKSHAYA and KISSAN Kerala. The study revealed that KISSAN Kerala beneficiaries had highest mean e- readiness score (3.04) followed by AKSHAYA beneficiaries (2.45) whereas, farmers scored the least (1.72). Majority of the KISSAN Kerala beneficiaries (85%) and two fifth of AKSHAYA beneficiaries (40%) belong to high e-readiness group, while most of the farmers exhibited low level of readiness. e-readiness was calculated by assessing six indicators namely, e-access, e-learning, e-society, e-business, e-governance and e-willingness. KISSAN Kerala beneficiaries secured highest scores for all indicators while farmers scored the least. All respondents had low scores for e-society, e-business and e-governance. In spite of their low scores for e-access and e-learning farmers expressed high e-willingness. There is a need to make beneficiaries aware about the ICT enabled services and care should be taken to make them adopt more such technologies.

Introduction

In recent years, there is a visible shift from the old ways of delivering information to the modern ways of information delivery systems. The global information society is undergoing a fast revolution. Integrating Information Technology with agriculture has increased the opportunities for effective management support. ICTs can empower rural communities and give them 'a voice' that permits them to contribute to the development process. A silent revolution is taking place in the communication systems in rural India. It is said that in the years to come digital divide will be one important factor in separating people in a society like caste and income. A huge sum of money is being pumped into this sector to reduce the so called digital divide. At the same time it raises a number of questions also. Whether our farmers and rural people are capable of facing this revolution?, What are the factors hindering ICT dissemination?, Are they really ready in terms of various ICT indices to use these technologies?. In this regard, an attempt is made through a study conducted in Kerala state to find out answers to the above mentioned questions with the following objectives.

1. To study the mean e-Readiness and its indicators among beneficiaries
2. To assess the e-Readiness levels of AKSHYA and KISSAN Kerala beneficiaries
3. To know the extent of participation of AKSHAYA and KISSAN Kerala beneficiaries
4. To document the information needs and the constraints of beneficiaries

Methodology

The study was conducted in three districts of Kerala namely Malappuram, Thrissur and Palakkad. Two Information Communication Technology projects that have more relevance to agriculture and rural development namely; KISSAN Kerala and AKSHAYA were selected. Forty respondents were selected from each projects making to a total of 80. Simple random sampling method was used to select the respondents. Data was collected using a well structured and pretested interview schedule.

e-Readiness is operationally defined as the readiness or preparedness of people to use various forms of ICTs in their daily life. The procedure developed by Harvard Universities Centre for International development was used to measure the e-readiness with slight modification (Anonymous, 2000). The tool prepared for the study has 6 major categories with sub-divisions. The six major categories assessed were, e-Access, e-Learning, e-Society, e-Business, e-Governance and e-Willingness. This tool provided a grid with four stages of development for each category and their sub-divisions. The respondents were asked to mark the stage they fit in according to their perception. They were asked to select only one stage. A score of 1 was given to first stage, 2 for second stage, 3 for third stage and 4 for fourth stage of development. The score obtained by the respondent in each category is the readiness of that category and the total e-Readiness is assessed by taking the average of all these six major categories. The maximum score of e-Readiness that can be attained by a respondent was 4 and the minimum score that can be attained was 1.

Findings and Discussion

Mean e-Readiness of ICT Beneficiaries

A critical glance of the table1 indicates that among the two categories of respondents KISSAN Kerala beneficiaries exhibited highest mean e-Readiness score followed by AKSHAYA beneficiaries.

Table 1 : Mean e-Readiness Scores of Beneficiaries (n=80)

Respondents	Mean e-Readiness score
AKSHAYA Beneficiaries	2.45
KISSAN Kerala Beneficiaries	3.04

The mean e-readiness score of KISSAN Kerala beneficiaries is 3.04, which is almost 25 per cent more than AKSHAYA beneficiaries who scored 2.45.

Mean levels of e-Readiness indicators

e-Readiness was calculated by assessing six indicators namely e-Access, e-Learning, e-Society, e-Business, e-Governance and e-Willingness to use. Each of these indicators were assessed and scored in such a way that maximum score that an individual can get is 4 and minimum score is 1. The mean score of these indicators obtained by the respondents presented in table 2.

Table 2 : Mean Levels of e-Readiness Indicators (n=80)

Respondents	e-Access	e-Learning	e-Society	e-Business	e-Governance	e-Willingness	Mean e-Readiness
AKSHAYA Beneficiaries	2.55	2.38	2.34	1.58	2.50	3.35	2.45
KISSAN Kerala Beneficiaries	3.35	3.29	2.44	2.83	2.65	3.68	3.04

It can be seen that AKSHAYA beneficiaries exhibited mean scores of 2.55, 2.38, 2.34, 1.58, 2.50 and 3.35 for e-Access, e-Learning, e-Society, e-Business, e-Governance and e-Willingness respectively. KISSAN Kerala beneficiaries had mean scores of 3.35, 3.29, 2.44, 2.83, 2.65 and 3.68 for the same indicators in that order. Further, it is observed that in case of all indicators, KISSAN Kerala beneficiaries had highest mean score among respondents. While comparing the scores of AKSHAYA and KISSAN Kerala beneficiaries, KISSAN Kerala beneficiaries had high scores in the case of e-Access, e-Learning and e-Business. Mean scores of e-Society and e-Willingness to use are almost equal to all the three groups.

e-Readiness Levels of AKSHAYA Beneficiaries

AKSHAYA beneficiaries were categorized into low, medium and high groups using mean and standard deviation of e-readiness scores. The results are presented in Table 3. It can be seen that majority of the AKSHAYA beneficiaries belong to high level of e-Readiness (40%) followed by medium (32.5%) and low (27.5%) categories. Each half of the respondents (50%) were found in the medium e-Access category followed by 37.5 per cent in the low and 12.5 per cent of the beneficiaries in the high e-Access category. In the case of e-Learning, it is interesting to quote that none of the beneficiaries were found in low level of e-Learning while 77.5 per cent and 22.5 per cent of the beneficiaries were found in medium and high categories. It is observed that none of the beneficiaries belong to low group of e-Society however, vast majority (92.5%) of beneficiaries belongs to high group and remaining 7.5 per cent of them comes under medium group for e-Society. As high as 45 per cent of the beneficiaries were observed in low group followed by 52.5 per cent in medium group and only 2.5 per cent in high group of e-Business. With regard to e-Governance it is seen that only 12.5 per cent of beneficiaries fall in low group as against 25 per cent of beneficiaries in the medium group and majority (62.5%) of beneficiaries in high group. The table further shows that majority of the beneficiaries (52.5%) exhibited high score followed by 30 per cent medium score and 17.5 per cent were found to have low score for e-Willingness.

Table 3 : e-Readiness Levels of AKSHAYA Beneficiaries (n=40)

Indicators	Category	No.	Per cent
e-Access	Low <2.36	15	37.50
	Medium 2.36-3.09	20	50.00
	High = 3.09	5	12.50
e-Learning	Low <.1.90	0	0.00
	Medium 1.90-2.84	31	77.50
	High = 2.84	9	22.50
e-Society	Low <1.99	0	0.00
	Medium 1.99-2.53	37	92.50
	High = 2.53	3	7.50
e-Business	Low =1	18	45.00
	Medium=2	21	52.50
	High =3	1	2.50
e-Governance	Low =1	5	12.50
	Medium=2	10	25.00
	High =3	25	62.50
e-Willingness	Low =1	7	17.50
	Medium=2	12	30.00
	High =3	21	52.50
e-Readiness	Low <2.06	11	27.50
	Medium 2.06-2.74	13	32.50
	High = 2.74	16	40.00

e-Readiness Levels of KISSAN Kerala Beneficiaries

The KISSAN Kerala beneficiaries were categorized into low, medium and high using mean and standard deviation of e-Readiness scores. Percent distribution of each category is presented in table 4. It can be observed that a vast majority of the KISSAN Kerala (85%) showed high level of e-Readiness and remaining 15 per cent had medium level of e-Readiness. None of the KISSAN Kerala beneficiaries are found in low e-Access category while 32.5 per cent and 67.5 per cent of the beneficiaries are found in medium and high e-Access category respectively. In the case of e-Learning, none of the KISSAN Kerala beneficiaries had low level of e-Learning whereas a greater proportion of them (80%) had high level of e-Learning followed by medium level (20%). From the table, it is observed that 20 per cent of beneficiaries exhibited low score of e-Society followed by 35 percent of beneficiaries in medium category and 45 per cent in high category for e-Society. Exactly half of the beneficiaries (50%) were observed in low group whereas 17.5 per cent of beneficiaries were found in medium group and only 32.5 per cent of beneficiaries were in the group with high score for e-Business. Exactly, 50 per cent of the beneficiaries in low group, 35 per cent in medium group and 15 per cent in the high group of e-Governance. It can be seen that none of the beneficiaries were found in low group based on their e-Willingness score whereas 32.5 per cent of beneficiaries were in the medium group and majority (67.5%) of beneficiaries were found in high group.

Table 4 : e-Readiness Levels of KISSAN Kerala Beneficiaries (n=40)

Indicators	Category	No.	Per cent
e-Access	Low <2.36	0	0.00
	Medium 2.36-3.09	13	32.50
	High = 3.09	27	67.50
e-Learning	Low <.1.90	0	0.00
	Medium 1.90-2.84	8	20.00
	High = 2.84	32	80.00
e-Society	Low <1.99	8	20.00
	Medium 1.99-2.53	18	45.00
	High = 2.53	14	35.00
e-Business	Low =1	20	50.00
	Medium=2	7	17.50
	High =3	13	32.50
e-Governance	Low =1	20	50.00
	Medium=1.1 to 2.9	14	35.00
	High =3	6	15.00
e-Willingness	Low =1	0	0.00
	Medium=2	13	32.50
	High =3	27	67.50
e-Readiness	Low <2.06	0	0.00
	Medium 2.06-2.74	6	15.00
	High = 2.74	34	85.00

Extent of Participation of Beneficiaries in ICT Projects

Extent of participation was calculated by assessing degree of participation, post participation and reasons for participation of beneficiaries in the programmes. The mean scores of AKSHAYA and KISSAN Kerala beneficiaries are presented in Table 5.

Table 5 : Extent of Participation of AKSHAYA and KISSAN Kerala Beneficiaries (n=80)

Beneficiary	Mean Participation Score
AKSHAYA	11.38
KISSAN Kerala	20.95

It can be observed that AKSHAYA beneficiaries are having a score of 11.38 for participation in ICT project and KISSAN Kerala beneficiaries are having a higher score (20.95) for the same.

Mean Scores of Degree of Participation and Post Participation

Table 6 highlights that AKSHAYA beneficiaries had a higher mean score (3.83) for degree of participation and KISSAN Kerala beneficiaries had a mean score of 2.73 for degree of participation. It can also be seen that mean post participation score of AKSHAYA beneficiaries is 5.15 while KISSAN Kerala beneficiaries obtained higher mean score (14.23) for post participation.

Table 6 : Mean Scores of Degree of Participation and Post Participation (n=80)

Beneficiaries	Mean Degree of Participation	Mean Post Participation
AKSHAYA	3.83	5.15
KISSAN Kerala	2.73	14.23

The low post participation score of AKSHAYA beneficiaries may be due to the lack of follow up by the authorities and this is the reason for their low score for extent of participation.

Extent of Participation of AKSHAYA Beneficiaries

The beneficiaries were categorized into three groups based on mean and standard deviation *viz.*, low, medium and high. The results are presented in the table 7. It is observed that majority (82.50%) of the respondents had high degree of participation and 17.5 per cent of the respondents had medium level of participation. It is also observed that none of the participants had low level of participation. In the case of post participation majority of the beneficiaries had low level (47.50%) followed by 30 per cent having medium post participation and 22.50 per cent having high level of participation. In overall majority of the beneficiaries (45.00%) had low participation while 32.50 per cent of the respondents had medium and 22.50 per cent of the respondents had high level participation.

Table 7 : Extent of Participation of AKSHAYA Beneficiaries (n= 40)

Indicators	Categories	No.	Per cent
Degree of participation	Low =1	0	0
	Medium = 2	7	17.50
	High = 3	33	82.50
Post participation	Low (<2.12)	19	47.50
	Medium (2.12-8.18)	12	30.00
	High (=8.18)	9	22.50
Overall participation	Low (<17.84)	12	30.00
	Medium (17.84-24.01)	14	35.00
	High (= 24.01)	14	35.00

Extent of Participation of KISSAN Kerala Beneficiaries

Based on their score of extent of participation the beneficiaries were categorized into low, medium and high groups. Table 8 shows that the beneficiaries are distributed equally in all the categories. Almost one third of the respondents had low (30%) participation followed by medium (35%) and high (35%) participation. From the same table, it can be observed that more than half of the respondents (55%) are having high degree of participation followed by 40 per cent in medium level and only 5 per cent with low level of participation. In the case of post participation almost equal per cent of the beneficiaries had low (30%), medium (32.5%) and high (37.5%) level of post participation.

Table 8 : Extent of Participation of KISSAN Kerala Beneficiaries (n = 40)

Indicators	Categories	No.	Per cent
Degree of participation	Low =1	2	5.00
	Medium = 2	16	40.00
	High = 3	22	55.00
Post participation	Low (<11.50)	12	30.00
	Medium (11.50-46.96)	13	32.50
	High (=46.96)	15	37.50
Overall participation	Low (<17.84)	12	30.00
	Medium (17.84-24.01)	14	35.00
	High (= 24.01)	14	35.00

Information Needs of ICT Beneficiaries

From table 9, it could be observed that market information was the main area of information needed by beneficiaries followed by input rice and availability, pests and disease management, crop insurance and others.

Table 9 : Information Needs of ICT Beneficiaries (n=80)

Sl. No.	Area of Information	Relevancy Ranking
1.	Market information	0.97
2.	Input price and availability	0.97
3.	Pests and disease management	0.95
4.	Crop insurance	0.90
5.	Plantation crops	0.88
6.	Spices and aromatic	0.87
7.	Organic farming	0.86
8.	Question and answer services	0.85
9.	Weather information	0.84
10.	General information	0.83
11.	Medicinal plants	0.83
12.	Post harvest technology	0.82

Constraints in using ICTs

The important constraints perceived by the beneficiaries using ICTs are listed in table 10. The results indicate that electricity problem ranked first while low affordability occupies the second position. Technology fear, Problem of foreign language, Lack of infrastructure facility and Technical problem were the other major constraints faced by the beneficiaries.

Table 10 : Constraints in Using ICTs (n=80)

Sl. No.	Constraints	Mean Rank AKSHAYA	Mean Rank KISSAN KERALA
1.	Electricity problem	8.44	6.53
2.	Low affordability	4.95	3.42
3.	Technology fear	5.70	6.50
4.	Problem of foreign language	6.93	7.30
5.	Lack of infrastructure facility	5.49	3.72
6.	Technical problem	5.61	5.38
7.	Cultural barriers	6.03	8.23
8.	Lack of local relevance to information	4.85	4.95
9.	Low e-literacy	3.41	4.55
10.	Lack of awareness about ICT	3.60	4.43
	Kendalls W	0.252	0.282
	X2	90.746**	101.596**

Summary and Implications

It was found that KISSAN Kerala beneficiaries were having more score for e-Readiness indicators. It was also found that both the set of beneficiaries were having low scores for e-Society, e-Business and e-Governance. KISSAN Kerala beneficiaries had more participation compared to Akshaya beneficiaries. More number of Akshaya beneficiaries exhibited low to medium level of participation (77.50%). Market information on input price and availability, pests and disease management, crop insurance and plantation crops were the most needed areas of information. Low e-Iliteracy, lack of awareness about ICT initiatives and low affordability were the biggest constraints in using ICTs. There is a need to make beneficiaries aware about the ICT enabled services and care should be taken to make them adopt more such technologies. It was also found that irrespective of the respondents all were having high score for e-Willingness, which indicates the scope for diffusion of ICTs and ICT enabled services to rural areas. Care should also be taken to concentrate more on technologies like mobile telephony, radio and television, where in the rural poor have more access. By exploiting the full potential of mobile telephony or land line telephony farmers and rural people can get more connected to the wired world. FM radio is a potential medium to reach the rural poor for providing them access to the outside world. The possibility of starting an FM channel for the farmers and rural people in all the states giving more relevance to local information can be explored and it will certainly help the rural poor in entering to the networked world.

Reference

Anonymous, 2000, Readiness for the Networked world - A Guide for Developing Countries, Centre for International Development, Harvard University. http://cyber.law.harvard.edu/readinessguide/test.html

27

ICT for Effective e-Extension and the Successof VRC Network in Assam

S. Borua and P. K. Neog

ABSTRACT

As population explosion continues and the shortage of extension personnel becomes acute, the wisest proposition is to utilize the latest information and communication technology (ICT) to make information available to the common man. The ICT tools with their recent advancements have become easily available in various forms such a mobile telephone, PDA, WLL phones, FM radio, community radio, television, wireless internet services, webcams, video conferencing facilities *etc.* Extension agencies need to harness the ICT tools usefully and improve the quality of extension services. As per new Extension Reforms, when the group approach is being promoted such as FIGs, CIGs *etc.*, it would be even easier to utilize ICTs, as they can have access to the information as groups and thereby make progress in their ventures. Public Private Partnership is being encouraged particularly in the field of providing extension services. There has been enough success in this regard with examples such as ITC-e-Choupals, Tata Kisan Kendras, Mahindra Krishi Vihar *etc.*, which have given considerable support to the farming community. Selected farmers need to be trained on few basic computer skills for enabling them to access such ICT centres and websites and get the benefit. The services being provided under the public-private partnership of ISRO and ABITA

(GKUP) through their VRC network can be highlighted as a success story of ICT in extension work in Assam. The VRCs are an unique facility which were devised to provide the information services to the rural areas of the country, especially to those engaged in agriculture. They are manned by the Agricultural Development Officers along with a supporting staff who serves as the VRC Operator and facilitates access to the farmers. VRCs were formally inaugurated on Nov 28, 2008 at Guwahati. The main components of VRC services to the rural areas are; Land and Water Resources Advisories, Interactive Farmers' Advisory Services/ Tele-agriculture, e-Governance services, Weather Information, Tele-education and Telemedicine. Thus, as a single window delivery mechanism, VRCs are also supposed to be involved in capacity building of the rural community to raise its awareness and knowledge regarding improved technologies. The VRCs provide information on a variety of aspects *viz.*, health, education, nutrition, gender issues, legal services and women empowerment. Upto 142 farmers enthusiastically participated in the video conferences on the various improved technologies. Eminent personalities also interact with farmers through the same facility. In order to make extension services effective, cyber extension has to be strengthened and utilized fully so that technology can be effectively transferred far and wide.

Introduction

The availability of effective and professional Extension services to the entire population living in every nook and corner of the country is essential for facilitating the process of development. More so, with the technical advances in agriculture and allied areas, extension services are seen as indispensable in order to disseminate the latest and improved technical know-how. Much of the extension work in the present day is the sharing of useful information in a prompt/timely manner. There are individuals who are even ready to pay for getting the information on time, as that information can give him the opportunity to earn better profit.

As agriculture is getting converted into a profitable commercial enterprise and agri-business ventures, marketing is becoming crucial. For profitable marketing, extension services are required, which in turn demand the availability of market related information.

As population explosion continues and the shortage of extension personnel becomes acute, the wisest proposition is to utilize the latest information and communication technology (ICT) to make information available to the common man.

The ICT tools with their recent advancements have become easily available in various forms such a mobile telephone, PDA, WLL phones, FM radio, community radio, television, wireless internet services, webcams, video conferencing facilities *etc.* Extension agencies need to harness the ICT tools usefully and improve the quality of extension services.

As per new Extension Reforms, when the group approach is being promoted such as Farmer Interest Groups (FIGs), Commodity Interest Groups (CIGs) *etc.*, it would be even easier to utilize ICTs, as they can have access to the information as groups and thereby make progress in their ventures.

Of late, the Government of India has been encouraging Public Private Partnerships in almost all sectors, particularly in the field of providing extension services. There has been enough success in this regard with examples such as ITC-e-Choupals, Tata Kisan Kendras, Mahindra Krishi Vihar *etc.*, which have given considerable support to the farming community. However, it is yet to make progress in North East India.

Concerned authorities can take steps to make the improved package of practices available to the farmers in the form of CDs or even VCDs by having few videos prepared on some critically important skills. The websites which have come up in recent years can be of great help to the farmers. Selected farmers need to be trained on few basic computer skills for enabling them to access such websites and get the benefit. 'Briddhi' the website of Assam Agricultural University, Jorhat can be of great help to the farmers not only in the state of Assam but entire North East India. The websites of Krishi Vigyan Kendras which are coming up one after the other, with ICAR support can provide location-specific information in agriculture and allied enterprises. However, concerned authorities need to make concerted efforts to develop locally relevant content.

The Common Service Centres (CSCs) which are being converted from the earlier existing centres *viz.*, Community Information Centres(CICs) can be utilised better by farmers, as they are conveniently located in the rural development blocks. However, a better coordination with development departments besides providing adequate publicity of the services available only can enable the villagers and others to come and avail the facility for improving their standard of life.

ISRO-ABITA (GKUP) VRC Network - A Success Story of ICT in Extension Work in Assam

In the present day context, it is said and realised by most people that 'information is power'. The space-based technology has made its advancements almost to its peak. However, it could be effectively utilized only when it can provide the some benefit to people in the grassroot. If the space technology can be utilized to provide the much required information to the people in the grassroots, it can appropriately empower them and enable people to taste prosperity.

With support from Indian Space Research Organisation (ISRO), the North Eastern Space Application Centre (NESAC) under Department of Space, Government of India launched the programme to set up Village Resource Centres (VRCs) in Assam in partnership with NGOs. Having adequate reach to the grassroots and also the capacity, the NGO which came forward to meet the requirements of establishing such a facility was ABITA-Gramin Krishi Unnayan Prakalpa sponsored by the Assam Branch of Indian Tea Association.

The VRCs are unique facilities which were devised to provide the benefit of space-based services to the rural areas of the country, especially to those engaged in agriculture. The mode of working is that of video conference being organised amongst two ends *i.e.,* expert end and learners'/ users' end. There can be useful discussions that can be organised amongst farmers sitting in some remote location with the experts in the University or NGO headquarter. Of course, the internet can also be accessed in the VRCs.

VRCs were formally inaugurated by the Hon'ble Chief Minister of Assam on Novermber 28, 2008 at Assam Branch of India Tea Association (ABITA)-Gramin Krishi Unnayan Prakalpa's (GKUP) office at Guwahati. However, a very good response has been seen from the farmers in all the 10 learners ends housed in the Agricultural Growth Centres of ABITA-Gramin Krishi Unnayan Prakalpa situated in places like Borgang, Moran, Margherita, Balipara *etc.* They are manned by the Agricultural Development Officers along with a supporting staff who serves as the VRC Operator and facilitates access to the farmers. There are 142 Farmers enthusiastically participated in the video conferences, particularly on the technology relating to System of Rice Intensification (SRI) and bamboo cultivation, Post Harvest Technologies in horticultural crops *etc.* Interaction sessions were organised for farmers also with the subject matter experts from agriculture and home science faculties. Eminent personalities who have been invited to interact with farmers included Vice-Chancellor, NEHU, DDG from IMD, Regional Meteorological Centre, Guwahati, Shri Hembhai, eminent Gandhian and social worker *etc.*

The main components of VRC services to the rural areas are; Land & Water Resources Advisories, Interactive Farmers' Advisory Services/ Tele-agriculture, e-Governance services, Weather Information, Tele-education and Telemedicine. Thus, as a single window delivery mechanism, VRCs are also supposed to be involved in capacity building of the rural community to raise its awareness and knowledge regarding improved technologies and be benefited from them, even though they may be residing in remote rural areas.

The VRCs provide information on a variety of aspects *viz.*, health, education, nutrition, gender issues, legal services and women empowerment.

In Assam, Assam Agricultural University, Jorhat was identified as the Expert agency and ABITA-GKUP as the Associating agency which was responsible for implementing and operationalizing the VRCs. ISRO provided the equipments *viz.*, the V-SAT along with necessary hardware and software as required.

The village resource centres proved as an effective channel for information dissemination even to the remotest parts of the state, which might otherwise be difficult to be accessed by extension workers. VRCs have been able to achieve the goal of utilizing the space and communication technology for development of the common man in the rural areas. Proper utilization and extension of VRC can result in reduction of rural poverty and improve the standard of life of the masses. The success of the VRC concept can be replicated elsewhere too by some other agencies even by utilizing the BSNL or other telecommunication network, as ICT gets further developed day after day.

In order to make extension services effective, cyber extension has to be strengthened and utilized fully so that technology can be effectively transferred far and wide. The power of ICT thus needs to be harnessed as a catalyzing agent in the process of development in the North East India region which has several other bottlenecks in accessibility due to a variety of factors such as improper roads, poor transport, insurgency *etc.* ICT utilization needs the initiative and affordable financial investment with the assurance of support services and power supply.

28

Internet Utilization Pattern in CommunityInformation Centre (CIC)

Bikash Borthakur and Mahesh Chander

ABSTRACT

The present study was conducted with 100 internet users in Lahowal CIC and Barbaruah CIC in Dibrugarh district of Assam to critically analyze the various aspects of internet utilization pattern. Results showed that majority (66%) of the respondents were between 18 to 26 years of age and were students (50%). Education qualification of majority of the respondents (32%) was above seventh standard and below graduation and seventy per cent were above graduation and below post graduation and also undergone training in computer. All of the respondents use internet for e-mail and information downloading followed by chatting (70%), friendship (70%), educational purpose (70%), job search (60%) and leisure and entertainment (30%). Results also showed that all the respondents use internet in CIC for low cost followed by only one café in the whole block (80%). Majority (50%) of the respondents were using internet in CIC for three to five years and used internet facility once in a week. Results also showed that majority of the respondents preferred Google (100%) followed by Gmail (80%), Orkut (80%), Naukri (60%), Monster (60%), Rediff (50%), Yahoo (50%), Bharatmatrimony (30%), Sify (20%) and others (10%). All the respondents expressed e-mail and abundance of information available as major positive aspects and Viruses and hackers and

Pornographic sites as major negative aspects of internet use. All the respondents expressed electricity failure was the major problem followed by slow connectivity and downloading (80%) in CIC.

Introduction

The North – Eastern region has traditionally been less developed due to its remoteness and difficult hilly terrain. A strong communication infrastructure is lacking. It was envisaged by the government that extending the reach of the IT revolution to this region would give a boost to all-round development. Satellite based communication overcomes the challenge of terrain and geographic remoteness. Prime Minister Shri Atal Behari Vajpayee, in February 2000, announced a plan for the establishment of IT facilities at the block level in the North – Eastern region of the country. The Minister of Communications and Information Technology launched a Pilot Project to establish Community Information Centres in 30 blocks in North – Eastern States in April 2000. On August 17, 2002, a landmark day for the eight North – Eastern states of the country, because Community Information Centres (CICs) set up by the government were launched and dedicated to the people of the North-East. The project aims to achieve the objectives of ICT Infrastructure at Block level, Web Access and Internet Services such as e-mail, Market Access and e-commerce, Access to Socio-Economic Databases, e-learning (Computer Aided Learning Processes) and e-education, e-medicine, e-consulting, Citizen Centric Services, Weather Information, IT awareness among local people, Computer Training Programmes, Tender Notification, e-employment Notification, Education and Entertainment. The project is a joint effort by Department of Information Technology (DIT) under Ministry of Communications and Information Technology (MoC&IT), National Informatics Centre (NIC) and the State Governments of the North-Eastern States. The project is managed centrally by DIT/ NIC, New Delhi and locally by active participation of NIC officials at State and District level. Each CIC has two CIC Operators for managing the centres and providing services to the public. Basic services provided by CICs include Internet access, e-mail, printing, data entry, word processing and training for the local populace. In addition, several citizen-centric or Government to Citizen (G2C) services are being delivered from the CICs. Internet is an acronym for international network that gives access to vast global information and as a substantial tool is characterized by information, versatility and interactivity. But, little is known about the perception and internet utilization patterns in CIC. So, it was decided to critically analyze the various aspects of the Internet utilization pattern in CIC and hence, the present study was conducted with the following specific objectives:

1. To study the Internet utilization pattern in CIC.
2. To study the positive and negative aspects of the Internet use.
3. To identify the problems faced by the internet users in CIC.

Materials and Methods

The present study was conducted in Lahowal and Barbaruah Development Block of Dibrugarh District of Assam. Persons who have visited Lahowal CIC and Barbaruah CIC at least once in a month and those who were belonging to Lahowal and Barabruah Development Block were considered as respondents. Fifty such respondents from each of the two CICs were randomly selected to make the total sample size one hundred. Data were collected through personal interview method.

Results and Discussions

Personal Information of the Internet Users

Findings of the study as presented in Table 1 indicated that the majority of the respondents (66%) were between 18 to 26 years of age. Majority (60%) of the respondents were male and unmarried (70%). Majority (50%) of them were students followed by service and unemployed (20%). The caste of the majority of the respondents (50%) were OBC/ MOBC followed by Gen (30%), SC (10%) and ST (10%). Majority of them (50%) were Assamese community followed by Bengali (10%), Marwari (10%), Bihari (10%), Nepali (10%) and Tea tribe (10%). The religion of the majority of the respondents were Hindu (80%) followed by Muslim (20%). Education qualification of majority of the respondents (32%) were above seventh standard and below graduation and 70 per cent had above graduation and below post graduation. Majority of the respondents (70%) had undergone training in computer and nearly half of the respondents possessed family income between Rs 10,000 - 25000 per month.

Internet Utilization Pattern of Respondents

The data in Table 2 expressed that eighty per cent of the respondents did not posses computer at home and among those who had computer at home, none of them had internet connection. All the respondents accessed internet in cyber café other than CIC followed by college (42%), Library (35%) and school (10%). All of the respondents use internet for e-mail and information downloading followed by chatting (70%), friendship (70%), educational purpose (70%), job search (60%) and leisure and entertainment (30%). Results also showed that all the respondents use internet in CIC for low cost followed by only one café in the whole block. Majority (50%) of the respondents were using internet in CIC for 3 to 5 years and used internet facility once in a week. Results also showed that majority of the respondents preferred Google (100%)

followed by Gmail (80%), Orkut (80%), Naukri (60%), Monster (60%), Rediff (50%), Yahoo (50%), Bharatmatrimony (30%), Sify (20%) and others (10%). About 80% of the respondents had family encouragement for using internet in CIC.

Table 1 : Personal Information of the Internet Users

Sl.No.	Item	Category	Frequency
1.	Age	Young = 13- 17 (mean-SD)	17
		Middle 18 - 26	66
		Elderly = 27-42(mean+SD)	17
2.	Marital status	Married	30
		Unmarried	70
3.	Sex	Male	60
		Female	40
4.	Occupation	Student	50
		Service	20
		Business	10
		Unemployed	20
5.	Caste	Gen	30
		OBC/MOBC	50
		SC	10
		ST	10
6.	Community	Assamese	50
		Bengali	10
		Marwari	10
		Bihari	10
		Nepali	10
		Tea tribe	10
7.	Religion	Hindu	80
		Muslim	20
8.	Education qualification	Below class X	6
		Up to class X	7
		Above X and below XII	8
		Above XII and below graduation	32
		Above graduation and below post graduation	32
		Post graduation and above	15
9.	Training in computer	Yes	70
		No	30
10.	Family income in Rs (per month)	1- 10000	25
		10000 -25000	50
		>25000	25

Table 2 : Internet Utilization Pattern of Respondents

Sl.No.	Item	Category	Frequency
1.	Possess computer at home	Yes	20
		No	80
2.	Place of access to internet other than CIC	Cyber café	100
		College	42
		School	10
		Library	35
3.	Reasons for Internet use	e-mail	100
		Information download	100
		Leisure and entertainment	30
		Job search	60
		Chatting	70
		Friendship	70
		Educational purpose	70
4.	Reasons for using internet in CIC	Low cost	100
		High rate in cyber café	70
		No other source for internet nearby	60
		Only one café in the b lock	80
5.	Years of using internet in CIC	Less than one year	20
		1 – 2 years	30
		3 -5 years	50
6.	Time of usage	Once in a week	50
		Once in a month	10
		More than once in a week	30
		When ever necessary	10
7.	Search engines/ Sites preferred	Google	100
		Rediff	50
		Yahoo	50
		Orkut	80
		Gmail	80
		Naukri	60
		Monster	60
		Bharat matrimony	30
		Sify	20
		others	10
8.	Family encouragement	Yes	80
		No	20

* Multiple responses were allowed

Positive and Negative Aspects of the Internet

Findings presented in Table 3 revealed that all of the respondents expressed e-mail and abundance of information available as major positive aspects followed by acquiring knowledge (90%), globalization possible (90%), communication with others (80%) and friendship (60%). Majority of the respondents (90%) perceived Viruses and hackers and Pornographic sites as major negative aspects of internet use followed by misuse (80%), harmful to Indian culture (70%), waste of time (60%) and Person may become addicted to web world (50%).

Table 3 : Positive and Negative Aspects of the Internet

Sl. No.	Items	Yes (%)	No (%)
1.	**Positive aspects**		
	Abundance of information	100	-
	e -mail	100	-
	Communication with others	80	20
	Acquiring knowledge	90	10
	Globalization possible	90	10
	Friendship	60	40
2.	**Negative aspects**		
	Viruses and hackers	90	10
	Pornographic sites	90	10
	Misuse	80	20
	Harmful to Indian culture	70	30
	Waste of time	60	40
	Person may become addicted to web world	50	50

* Multiple responses were allowed

Problems Faced by the Internet Users in CIC

Regarding the problems faced by the internet users in CIC, all the respondents expressed electricity failure was the major problem followed by slow connectivity and downloading (80%), less number of computers (70%), Sunday closed (70%), service available only at office hours (60%), most of the time computer were busy due to ongoing training (50%) and improper services provided in CIC (30%) (Table 4).

Conclusion

CIC is a very noble project offering service to the remotest part of India. People of North – East India are highly benefited from this project. From the above findings, it can be interpreted that the project can be made much more useful to the users by providing battery back-up to overcome power failure, more number of computers, increase of service hours especially in evening, opening on Sunday and broad band connection for faster connectivity and downloading. Better connectivity with the world using internet will ultimately help the people of North -East to go hand in hand with the fast moving world with greater competencies.

Table 4 : Problems Faced by the Internet Users in CIC

Sl. No.	Problems	Frequency
1.	Electricity failure	100
2.	Slow connectivity and downloading	80
3.	Less number of computers	70
4.	Sunday closed	70
5.	Service available only at office hours	60
6.	Most of the time computer were busy due to ongoing training	50
7.	Improper services provided in CIC	30

* Multiple responses were allowed

29

Application of ICT in Agricultural Extension An Evaluation Study of Kissan Call Centres(KCCs)

Hemnath Rao Hanumankar

ABSTRACT

The concept of the Kissan Call Centres-KCCs (Farmers' Call Centres-FCCs)) was a logical outcome of the Government of India's (GOI's) commitment to leverage the ICT for overcoming the constraints of distance and time in providing new generation extension services to the farmers of the country by facilitating their direct virtual contact with the agricultural extension and expert personnel. The study on KCC has shown that farmers in the age group of 29 to 48 years are the biggest user group of KCC helpline. In terms of academic background, farmers with a minimum exposure of six to seven years to formal schooling (86%) seek to utilize the KCC support and relatively inexperienced farmers with less than 10 years of experience are more likely (47.5%) to call the KCC. The scope to enhance awareness and access among women farmers to the KCC helpline is huge as their current levels of participation are negligible. As regards the performance of KCCs, 84% of the farmers who called the KCCs expressed overall satisfaction from the advice provided through the helpline. Of all the supply side factors like promptness of response, patient hearing and completeness of answer, farmers who received a complete response are 47% more satisfied than those who did not receive a complete response.

Introduction

Indian agriculture has made significant gains during the post independence period in pushing the frontiers of production and productivity for major food and non-food crops in the country. The 'Green Revolution', now referred often as the 'First Revolution in Indian Agriculture' was ushered in during the 1970's and 80's, contributing to the country's self-sufficiency in food grain production. However, the agricultural sector has been confronted during the last decade with new problems of declining farm income and an uneasy picture of weak competitiveness of many Indian agricultural products as shown by recent studies. The transition from a subsistence based agricultural economy to a market oriented farming system has begun and the burden of managing the challenge of change seems manifold for the farmers and agriculture development institutions.

The pace of agricultural growth had slipped from 3.4% (1985-95) to 1.8% (1995-96 to 2004-05). The subsequent recovery has not been too comforting either. With the rest of the economy growing briskly propelled by the services sector, the share of agriculture in the country's GDP has declined to around 15%. The relative decline of agriculture component of the GDP is not a cause for concern in itself. But the fact remains that 60% of the country's population continues to depend on agriculture, directly or indirectly for their livelihoods. Unlike the linear thrust of the past on increasing agricultural production, the current challenge is to ensure accelerated growth of agriculture with appreciable gains in farm income inclusively spread out for the benefit of all actors across the sector value chain. This calls for a second revolution and a new paradigm of market orientation and competitiveness would need to be built in to the renewed framework for agriculture development.

Leveraging ICT to Reach Farmers

The information asymmetry at all levels of the agriculture value chain – input suppliers, farmers, traders, agro-processing units, wholesalers, retailers *etc.*, needs to be corrected and here, among many other changes, the agricultural extension system needs to be reinvented with new tools of communication, knowledge of markets and infrastructure to support the farmers with new generation services. Fortunately, the developments in the field of Information and Communication Technology (ICT) in India make it possible for the agricultural extension system to make full use of the growing communication bandwidth to serve the farming community. The ICT is an important value multiplier as it will enable establishment of closer linkages and seamless communication among the key stakeholders in the extension system namely – Agricultural Scientists, Extension Functionaries, Farmers and Marketing Agencies. Where the existing extension system finds it difficult (in terms of infrastructure and finances) to reach their targeted clientele, particularly in the remote hilly and difficult to access areas, the ICT can be deployed productively to connect the disadvantaged farmers with the best of the agricultural scientists.

The concept of the Kissan Call Centres (KCCs) was a logical outcome of the Government of India's (GOI's) commitment to leverage the ICT for overcoming the constraints of distance and time in providing new generation extension services to the farmers of the country by facilitating their direct though virtual contact with the agricultural extension and expert personnel. The Department of Agriculture and Cooperation (DoAC), MoA, GoI initiated a centrally sponsored scheme to set up the KCCs in January, 2004. The Kissan Call Center (KCC) synthesises two separate technologies namely, the Information & Communication Technology (ICT) and the agricultural Technology to deliver online extension support to farmers. Both technologies have their specialized domains and work cultures. KCCs optimally utilize the strengths of both these systems by channeling the specialized knowledge of Subject Matter Specialists (SMSs) from among the Agricultural Scientists and Extension Officers through a professionally managed 'Call Center' model so as to empower the farming community with information based extension services in Agriculture, Horticulture, Animal Husbandry, Marketing and other related areas.

The KCC Operation

The KCC operates at three levels – Level - 1 (the basic Call Center interface, with high quality bandwidth and Agriculture Graduates proficient in local language), Level - 2 (Subject Matter Specialists in agriculture and allied areas connected through good bandwidth telecom and computer connectivity) and Level - 3 (the Management Group connected off line that ensures ultimate answering and resolution of all the farmers' queries not resolved at Level - 2). The flow of activity in a typical KCC is shown in Figure 1.

Level –1: The call coming to the call center is picked up by a level – 1 functionary who after a brief welcome message, solicits and records the basic information and the query of the caller. These details are fed into a computer at the Level – 1. The Level -1 functionary is an agricultural graduate familiar with the local language, soil/ agri-climatic conditions and cropping patterns. The farmers' toll free call lands on a switch in the telecom exchange and the same is transferred to one of the 4 hunting lines (on first come first served basis) at the Call Center premises. One of the agriculture graduates available in the shift duty picks up and processes the call.

Level-2 : The Level –2 consists of Subject Matter Specialists (SMSs) who are in the regular service at State and Central Agricultural Research Stations, Agriculture Technology Information Centres (ATICs), State Agricultural Universities (SAUs) and agriculture and related departments of GOI/ State Governments. Where the Level - 1 agent is not able to answer the farmer's query, the call is transferred in a call sharing mode to the concerned SMS available to the Level -1. The data relating to the caller including the question asked is also transferred to the Level -2 functionary on the latter's computer along with the call. In case, the Level - 2 resource cannot find an appropriate answer on-line, the farmer is advised to call back at a particular time or the Level - 2 resource reverts back to the caller.

Level– 3 : The Level –3 consists of a dedicated cell located at the nodal agency which is invariably a commodity directorate or an agriculture development institution under the administrative control of the Ministry of Agriculture (MOA) in the GOI. The questions not resolved at Level -2 are escalated to Level -3 where appropriate replies are framed in consultation with the concerned specialists available within or outside the State. The replies would be sent to the farmers promptly by post/ e-mail/ fax/ telephone *etc.* within 72 hours of the question being registered at the KCC.

Who Accesses KCC?

Evaluating KCCs Performance

An evaluation study was carried out by the Administrative Staff College of India (ASCI) at the instance of the Ministry of Agriculture, Government of India to assess the performance of the KCCs. The study covered five States in the country namely Haryana, Madhya Pradesh, Maharashtra, West Bengal, and Andhra Pradesh. In each state, three districts were sampled to evaluate the impact of Kissan Call Centres (KCCs). Further, three districts in each state were sampled based on the density of calls received in the districts. The sample plan envisaged selection of three districts in each state on the basis of the total number of calls registered at the Kissan Call Centres (KCC's) during the two successive financial years- April, 2004 to March, 2005 and April, 2005 to March, 2006. One district with the highest cumulative number of calls over the two years, one district with the lowest number of calls and one district that recorded the least growth in the number of calls between the financial years, 2004-'05 and 2005-'06 were sampled in each state.

Sampling Caller and Non-Caller Farmers

Assessing the satisfaction among users of KCC's- the farmers, is vital to understanding how the KCCs have performed. A sample size of 350 caller farmers was determined as adequate for each district. Within the district, six blocks with the highest volume of calls were selected and three to four villages with high volume of calls were sampled within each block in proportion to the share of the village/ block in the total number of calls. An element of convenience in terms of access to the villages was, however, a significant influence on the above multi-stage sampling process.

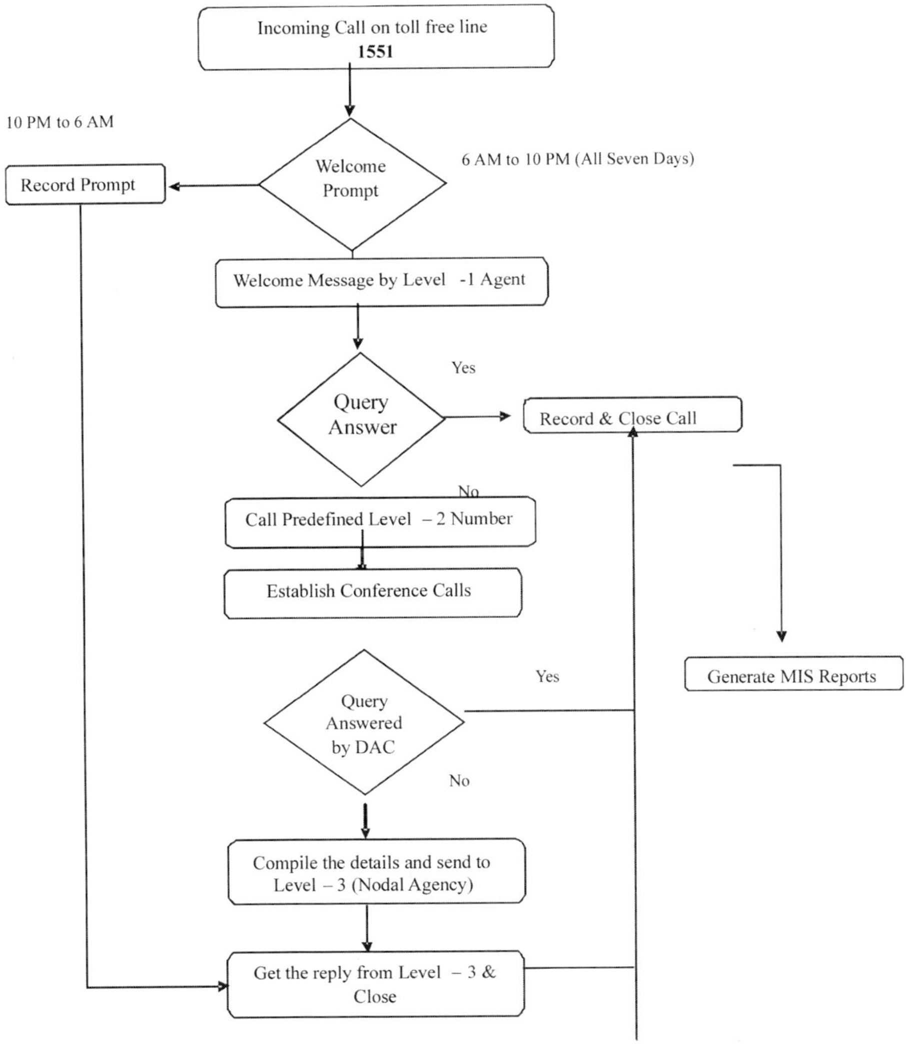

Source: KCC manual of MOA, GOI and Telecom Consultants of India Limited (TCIL), Technical Consultants to the Project.

Figure 1 : Activity Flow at the KCCs

Callers (farmers who called the KCC) were cluster sampled in the selected villages for the survey. Further, a base sample of 30 non-callers (farmers who did not call the KCCs) was surveyed from one village each in the same block from where the sample of caller-farmers was drawn. The non-callers' survey was aimed at eliciting the reasons why farmers have not utilized the services of KCCs. The Tables 1 to 5 show the details of the farmers sampled for the study and responses received in each state. In addition to surveying callers or user farmers, the effectiveness of the functioning of the KCC's was further evaluated through interviews/ focus discussions among various stakeholders including the primary level agents of Kissan Call Centers (KCCs).

Table 1 : Andhra Pradesh

Name of the District	Caller Farmers Contacted	Caller Farmers Responded	Farmers Not Available / Not Responded	Non Callers
Krishna	373	219	154	58
Kadapa	229	130	99	68
Nizamabad	167	70	97	54
Total	769	419	350	180

Table 2 : Haryana

Name of the District	Caller Farmers Contacted	Caller Farmers Responded	Farmers Not Available / Not Responded	Non Callers
Kurukshetra	350	195	455	50
Hissar	350	277	73	50
Jhajjar	300	128	172	50
Total	1000	600	400	150

Table 3 : Maharashtra

Name of the District	Caller Farmers Responded	Non Callers	Total
Ahmednagar	301	52	353
Nandurbar	148	51	199
Amaravati	238	50	288
Total	687	153	940

Table 4 : West Bengal

Name of the District	Caller Farmers Contacted	Caller Farmers Responded	Farmers Not Available /Not Responded	Non Callers
Nadia	360	228	132	53
Purulia	107	35	72	61
Jalpaiguri	200	58	142	60
Total	667	321	346	174

Table 5 : Madhya Pradesh

Name of the District	Caller Farmers Contacted	Caller Farmers Responded	Farmers Not Available /Not Responded	Non Callers
Ujjain	350	211	139	41
Datia	300	225	75	38
Jabalpur	350	211	139	40
Total	1000	647	353	119

Profile of Farmers' by Age and Gender

An understanding of the demographic and socio-economic profile of the farmers calling the KCCs gives an insight into who are most likely to access ICT based support services and how these groups could be targeted effectively with extension services. Majority of farmers (59%) who called the KCCs are in the age group, 29 to 48 years which can be described as the middle age group. This age group accounted for around 55% of callers each in AP, Haryana and Madhya Pradesh and around 65% each in Maharashtra and West Bengal. Table 6 shows the distribution of caller farmers by age groups in all the five states included in the study.

Table 6 : Distribution of Callers by Age

Age	A.P	Haryana	Maharashtra	M.P	W.B	Total
8 to 28 yrs	67 (8.9) (15.7)	206 (27.4) (35.6)	143 (19.0) (20.5)	259 (19.0) (40.0)	78 (10.4) (24.5)	753 (100) (24.5)
29 to 48 yrs	240 (15.3) (56.3)	317 (20.2) 54.8)	446 (28.4) (64.1)	358 (22.8) (55.2)	207 (13.2) (65.1)	1568 (100) (58.8)
49 to 68 yrs	108 (34.5) (25.4)	52 (16.6) (9.0)	99 (31.6) (14.2)	27 (8.6) (4.2)	27 (8.6) (4.2)	313 (100) (11.7)
69 and above	11 (34.4) (2.6)	3 (9.4) (0.5)	8 (25.0) (1.1)	4 (12.5) (0.6)	6 (18.8) (1.9)	32 (100) (1.2)
Total	426 (16.0) (100)	578 (21.7) (100)	696 (26.1) (100)	648 (24.3) (100)	318 (11.9) (100)	2666 (100) (100)

Farmers in the early age group of less than 28 years constitute the second highest group of callers in four out of five states excluding Andhra Pradesh. Further, almost 99% of the farmers who called the KCC helpline in the five sampled states are from the male population.

Academic Profile of Callers

Interestingly, a clear pattern has emerged in regard to the academic background of most callers sampled for the study. Except in West Bengal where the farmers with middle to secondary school background (class 6th to 10th) accounted for the highest percentage of callers (52%), the farmers with post matriculation education up to graduation are the most dominant group (46.5) among the callers in other four states. As may be seen in Figure 2, this group accounts for 44.5% in AP, 47% in Haryana, 41% in Madhya Pradesh and 56% in Maharashtra. In all the five states, 86% of farmers calling the KCCs bear an academic achievement ranging between class 6th and graduation.

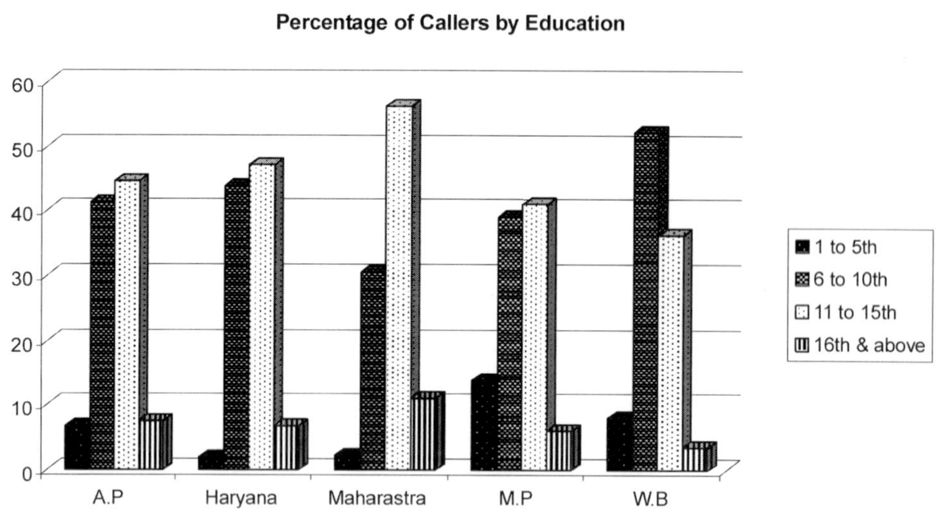

Figure 2 : Percentage of Callers by Education

Farming Experience & Farm Size

Yet another interesting feature of the farmers' profile who called the KCCs is that farmers with less than ten years experience accounted for the highest number (47.5%) of callers- 52% in Haryana, 59% in MP, 45% in Maharashtra and 40% in West Bengal followed by those with 11 to 20 years of experience. AP is an exception here as the farmers with 11-20 years of farming experience accounted for the highest percentage

(34.3) of callers followed by those with less than 10 years experience (31.5%). In the context of land holding size of farmers who access KCCs, the study shows that the callers in all the four states are farmers with an average land holding of nine acres or more except in West Bengal where it is three acres as may be observed from Figure 3.

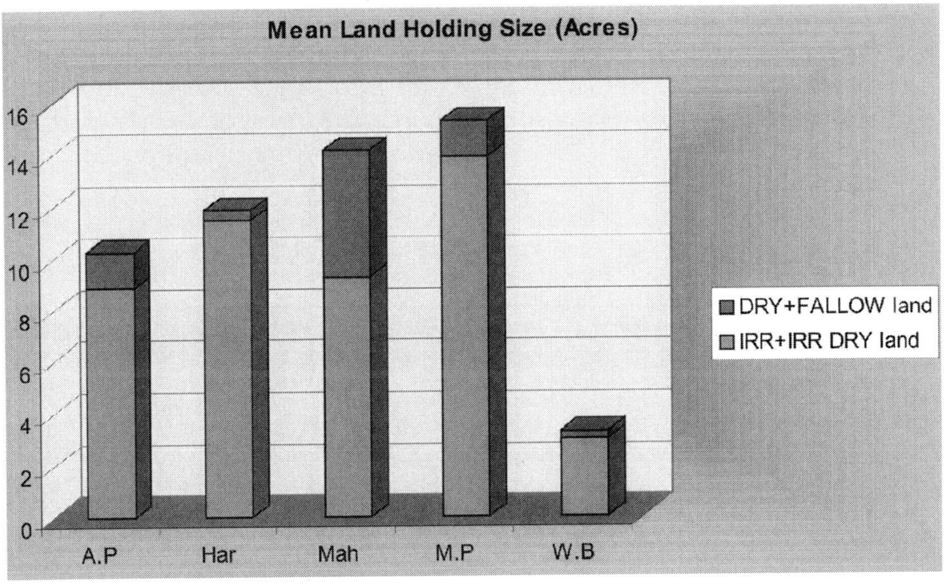

Figure 3 : Mean Land Holding Size (Acres)

Income Levels

It is observed that there is much variation across the states in the income levels of farmers accessing the KCC helpline. 83% of farmers who called the KCC in West Bengal reported annual income levels below Rs.30,000/- while farmers with this level of income accounted for only 7% of the callers in MP. As shown in Table 7, farmers with annual income levels between Rs.31,000/- and 60,000/- constituted the highest segment of callers in the three states of AP (54.2%), Haryana (42.4%) and MP (48.9%) while Maharashtra has drawn the highest percentage (36.2) from those with annual income in excess of Rs. 91,000/-.

Farmers' Satisfaction with KCCs

Almost 84% of the farmers who responded to the question about their satisfaction with the answer provided by KCCs, confirmed having been satisfied. In Haryana and Maharashtra, 91% and 90% of the farmers who made their first call to the KCCs were satisfied, respectively while the percentage of satisfied farmers making the first call was around 75% in the other three states. Table 8 shows the distribution of farmers satisfied with the KCC response to their questions. Further analysis to establish linkages between the farmers' satisfaction with KCC answers and their demographic profiles showed that 66% of the satisfied callers belonged to the younger and middle age groups up to 38 years. Again, in terms of the relationship between farmers' academic background and their satisfaction with KCC response, the study showed that almost 80% of the satisfied farmers were those with a minimum formal educational exposure of 6 years and above. The largest segment of satisfied farmers were those with a minimum qualification of matriculation and who had studied up to graduation.

Table 7 : Total Annual Income from All Sources (State-wise)

Annual income	A.P	Haryana	Maharashtra	M.P	W.B	Total
Rs. 500 to 30000	119 (18.0) (27.9)	84 (12.7) (14.5)	145 (21.9) (20.8)	47 (7.1) (7.3)	266 (40.2) (83.6)	661 (100) (24.8)
Rs.31000 to 60000	231 (22.3) (54.2)	245 (23.6) (42.4)	202 (19.5) (29.0)	317 (30.6) (48.9)	42 (4.1) (13.2)	1037 (100) (38.9)
Rs. 61000 to 90000	52 (12.7) (12.2)	143 (25.0) (24.7)	97 (23.7) (13.9)	111 (27.1) (17.1)	6 (1.5) (1.9)	409 (100) (15.3)
Rs. 91000 and above	24 (4.3) (5.6)	106 (19.0) (18.3)	252 (45.1) (36.2)	173 (30.9) (26.7)	4 (0.7) (1.3)	559 (100) (21.0)
Total	426 (16.0) (100)	578 (21.7) (100)	696 (26.1) (100)	648 (24.3) (100)	318 (11.9) (100)	2666 (100) (100)

Table 8 : Farmers Satisfied with the Answer

	A.P	Haryana	Maharashtra	M.P	W.B	Total
First	N=420 329 (78.3)	N=530 486 (91.7)	N=675 612 (90.7)	N=625 474 (75.8)	N=220 163 (74.1)	N=2470 2064 (83.6)
Second	N=242 188 (77.7)	N=170 149 (87.6)	N=284 249 (87.7)	N=148 110 (74.3)	N=70 56 (80.0)	N=914 752 (82.3)
Third	N=85 70 (82.4)	N=89 79 (88.8)	N=152 136 (89.5)	N=63 47 (74.6)	N=73 66 (90.4)	N=462 398 (86.1)
Last	N=35 32 (91.4)	N=73 67 (91.8)	N=128 116 (90.6)	N=229 177 (77.3)	N=82 74 (90.2)	N=547 466 (85.2)

Where the farmers were not satisfied (16%), an attempt was made to identify the causes for their dissatisfaction. The most common causes of dissatisfaction reported by the farmers who called the KCCs related to the accuracy and credibility of advice offered by the KCC agents at Level - 1 as not being satisfying enough. The caller farmers were often not sure whether their problem has been fully and rightly appreciated by the KCC agents who did not have physical access to the problem situation of the farmers and hence the latter's skepticism of the Level - 1 advice. Further, differences in accent and dialect coupled with delays in connecting farmers to Level - 2 resources have also been found to be among the main causes of dissatisfaction among the caller farmers not satisfied with KCCs.

The survey of the non-caller farmers shows that there is no substantive difference in the demographic profile of farmers who did not call the KCCs (Non-Callers) and the farmers who made the calls (Caller-Farmers), by age, gender, education and length of experience in farming. Understandably, lack of awareness (71%) has been the main reason for farmers not calling the KCCs. There is a need to step up the publicity campaign with the active involvement of the state departments of agriculture. The media mix must be customised to address the media preferences of local farming populations such as harnessing radio as the medium of preference in states like Haryana and West Bengal. A small percentage of farmers (20%) were aware of KCCs but did not make the call for many reasons including difficulty in accessing the KCC from PCOs.

Conclusion

There is considerable interest and acceptance among farmers for ICT based agricultural extension services. The study has shown that farmers in the age group of 29 to 48 years are the biggest user group of KCC helpline. In terms of academic background, farmers with a minimum exposure of six to seven years to formal schooling (86%) seek to utilize the KCC support and relatively inexperienced farmers with less than 10 years of experience are more likely (47.5%) to call the KCC. The scope to enhance awareness and access among women farmers to the KCC helpline is huge as their current levels of participation is negligible.

As regards the performance of KCCs, 84% of the farmers who called the KCCs expressed overall satisfaction from the advice provided through the helpline. Of all the supply side factors like promptness of response, patient hearing and completeness of answer, farmers who received a complete response are 47% more satisfied than those who did not receive a complete response. There is also evidence of farmers' satisfaction growing with repeat calls made by them to the KCCs. Analysis of satisfied farmers by the demographic profile suggests that younger farmers below 38 years of age with post matriculate academic background and less than 10 years of farming experience expressed the highest levels of satisfaction.

Where the farmers were dissatisfied the common reasons were found to be the impracticality of advice provided by KCC agents at level -1; their obsolete knowledge and inability to comprehend local accents and dialects. Delayed access to Level - 2 experts was also found to be an irritant by the farmers. As regards the reasons for farmers not calling the KCCs, the study unambiguously established that nothing short of a professionally planned and promoted publicity blitz can crack the lack of awareness about KCCs among the farming community. There is also a need to review issues of telecom connectivity particularly in remote and far-flung villages.